Remediation of Heavy Metals

Remediation of Heavy Metals

Sustainable Technologies and Recent Advances

Edited by

Rangabhashiyam Selvasembian
SRM University-AP, Andhra Pradesh, India

Binota Thokchom
Indian Institute of Technology Guwahati, Assam, India

Pardeep Singh
University of Delhi, New Delhi, India

Ali H. Jawad
Universiti Teknologi MARA, Selangor, Malaysia

Willis Gwenzi
Leibniz Institute of Agricultural Engineering and Bio-economy e.V. (ATB), Potsdam, Germany

Universität Kassel, Witzenhausen, Germany

WILEY

This edition first published 2024

Registered Office(s)
John Wiley & Sons, Inc., 111 River Street, Hoboken, NJ 07030, USA
John Wiley & Sons Ltd, The Atrium, Southern Gate, Chichester, West Sussex, PO19 8SQ, UK

For details of our global editorial offices, customer services, and more information about Wiley products visit us at www.wiley.com.

Wiley also publishes its books in a variety of electronic formats and by print-on-demand. Some content that appears in standard print versions of this book may not be available in other formats.

Library of Congress Cataloging-in-Publication Data Applied For
[ISBN: 9781119853558; ePDF: 9781119853565; epub: 9781119853572; oBook: 9781119853589]

Cover Design: Wiley
Cover Image: © Ichy WHN/Shutterstock

Set in 9.5/12.5pt STIXTwoText by Straive, Pondicherry, India

Printed and bound by CPI Group (UK) Ltd, Croydon, CR0 4YY
C9781119853558_120124

Contents

List of Contributors

Laxmi R. Adil
Department of Chemistry
Indian Institute of Technology Guwahati
Guwahati, Assam
India

Pinki R. Agrawal
Environmental Sciences and
Biomedical Metrology
CSIR-National Physical Laboratory
New Delhi, Delhi
India

Deeksha Aithani
School of Environmental Sciences
Jawaharlal Nehru University
New Delhi
India

V. Alagesan
School of Chemical and Biotechnology
SASTRA Deemed University
Thanjavur, Tamil Nadu
India

C. Arun
Department of BioTechnology
Periyar Maniammai Institute of Science &
Technology (Deemed to be University)
Thanjavur, Tamil Nadu
India

Debika Barman
Department of Chemistry
Indian Institute of Technology Guwahati
Guwahati, Assam
India

Rinkumoni Barman
Central Ground Water Board, North
Eastern Region
Department of Water Resources, River
Development and Ganga Rejuvenation
Ministry of Jal Shakti
Guwahati, Assam
India

Soumalya Bhowmik
Centre for Nanotechnology
Indian Institute of Technology Guwahati
Guwahati, Assam
India

Chanchal
Department of Chemistry
DCRUST
Murthal, Haryana
India

Chankit
Department of Chemistry
DCRUST
Murthal, Haryana
India

Suparna Datta
Central Ground Water Board,
Eastern Region
Department of Water Resources, River
Development and Ganga Rejuvenation
Ministry of Jal Shakti
Kolkata, West Bengal
India

Ngangbam R. Devi
Department of Chemistry
Indian Institute of Technology Guwahati
Guwahati, Assam
India

Snigdha Dutta
Central Ground Water Board, North Eastern Region
Department of Water Resources, River Development and Ganga Rejuvenation
Ministry of Jal Shakti
Guwahati, Assam
India

Dison S. P. Franco
Department of Civil and Environmental
Universidad de la Costa, CUC
Barranquilla, Colombia

Anadi Gayen
Central Ground Water Board, Eastern Region, Department of Water Resources
River Development and Ganga Rejuvenation
Ministry of Jal Shakti
Kolkata, West Bengal
India

Jordana Georgin
Department of Civil and Environmental
Universidad de la Costa, CUC
Barranquilla, Colombia

Rajdikshit Gogoi
Department of Chemistry
Indian Institute of Technology Guwahati
Guwahati, Assam
India

Willis Gwenzi
Faculty of Organic Agricultural Sciences
Universität Kassel, Grassland Science and Renewable Plant Resources
Witzenhausen, Germany

Department of Technology Assessment
Leibniz Institute for Agricultural Engineering and Bioeconomy (ATB)
Potsdam, Germany

Ziaul Hasan
Department of Biosciences
Jamia Millia Islamia
New Delhi, Delhi
India

Centre for Interdisciplinary Research in Basic Sciences
Jamia Millia Islamia
New Delhi, Delhi
India

Tauseef Hassan
Center for Nanoscience and Nanotechnology
Jamia Millia Islamia
New Delhi, Delhi
India

R.V. Hemavathy
Department of Biotechnology
Rajalakshmi Engineering College
Thandalam, Chennai, Tamil Nadu
India

Maimur Hossain
Department of Chemistry
Indian Institute of Technology Guwahati
Guwahati, Assam
India

Ittishree
Department of Pharmacognosy
Hindu College of Pharmacy
Sonipat, Haryana
India

Parameswar K. Iyer
Department of Chemistry
Indian Institute of Technology Guwahati
Guwahati, Assam
India

Centre for Nanotechnology
Indian Institute of Technology Guwahati
Guwahati, Assam
India

Arif Jamal
Department of Mechanical Engineering
Jamia Millia Islamia
New Delhi, Delhi
India

Vinod Kashyap
Department of Chemistry
National Institute of Technology
Tiruchirappalli, Tamil Nadu
India

Raju Khan
CSIR – Advanced Materials and Processes Research Institute (AMPRI)
Bhopal, Madhya Pradesh
India

Academy of Scientific and Innovative Research (AcSIR)
Ghaziabad, Uttar Pradesh
India

Ruhima Khan
Department of Chemistry
Gauhati University
Guwahati, Assam
India

Vedika Khare
School of Nanotechnology
UTD, RGPV Campus
Bhopal, Madhya Pradesh
India

Jyoti Kushawaha
School of Environmental Sciences
Jawaharlal Nehru University
New Delhi
India

Zakio Makuvara
Department of Physics, Geography and Environmental Sciences
School of Natural Sciences
Great Zimbabwe University
Masvingo, Zimbabwe

Department of Life and Consumer Sciences, School of Agriculture and Life Sciences
College of Agriculture and Environmental Sciences
University of South Africa
Pretoria, South Africa

Jerikias Marumure
Department of Physics, Geography and Environmental Sciences, School of Natural Sciences
Great Zimbabwe University
Masvingo, Zimbabwe

Department of Life and Consumer Sciences, School of Agriculture and Life Sciences
College of Agriculture and Environmental Sciences
University of South Africa
Pretoria, South Africa

Ching T. Moi
Department of Chemistry
Indian Institute of Technology Guwahati
Guwahati, Assam
India

Geetmani S. Nongthombam
Department of Chemistry
Indian Institute of Technology Guwahati
Guwahati, Assam
India

Arpana Parihar
CSIR – Advanced Materials and Processes Research Institute (AMPRI)
Bhopal, Madhya Pradesh
India

Retwik Parui
Department of Chemistry
Indian Institute of Technology Guwahati
Guwahati, Assam
India

Keisham Radhapyari
Central Ground Water Board, North Eastern Region
Department of Water Resources, River Development and Ganga Rejuvenation
Ministry of Jal Shakti
Guwahati, Assam
India

Chandrasekaran Ramprasad
School of Civil Engineering
Centre for Advanced Research in Environment (CARE)
SASTRA Deemed to be University
Thanjavur, Tamil Nadu
India

Sathya Albert Manoharan
School of Chemical and Biotechnology
SASTRA University
Thanjavur, Tamil Nadu
India

A. Sethupathy
Department of Chemical Engineering
Anjalai Ammal Mahalingam Engineering College Kovilvenni
Thiruvarur, Tamil Nadu
India

Ashok K. Sharma
Department of Chemistry
DCRUST
Murthal, Haryana
India

Rahul Sharma
Environmental Sciences and Biomedical Metrology
CSIR-National Physical Laboratory
New Delhi, Delhi
India

Tinoziva T. Simbanegavi
Department of Soil Science and Environment
Faculty of Agriculture Environment and Food Systems
University of Zimbabwe
Harare, Zimbabwe

Ayushi Singhal
CSIR – Advanced Materials and Processes Research Institute (AMPRI)
Bhopal, Madhya Pradesh
India

Academy of Scientific and Innovative Research (AcSIR)
Ghaziabad, Uttar Pradesh
India

Gagan Kant Tripathi
School of Nanotechnology
UTD, RGPV Campus
Bhopal, Madhya Pradesh
India

Priya Dharshini Veeraragavan
School of Chemical and Biotechnology
SASTRA University
Thanjavur, Tamil Nadu
India

Editors' Biography

Dr. Rangabhashiyam Selvasembian is currently working as an associate professor in the Department of Environmental Science and Engineering, School of Engineering and Sciences, SRM University-AP, India. Previously, he worked in the Department of Biotechnology, School of Chemical and Biotechnology, SASTRA Deemed University, India. He received his Doctor of Philosophy degree from the National Institute of Technology, Calicut, India. He was awarded a Post-Doctoral Fellowship by the Max Planck Institute for Dynamics of Complex Technical Systems, 2015, Germany. He was further awarded the National Post-Doctoral Fellowship from SERB-DST, 2016–2018, India. Dr. Selvasembian was awarded as Young Scientist by DST, India, for the BRICS Conclave held in Durban, South Africa, in 2018, and was the recipient of the Hiyoshi Young Leaf Award from Hiyoshi Ecological Services, Hiyoshi Corporation, Japan, in 2018. His major research interests are bioremediation and wastewater treatment. He is an Editorial Board Member in *Separation and Purification Reviews*, *Scientific Reports*, *Biomass Conversion and Biorefinery*, and *Environmental Management*. He is also serving as Associate Editor in the International Journal of Environmental Science and Technology, IET Nanobiotechnology, Frontiers in Environmental Chemistry, and as an Academic Editor for Adsorption Science and Technology. He has published more than 100 peer-reviewed international research articles and contributed several book chapters. He is editing books from prestigious publishers such as Elsevier, Wiley, CRC, and Springer. Recently, Dr. Selvasembian was listed in the top 2% of most cited research scientists in the world, as per data published by Stanford University USA.

Dr. Binota Thokchom is a DST-Inspire faculty at the Centre of Nanotechnology, Indian Institute of Technology, Guwahati. She obtained her PhD from Korea University, Seoul, in 2016. Her working area varies from combating environmental issues, especially water-related crises, to nanomaterial synthesis, simulation, and prediction of works through various software. She has published many research articles and technical reports while participating in various national and international conferences, workshops, and seminars.

Dr. Pardeep Singh is presently working as an assistant professor (Department of Environmental Science, PGDAV College, University of Delhi, New Delhi, India). He obtained his master degree from the Department of Environmental Science, Banaras Hindu University, Varanasi, India, in 2011. He obtained his doctorate degree from the Indian Institute of Technology (Banaras Hindu University) in Varanasi in 2017. The area of his doctoral research is the degradation of organic pollutants through various indigenously isolated microbes and by using various types of photocatalysts. He has published more than 35 papers in international journals in the field of waste management.

Dr. Ali H. Jawad is an associate professor at the Faculty of Applied Sciences, Universiti Teknologi MARA, Malaysia. He received his Ph.D. in environmental chemistry from Universiti Sains Malaysia in 2011. He has over 100 publications in reputed and high-impact factor international journals with total citations (1938) and a Hirsch (H) index of 27. He is currently the Chief Editor of *Science Letters* journal, and he also serves as Editor for several international journals. He received a Top Peer Review award from Publons (Web of Science) in 2019. So far, he has served as an active reviewer for 72 high-impact international journals with a total number of completed reviews of 284. He is a principal investigator for several international and national research grants. His current research is focused on the synthesis of chitosan-based nanocomposite biopolymers and the development of activated carbon from biomass waste for waste water treatment applications.

Prof. Willis Gwenzi is an Alexander von Humboldt Fellow, and Guest Full Professor of Biosystems and Environmental Engineering co-hosted by Leibniz Institute of Agricultural Engineering and Bio-economy e.V. (ATB), Potsdam, and Universität Kassel, Germany. Willis has published over 150 articles in high-impact international peer-reviewed journals, one edited book on merging contaminants by Elsevier, and over 30 edited book chapters. Four of his seminal papers are rated as Highly Cited and Hot papers by Clarivate/Web of Science. He has over 15 experience in university research and development, teaching, community service and consulting. His research focuses on environmental remediation, water/wastewater treatment, emerging contaminants, bioenergy, circular bioeconomy, life cycle assessment, health risk assessments, and industrial symbiosis. He has mentored over 100 early-career researchers. Willis is a regular reviewer for top international journals, and international funding agencies. His academic and research excellence has be recognized internationally by over 20 merit awards, honours, scholarships and research grants. Willis earned: (1) a PhD in Environmental Systems Engineering from the University of Western Australia, (2) a M.Sc. in Water Resources Engineering and Management with merit, (3) a B. Sc. Honours with a first class/distinction, and (4) a Postgraduate Certificate in Applied Groundwater Modelling (UNESCO-IHE, Netherlands).

Preface

1 Motivation and Novelty

Environmental pollution by heavy metals is a global problem posing environmental and human health risks. Heavy metals are highly toxic, bioaccumulate, and persistent, thus, their health risks may last for a long period following their release into the environment. Traditionally, several insitu and ex-situ conventional technologies have been used for the remediation of heavy metals, but most of them have limitations, including high cost. Thus, recent years have witnessed the rapid advances in our understanding of heavy metals. These advances cover the following aspects: (i) ecological and health risks, including toxicology and phytotoxicity; (ii) detection tools; (iii) sustainable and novel remediation technologies; and (iv) final disposal of heavy metal-laden waste materials such as spent adsorbents. However, on the one hand, existing books tend to pay particular attention to conventional remediation methods. On the other hand, information on recent advances on heavy metals remains scattered in individual articles. Therefore, the present 12-book chapter addresses this gap by providing an up-to-date comprehensive synthesis of recent advances in heavy metals.

The 12 chapters address aspects that can be classified under the following main categories:

- The environmental release, occurrence, and behavior of heavy metals
- The detection methods for heavy metals in environmental matrices
- Ecological risk of heavy metals and metalloids, including phytotoxicity in food crops
- Human health risks of heavy metals
- Remediation of heavy metals using chemical methods
- Novel carbon-based adsorption (bio)materials for metal remediation
- Application of novel adsorbents from industrial wastes to remove heavy metals
- Phytoremediation of heavy metals, including microbe-assisted removal
- Bioremediation of heavy metals in acid mine drainage
- Recycling and disposal of spent metal(loid)-laden adsorbents

Compared to earlier books on related topics that often focus on one aspect, the present book tracks heavy metals from their release into the environment, behavior and fate, ecological and human health risks, remediation using various methods, and disposal of metal-laden spent adsorbents. Overall, the book provides a one-stop comprehensive discussion of heavy metals in the environment.

2 Layout/Structure of the Book

- The book has a total 12 chapters, each one structured in the format of a comprehensive review paper based on critical analysis of the evidence. In the chapters, a state-of-the art on the topic is presented, followed by future perspectives, including knowledge gaps.
- The chapters are presented in a logical sequence, starting with those on release and occurrence of heavy metals, followed by health risks, mitigation, and then disposal of spent metal-laden adsorbents.
- The book is well-illustrated with several figures and tables.
- The presentation in terms of structure, language, and style is approachable, making it easy for the reader to understand.

3 Target Audience

This book is a must-have resource and indispensable for those interested in understanding recent advances in the environmental occurrence, behavior, health risk, mitigation of heavy metals, and disposal of metal-laden wastes. Specific target groups for the book include:

- Undergraduate students in environmental sciences, environmental engineering, and mining-related disciplines such as environmental geology and metallurgy.
- Postgraduate students in environmental sciences, environmental engineering, and related disciplines.
- Academics and researchers in environmental sciences, environmental engineering, and health risk assessment.
- Practitioners in environmental management, public health, and occupational health and safety.
- Environmental regulatory agencies and decision- and policy-makers.

Editors
Assoc. Prof. Dr. Rangabhashiyam S
Dr. Binota Thokchom
Dr. Pardeep Singh
Assoc. Prof. Dr. Ali H. Jawad
Prof. Dr. Willis Gwenzi

1

Release, Detection, and Toxicology of Heavy Metals: A Review of the Main Techniques and Their Limitations in Environmental Remediation

Dison S. P. Franco[1], Jordana Georgin[1], and Chandrasekaran Ramprasad[2]

[1] *Department of Civil and Environmental, Universidad de la Costa, CUC, Barranquilla, Colombia*
[2] *School of Civil Engineering, Centre for Advanced Research in Environment (CARE), SASTRA Deemed to be University, Thanjavur, Tamil Nadu, India*

1.1 Introduction to Heavy Metals: An Overview

The group of heavy metals has been the focus of much research and scientific studies for two reasons: they are released into the environment on a daily basis in different regions of the world and are highly harmful to human and animal health (Proshad et al. 2021). They are classified as heavy due to their high molecular weight being detected in biological and environmental samples in various source compositions (Proshad et al. 2021). When they are harmful to health, regardless of the concentration, they are classified as heavy metals, namely: zinc (Zn), chromium (Cr), copper (Cu), mercury (Hg), thallium (Tl), cadmium (Cd), arsenic (As), vanadium (V), nickel (Ni), cobalt (Co), iron (Fe), and lead (Pb) (Li et al. 2014). Both physiological and biological functioning depend on these elements (Lien et al. 2014; Kumar et al. 2018). Several activities in society are responsible for the release of these ions (Kim and Kim 2020). The source of release in food crops varies greatly by region of the world. In developed countries, the use of fertilizers such as sewage sludge and industrial effluents is the activity that mostly releases metals into the environment. In developing countries, irrigation with incorrectly treated sludge and effluents becomes the main input of heavy metals into the soil (Rai and Singh 2018). However, the deposition of these contaminants in the environment still involves several mechanisms. Therefore, these ions are released in different compartments present in society, namely: urban systems with industries, agricultural activities, watersheds, mining activities, public transport vehicles, and estuaries, among others (Figure 1.1). Distribution can occur on a regional, local, or global scale. Studies on the presence of heavy metals in watersheds are complex because they involve several sources such as soil, atmospheric deposition, river water, sediment, and biosphere. Physical–chemical processes, such as water-rock interactions from riverside and flood plains, atmospheric deposition (Nickel et al. 2014), and soil erosion and leaching (Li et al. 2020) are classified as carriers of contaminants. In this respect, river water acts as

Remediation of Heavy Metals: Sustainable Technologies and Recent Advances, First Edition. Edited by Rangabhashiyam Selvasembian, Binota Thokchom, Pardeep Singh, Ali H. Jawad, and Willis Gwenzi.

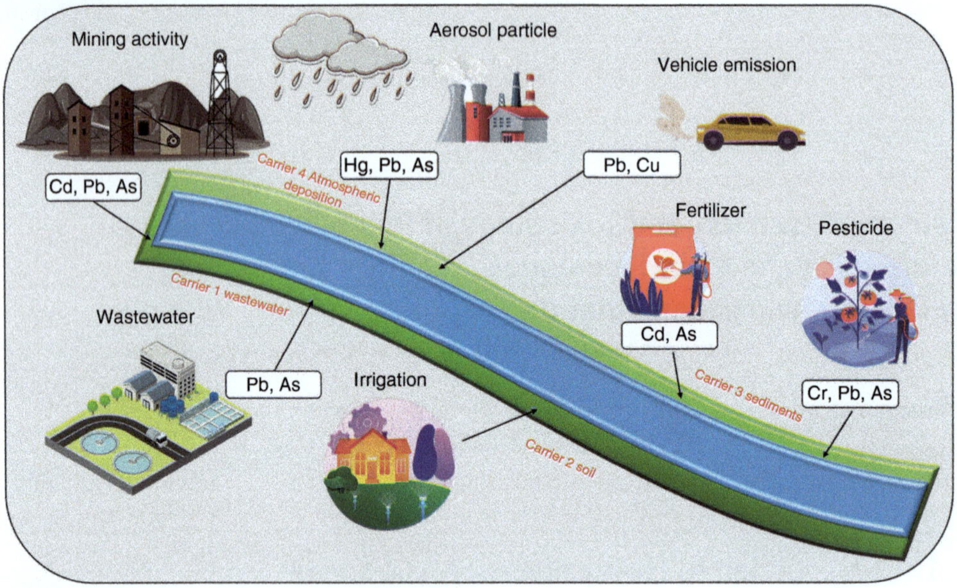

Figure 1.1 Mechanisms and sources of release of heavy metals in water resources.

a sink for contaminants within the watershed. In the image, we can see that river water is the primary carrier, whereas the secondary carriers are sediments, atmospheric deposition, and soil.

The degree of toxicity of these ions in food crops requires specific attention to define the real damage to health that each metal can generate. Some metals such as iron, copper, and zinc are essential in metabolic processes in biota (Zhuang et al. 2009). Uréase has nickel in its composition, and it is known that in large concentrations it can cause problems for human health (Zhuang et al. 2009). These interactions between food, soil, and plants are a model of abiotic-biotic linkages in the environment. Maintaining a balanced soil in relation to the presence of metals is essential, as sustenance and the food source depend on it. However, the unbalanced presence of these elements can cause a disturbance; an example is industrial activities such as the manufacture of chemicals (alkalis and chlorine), thermoelectric plants, energy industries, foundries, and textiles, among others, which are considered punctual sources of metal release. Agricultural runoff and soil erosion are classified as non-point sources (Figure 1.1). In addition to the human health problem, these elements alter the activities and microbial interactions of the soil, directly affecting its fertility and quality (Gadd 2010; Gall et al. 2015; Rai and Singh 2018). Other living organisms such as insects and even mammals that are needed by the soil are harmed (Gall et al. 2015; Bartrons and Peñuelas 2017; Rai and Singh 2018). Some cultures use various plants for medicinal purposes; therefore, determining the presence of metals in these plants is highly necessary since it can bring other damage to the health of these people (Shen et al. 2017). Another concern is that some species can bioaccumulate in their tissues metals such as cadmium, chromium, arsenic, lead, and iron; this situation is aggravated when these plants are close to areas of point sources (El Hamiani et al. 2015; Bolan et al. 2017; Kim et al. 2017; Kohzadi et al. 2019). Not only are species grown in the open air subject to contamination, but those grown in greenhouses are also subject to contamination; however, iron is not present in these crops due to lower lighting (Ling et al. 2017). Knowing the mechanisms that involve the release of metals in soil and food is essential for planning remediation technologies.

Due to the high toxicity of some metals, public bodies such as the World Health Organization (WHO) and the Environmental Protection Agency (EPA) have established release limits into the environment (Carolin et al. 2017). The permitted range (10–250 $mg\,l^{-1}$) varies depending on the metal and the regulatory agency; however, if the value is exceeded, it can cause serious damage to different regions of the human body (Vinod Kumar et al. 2014). Due to physical and chemical interactions, ions are rapidly absorbed by biological matrices through different pathways (Ray and Shipley 2015; Tan et al. 2015; Priyadarshini and Pradhan 2017; Ragab et al. 2017). The excessive release of these ions, added to consumption by the body, leads to serious health problems, making their detection and removal from the environment a major social challenge, as illustrated in Figure 1.2.

In order to perform the detection of toxic ions even in small concentrations, highly sensitive techniques were developed (Figure 1.2), with the most commonly applied being inductively coupled plasma mass spectrometry (ICP-MS) and atomic absorption spectrometry (AAS) (Sener et al. 2014; Oliveira et al. 2015). For removal and remediation, it is possible to observe different analytical methods that have been created and improved in recent times (chemical precipitation, electrochemical methods, ion exchange, and bioremoval); however, they are little applied due to cost (Kiatkumjorn et al. 2014; Swain et al. 2015). With the application of nanomaterials, the colorimetric technique has shown high selectivity, being efficient in environmental monitoring. Based on color change (plasmon resonance), detection by visible UV absorption spectrophotometry can also be applied. However, bringing together all the advantages such as selectivity, low cost, and environmental friendliness, there are still barriers that can be overcome with advanced technologies in optics, electronics, and electrochemistry (Zhao et al. 2017). With a greater number of desirable variables, detection in nanomaterials has been gaining ground in recent years (Hung et al. 2010; Anwar et al. 2018; Rossi et al. 2021). In this area, sensitive and efficient nanoprobes have also been developed for detecting metal ions (Liu et al. 2019). By causing a change in the surface and color plasmon resonance absorption peak, colorimetric sensors modify metal-induced particle segregation. The target ion provides the agglomeration of nanoparticles by changing the color of the solution (Wang et al. 2020).

Aiming to prevent damage to human and environmental health, calorimetric probes based on nanomaterials using plasmon resonance were applied to detect metals in low concentrations present in food samples and water resources (Shrivas and Wu 2007; Shrivas et al. 2015). For the same purpose, Khalkho et al. (2021) used Fourier transform infrared spectroscopy (FTIR) coupled with liquid/liquid solvent extraction, single drop microextraction (SDME), and cloud point microextraction techniques. By focusing on metals harmful to health, studies prove that the method of preconcentration of ions using sodium diethyldithiocarbamate immobilized in polyurethane foam (PU-NaDDC) successfully detected the target pollutant both in biological environmental samples and in liquid environmental samples (dos Santos et al. 2009). Using the same method, it was also possible to detect mercury ions in sedimented areas (dos Santos et al. 2009). Seeking to determine the presence of copper in organic food samples,Shrivas and Jaiswal (2013) used the method based on the separation and preconcentration of the target present in an extraction solvent

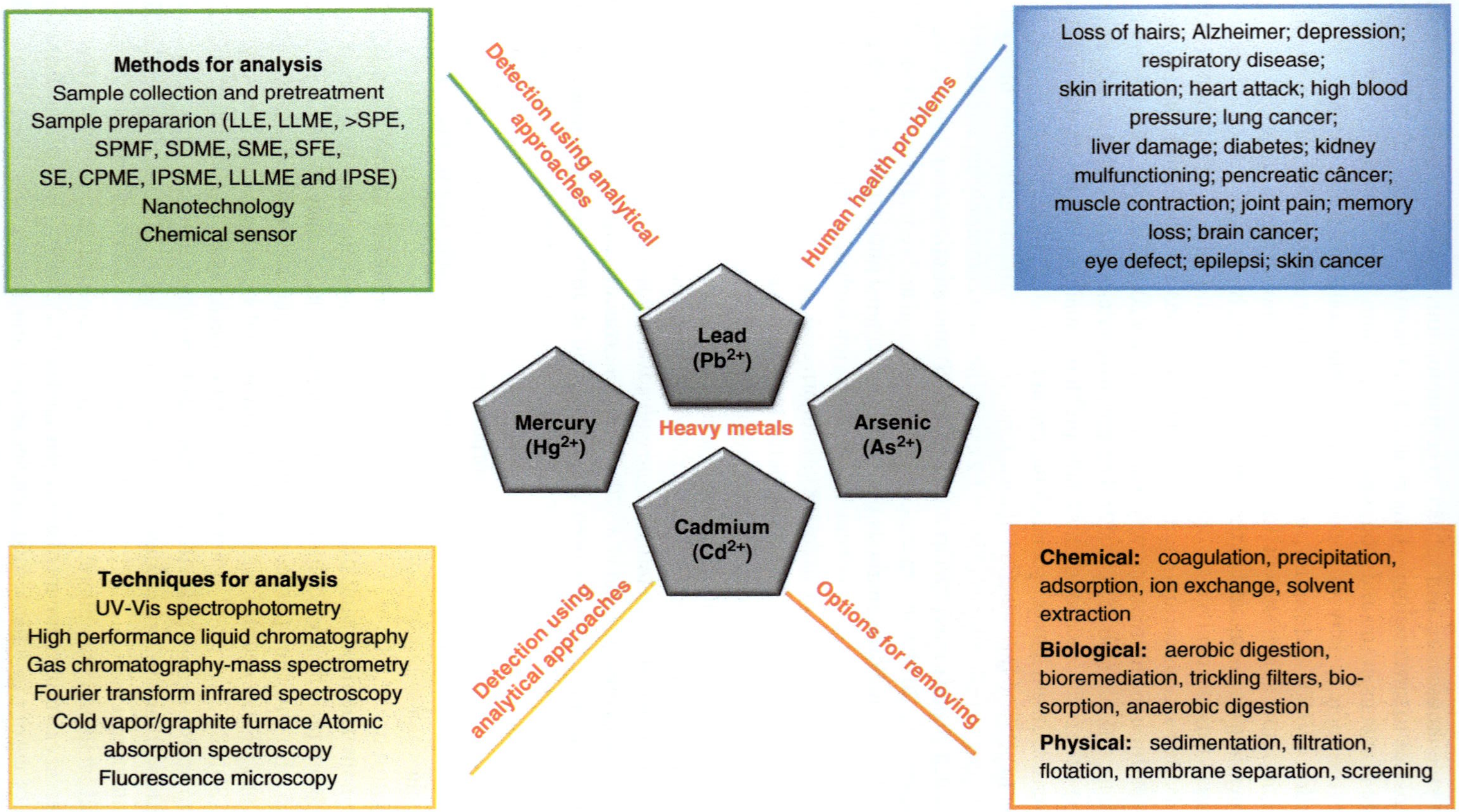

Figure 1.2 Impact of hazardous heavy metals on human health and analysis, detection, and treatment technologies.

containing PBITU together with a dispersion solvent. Shrivas and Wu (2007) jointly used SDME with atmospheric pressure mass spectrometry (AP)-MALDI to analyze the presence of metal ions in real river water samples in gas chromatography (GC), high-performance liquid chromatography (HPLC), and capillary electrophoresis (CE).

The analytical methods improved and developed so far are soxhlet extraction, supercritical fluid extraction (SFE), solvent-based liquid phase microextraction (LLME), solid phase extraction (SPE), solvent microextraction (SME), dispersive microextraction (DME), solid phase microextraction (SPME), SDME, and liquid–liquid microextraction (SLM), all focused on preconcentration and isolation of metals in food, biological, and environmental samples (Liang and Sang 2008; Kocot and Sitko 2014; Vessally et al. 2018). The great advantage of these analytical methods is the speed of the analysis, the ease of operation, and the high sensitivity to the presence of metallic ions; all these points provide high performance in analysis with different sophisticated instruments.

1.2 Industrial Application of Different Metal Ions

Heavy metals and metalloids have various applications in the industrial arena for manufacturing processes, coating purposes, and finishing purposes. The elements such as cadmium (Cd), copper (Cu), lead (Pb), zinc (Zn), mercury (Hg), arsenic (As), silver (Ag), chromium (Cr), iron (Fe), and the platinum group elements are considered heavy metals because their specific densities are at least five times that of water. Since the beginning of recorded history, metals have influenced human communities and history. Several cultures have transitioned between the bronze and iron ages, and various metals have been widely tested in their tools and weaponry. Although the ancients were aware of a variety of metals and non-metals, the greatest number of elements from raw ore was crucial throughout the industrial revolution (1700–1900) (Buyst 2018; Wångmar 2022). Several heavy metals discovered in ancient times came into use in the late 1990s. For instance, scandium, found in the year 1879, has a niche usage in the modern day (Benvenuto 2016). Researchers nowadays are focusing on the use of recovered heavy metals and rare earth metals from waste for various applications (Selvi et al. 2019; Tang et al. 2019; Ramprasad et al. 2022).

Heavy metals have several applications in various industries and other sectors that are discussed in Table 1.1. There are limited reports on the application of metal and derived metal oxides in various industries like textile, tanning, leather, rubber, electrical, electronics, plating, paint, petroleum, mining, and biomedical. Raw heavy metals and metallic composites, as well as nanoparticles made of lead, gold, silver, platinum, zinc, copper, and palladium, for instance, have been widely employed in a variety of goods, from cosmetics to pharmaceuticals, electrical to automobiles, and medical industries.

1.3 Conclusion

The environment and ecosystem have been harmed by the rapid industrialization and increasing pace of population expansion, which is a very serious worry on a global scale. Solitarily, if the three sphere's such as atmosphere, hydrosphere, and lithosphere are

Table 1.1 Heavy metal and metalloids applications in industrial and other sectors.

Heavy metals or metalloids	Industry applied	The function of heavy metals in industry	Occupational health risks and effects	Permissible limits (mg l^{-1})	References
Arsenic (As)	Rubber	Additive in catalyst	Internal cancer, skin lesions, and death	0.01	Yahaya and Don (2014), Kim et al. (2015), Pronk et al. (2020), and Orosun (2021)
	Steel	Metal processing plants			
	Dye and paint	Additive and preservative			
Chromium (Cr)	Chrome plating	Raw material for processing	Ulcer, skin irritation, liver, and kidney damage	0.05	Jobby et al. (2018), Vaiopoulou and Gikas (2020), and Alvarez et al. (2021)
	Leather	Tanning process			
	Textile and pulp	Manufacturing process			
Copper (Cu)	Plastic	Coating and refining	Hypochromic anemia, leukopenia, and osteoporosis	0.1	Lipowsky and Arpaci (2007), ur Rehman and Lee (2014), Siemon et al. (2020), Zamindar et al. (2020), and Li et al. (2021)
	Electroplating	Metal refining and solar cell preparation			
	Pesticide and fertilizer	Raw material for processing			
	Automotive industry	Plain bearings, bushes, valve guides, and brake pipes			
	Electrical industry	Wires, ignition solenoid, resistors, relays, and connectors			
Nickel (Ni)	Painting	Raw material for painting	Dermatitis, myocarditis, fibrosis, and nasopharyngeal tumors	0.2	Scheppe et al. (2002), Salihoglu and Salihoglu (2016), and Wahba et al. (2017)
	Automobile and electrical	Galvanization and paint processing			
Lead (Pb)	Rubber and leather	Catalyst additive	Vomiting, immediate abortion, kidney, brain, and nervous system damage	0.1	Chen et al. (2009), Rocha et al. (2012), Drath and Horch (2014), Tian et al. (2015), and Varshney et al. (2019)
	Electroplating	Metal coating			
	Electrical and electronics	Mobile batteries			
	Petroleum	Leaded gasoline, petrol-based materials			

Table 1.1 (Continued)

Heavy metals or metalloids	Industry applied	The function of heavy metals in industry	Occupational health risks and effects	Permissible limits (mg l^{-1})	References
Mercury (Hg)	Metallurgy and metal plating	Raw material and additive	Nausea, hypersensitivity, memory problems, kidney and heart diseases	0.01	Brooks (2012), Izatt et al. (2015), Bradberry (2016), and Matta and Gjyli (2016)
	Chemical manufacturing	Processing and synthesis of chemicals			
Zinc (Zn)	Rubber and metallurgical	Raw material processing	Vomiting, nausea, immunity disorders, and respiratory effects	0.5	Yin et al. (2015), Uikey and Vishwakarma (2016), and Mirzaei and Darroudi (2017)
	Electrical and electronics	Batteries and wires			
	Textile and paint	Additive and pigments			
	Biomedical	Ointments and preservatives			
Metal–organic framework (MOF)	Electrical and electronic	Energy storage devices mainly include supercapacitors, batteries, and electrode materials	NA	NA	Kinik et al. (2017), Zhou et al. (2021), Ryu et al. (2021), Dharmalingam et al. (2022), Mujahid et al. (2022), and Lin et al. (2023)
	Biomedical	Carrier in anticancer and antimicrobial drugs			
	Chemical industry	Gas storage			
Liquid metals (LMs)	Biomedical industry	Self-healant, electrical and thermal conductor	NA	NA	Yan et al. (2018) and Deng et al. (2021)
	Electrical and electronics	Solar power panels			

devoid of harmful substances, only then we can attain a sustainable ecosystem. All types of ecosystems have been greatly impacted by these dangerous components, which have disrupted the various interconnected food chains of ecosystems. The present study concludes that there are various sources for the heavy metals released into the environment such as industries, domestic utilities, agricultural fertilizers, vehicular emissions, and mining activities. The released heavy metals have many toxicological effects, such as bioaccumulation and biomagnification, leading to cancer and other health-related issues.

The detection methods of heavy metals in the various environmental compartments like soil, water, air, and plants were discussed like UV spectrometer, inductively coupled plasma mass spectrometry (ICPMS), gas chromatography–mass spectrometry (GC MS), and HPLC. The various physical, chemical, and biological remediation techniques available for cleaning heavy metals from the environment were elaborately discussed. The study concludes that heavy metals and metalloids have various industrial applications that are highlighted. Heavy metals have their own occupational health risk if they exceed the permissible limits.

Advanced functional materials, metal-organic frameworks, graphene oxides, carbon nanotubes, and metalloids were being developed and used predominantly in many industries. Further research is needed to address the removal of such emerging pollutants using novel and sustainable technologies. Additionally, the emerging removal technology should address issues of efficiency, effectiveness, cost of operation and maintenance, toxicity of by-products, and proper handling of post-remediation waste. Effective enforcement of public policies and legal requirements governing environmental protection is also necessary, in addition to the use of engineering and technological methods.

References

Alvarez, C.C., Bravo Gómez, M.E., and Hernández Zavala, A. (2021). Hexavalent chromium: regulation and health effects. *Journal of Trace Elements in Medicine and Biology* 65: 126729. https://doi.org/10.1016/j.jtemb.2021.126729.

Anwar, A., Minhaz, A., Khan, N.A. et al. (2018). Synthesis of gold nanoparticles stabilized by a pyrazinium thioacetate ligand: a new colorimetric nanosensor for detection of heavy metal Pd(II). *Sensors and Actuators B: Chemical* 257: 875–881. https://doi.org/10.1016/j.snb.2017.11.040.

Bartrons, M. and Peñuelas, J. (2017). Pharmaceuticals and personal-care products in plants. *Trends in Plant Science* 22: 194–203. https://doi.org/10.1016/j.tplants.2016.12.010.

Benvenuto, M.A. (2016). *Industrial Applications*. Berlin, Boston: De Gruyter.

Bolan, S., Kunhikrishnan, A., Seshadri, B. et al. (2017). Sources, distribution, bioavailability, toxicity, and risk assessment of heavy metal(loid)s in complementary medicines. *Environment International* 108: 103–118. https://doi.org/10.1016/j.envint.2017.08.005.

Bradberry, S.M. (2016). Metals (cobalt, copper, lead, mercury). *Medicine* 44: 182–184. https://doi.org/10.1016/j.mpmed.2015.12.008.

Brooks, W.E. (2012). Chapter 2. Industrial use of mercury in the ancient world. In: *Pattern and Process* (ed. M.S. Bank), 19–24. Berkeley: University of California Press.

Buyst, E. (2018). The causes of growth during Belgium's industrial revolution. *Journal of Interdisciplinary History* 49: 71–92. https://doi.org/10.1162/jinh_a_01232.

Carolin, C.F., Kumar, P.S., Saravanan, A. et al. (2017). Efficient techniques for the removal of toxic heavy metals from aquatic environment: a review. *Journal of Environmental Chemical Engineering* 5: 2782–2799. https://doi.org/10.1016/j.jece.2017.05.029.

Chen, H., Cong, T.N., Yang, W. et al. (2009). Progress in electrical energy storage system: a critical review. *Progress in Natural Science* 19: 291–312. https://doi.org/10.1016/j.pnsc.2008.07.014.

Deng, Y., Jiang, Y., and Liu, J. (2021). Liquid metal technology in solar power generation - Basics and applications. *Solar Energy Materials and Solar Cells* 222: 110925. https://doi.org/10.1016/j.solmat.2020.110925.

dos Santos, J.S., de la Guárdia, M., Pastor, A., and dos Santos, M.L.P. (2009). Determination of organic and inorganic mercury species in water and sediment samples by HPLC on-line coupled with ICP-MS. *Talanta* 80: 207–211. https://doi.org/10.1016/j.talanta.2009.06.053.

Dharmalingam, P., Palani, G., Apsari, R. et al. (2022). Synthesis of metal oxides/sulfides-based nanocomposites and their environmental applications: a review. *Materials Today Sustainability* 20: 100232. https://doi.org/10.1016/j.mtsust.2022.100232.

Drath, R. and Horch, A. (2014). Industrie 4.0: hit or hype? [industry forum]. *IEEE Industrial Electronics Magazine* 8: 56–58. https://doi.org/10.1109/MIE.2014.2312079.

El Hamiani, O., El Khalil, H., Sirguey, C. et al. (2015). Metal concentrations in plants from mining areas in South Morocco: health risks assessment of consumption of edible and aromatic plants. *Clean: Soil, Air, Water* 43: 399–407. https://doi.org/10.1002/clen.201300318.

Gadd, G.M. (2010). Metals, minerals and microbes: geomicrobiology and bioremediation. *Microbiology* 156: 609–643. https://doi.org/10.1099/mic.0.037143-0.

Gall, J.E., Boyd, R.S., and Rajakaruna, N. (2015). Transfer of heavy metals through terrestrial food webs: a review. *Environmental Monitoring and Assessment* 187: 201. https://doi.org/10.1007/s10661-015-4436-3.

Hung, Y.L., Hsiung, T.M., Chen, Y.Y. et al. (2010). Colorimetric detection of heavy metal ions using label-free gold nanoparticles and alkanethiols. *Journal of Physical Chemistry C* 114: 16329–16334. https://doi.org/10.1021/jp1061573.

Izatt, R.M., Izatt, S.R., Izatt, N.E. et al. (2015). Industrial applications of molecular recognition technology to separations of platinum group metals and selective removal of metal impurities from process streams. *Green Chemistry* 17: 2236–2245. https://doi.org/10.1039/C4GC02188F.

Jobby, R., Jha, P., Yadav, A.K., and Desai, N. (2018). Biosorption and biotransformation of hexavalent chromium [Cr(VI)]: a comprehensive review. *Chemosphere* 207: 255–266. https://doi.org/10.1016/j.chemosphere.2018.05.050.

Khalkho, B.R., Kurrey, R., Deb, M.K. et al. (2021). A simple and convenient dry-state SEIRS method for glutathione detection based on citrate functionalized silver nanoparticles in human biological fluids. *New Journal of Chemistry* 45: 1339–1354. https://doi.org/10.1039/d0nj04065g.

Kiatkumjorn, T., Rattanarat, P., Siangproh, W. et al. (2014). Glutathione and l-cysteine modified silver nanoplates-based colorimetric assay for a simple, fast, sensitive and selective determination of nickel. *Talanta* 128: 215–220. https://doi.org/10.1016/j.talanta.2014.04.085.

Kim, Y. and Kim, J. (2020). Bioinspired thiol functionalized carbon dots for rapid detection of lead (II) ions in human serum. *Optical Materials* 99: 109514. https://doi.org/10.1016/j.optmat.2019.109514.

Kim, H.S., Kim, Y.J., and Seo, Y.R. (2015). An overview of carcinogenic heavy metal: molecular toxicity mechanism and prevention. *Journal of Cancer Prevention* 20: 232–240. https://doi.org/10.15430/JCP.2015.20.4.232.

Kim, S.W., Chae, Y., Moon, J. et al. (2017). In situ evaluation of crop productivity and bioaccumulation of heavy metals in paddy soils after remediation of metal-contaminated soils. *Journal of Agricultural and Food Chemistry* 65: 1239–1246. https://doi.org/10.1021/acs.jafc.6b04339.

Kinik, F.P., Uzun, A., and Keskin, S. (2017). Ionic liquid/metal–organic framework composites: from synthesis to applications. *ChemSusChem* 10: 2842–2863. https://doi.org/10.1002/cssc.201700716.

Kocot, K. and Sitko, R. (2014). Trace and ultratrace determination of heavy metal ions by energy-dispersive X-ray fluorescence spectrometry using graphene as solid sorbent in dispersive micro solid-phase extraction. *Spectrochimica Acta Part B: Atomic Spectroscopy* 94–95: 7–13. https://doi.org/10.1016/j.sab.2014.02.003.

Kohzadi, S., Shahmoradi, B., Ghaderi, E. et al. (2019). Concentration, source, and potential human health risk of heavy metals in the commonly consumed medicinal plants. *Biological Trace Element Research* 187: 41–50. https://doi.org/10.1007/s12011-018-1357-3.

Kumar, A., Singh, N., Pandey, R. et al. (2018). Biochemical and molecular targets of heavy metals and their actions. In: *Biomedical Applications of Metals*, 297–319. Springer https://doi.org/10.1007/978-3-319-74814-6_14.

Li, K., Zhang, Y., Dang, Y. et al. (2014). Removal of Cr(VI) from aqueous solutions using buckwheat (*Fagopyrum esculentum* Moench) Hull through adsorption-reduction: affecting factors, isotherm, and mechanisms. *Clean: Soil, Air, Water* 42: 1549–1557. https://doi.org/10.1002/clen.201300399.

Li, J., Li, Z., Brandis, K.J. et al. (2020). Tracing geochemical pollutants in stream water and soil from mining activity in an alpine catchment. *Chemosphere* 242: 125167. https://doi.org/10.1016/j.chemosphere.2019.125167.

Li, H., Qi, H., Yin, M. et al. (2021). Carbon tubes from biomass with prominent adsorption performance for paraquat. *Chemosphere* 262: 127797. https://doi.org/10.1016/j.chemosphere.2020.127797.

Liang, P. and Sang, H. (2008). Determination of trace lead in biological and water samples with dispersive liquid-liquid microextraction preconcentration. *Analytical Biochemistry* 380: 21–25. https://doi.org/10.1016/j.ab.2008.05.008.

Lien, C.W., Tseng, Y.T., Huang, C.C., and Chang, H.T. (2014). Logic control of enzyme-like gold nanoparticles for selective detection of lead and mercury ions. *Analytical Chemistry* 86: 2065–2072. https://doi.org/10.1021/ac4036789.

Lin, G., Zeng, B., Li, J. et al. (2023). A systematic review of metal organic frameworks materials for heavy metal removal: synthesis, applications and mechanism. *Chemical Engineering Journal* 460: 141710.

Ling, L.L., Liu, W.J., Zhang, S., and Jiang, H. (2017). Magnesium oxide embedded nitrogen self-doped biochar composites: fast and high-efficiency adsorption of heavy metals in an aqueous solution. *Environmental Science & Technology* 51: 10081–10089. https://doi.org/10.1021/acs.est.7b02382.

Lipowsky, H. and Arpaci, E. (2007). *Copper in the Automotive Industry*. Wiley.

Liu, X., Yao, Y., Ying, Y., and Ping, J. (2019). Recent advances in nanomaterial-enabled screen-printed electrochemical sensors for heavy metal detection. *TrAC, Trends in Analytical Chemistry* 115: 187–202. https://doi.org/10.1016/j.trac.2019.03.021.

Matta, G. and Gjyli, L. (2016). Mercury, lead and arsenic: impact on environment and human health. *Journal of Chemical and Pharmaceutical Sciences* 9: 718–725.

Mirzaei, H. and Darroudi, M. (2017). Zinc oxide nanoparticles: biological synthesis and biomedical applications. *Ceramics International* 43 (1): 907–914.

Mujahid, M.H., Upadhyay, T.K., Khan, F. et al. (2022). Metallic and metal oxide-derived nanohybrid as a tool for biomedical applications. *Biomedicine & Pharmacotherapy* 155: 113791. https://doi.org/10.1016/j.biopha.2022.113791.

Nickel, S., Hertel, A., Pesch, R. et al. (2014). Modelling and mapping spatio-temporal trends of heavy metal accumulation in moss and natural surface soil monitored 1990–2010 throughout Norway by multivariate generalized linear models and geostatistics. *Atmospheric Environment* 99: 85–93. https://doi.org/10.1016/j.atmosenv.2014.09.059.

Oliveira, E., Núñez, C., Santos, H.M. et al. (2015). Revisiting the use of gold and silver functionalised nanoparticles as colorimetric and fluorometric chemosensors for metal ions. *Sensors and Actuators B: Chemical* 212: 297–328. https://doi.org/10.1016/j.snb.2015.02.026.

Orosun, M.M. (2021). Assessment of arsenic and its associated health risks due to mining activities in parts of North-Central Nigeria: probabilistic approach using Monte Carlo. *Journal of Hazardous Materials* 412: 125262. https://doi.org/10.1016/j.jhazmat.2021.125262.

Priyadarshini, E. and Pradhan, N. (2017). Gold nanoparticles as efficient sensors in colorimetric detection of toxic metal ions: a review. *Sensors and Actuators B: Chemical* 238: 888–902. https://doi.org/10.1016/j.snb.2016.06.081.

Pronk, M.E.J., Woutersen, M., and Herremans, J.M.M. (2020). Synthetic turf pitches with rubber granulate infill: are there health risks for people playing sports on such pitches? *Journal of Exposure Science & Environmental Epidemiology* 30: 567–584. https://doi.org/10.1038/s41370-018-0106-1.

Proshad, R., Islam, S., Tusher, T.R. et al. (2021). Appraisal of heavy metal toxicity in surface water with human health risk by a novel approach: a study on an urban river in vicinity to industrial areas of Bangladesh. *Toxin Reviews* 40: 803–819. https://doi.org/10.1080/15569543.2020.1780615.

Ragab, M.A.A., Korany, M.A., Ibrahim, H.Z. et al. (2017). Adsorption behavior of some metal ions on nanoparticles used in pharmaceutical matrices: application to laboratory made drug formulation. *Bulletin of Faculty of Pharmacy, Cairo University* 55: 155–162. https://doi.org/10.1016/j.bfopcu.2017.01.002.

Rai, P. and Singh, K.P. (2018). Valorization of Poly (ethylene) terephthalate (PET) wastes into magnetic carbon for adsorption of antibiotic from water: characterization and application. *Journal of Environmental Management* 207: 249–261. https://doi.org/10.1016/j.jenvman.2017.11.047.

Ramprasad, C., Gwenzi, W., Chaukura, N. et al. (2022). Strategies and options for the sustainable recovery of rare earth elements from electrical and electronic waste. *Chemical Engineering Journal* 442: 135992. https://doi.org/10.1016/j.cej.2022.135992.

Ray, P.Z. and Shipley, H.J. (2015). Inorganic nano-adsorbents for the removal of heavy metals and arsenic: a review. *RSC Advances* 5: 29885–29907. https://doi.org/10.1039/c5ra02714d.

Rocha, J.H.B., Gomes, M.M.S., Fernandes, N.S. et al. (2012). Application of electrochemical oxidation as alternative treatment of produced water generated by Brazilian petrochemical industry. *Fuel Processing Technology* 96: 80–87. https://doi.org/10.1016/j.fuproc.2011.12.011.

Rossi, A., Zannotti, M., Cuccioloni, M. et al. (2021). Silver nanoparticle-based sensor for the selective detection of nickel ions. *Nanomaterials* 11: 1–16. https://doi.org/10.3390/nano11071733.

Ryu, U., Jee, S., Rao, P.C. et al. (2021). Recent advances in process engineering and upcoming applications of metal–organic frameworks. *Coordination Chemistry Reviews* 426: 213544.

Salihoglu, G. and Salihoglu, N.K. (2016). A review on paint sludge from automotive industries: generation, characteristics and management. *Journal of Environmental Management* 169: 223–235. https://doi.org/10.1016/j.jenvman.2015.12.039.

Scheppe, F., Sahm, P.R., Hermann, W. et al. (2002). Nickel aluminides: a step toward industrial application. *Materials Science and Engineering A* 329–331: 596–601. https://doi.org/10.1016/S0921-5093(01)01587-8.

Selvi, A., Rajasekar, A., Theerthagiri, J. et al. (2019). Integrated remediation processes toward heavy metal removal/recovery from various environments—a review. *Frontiers in Environmental Science* 7: https://doi.org/10.3389/fenvs.2019.00066.

Sener, G., Uzun, L., and Denizli, A. (2014). Lysine-promoted colorimetric response of gold nanoparticles: a simple assay for ultrasensitive mercury(II) detection. *Analytical Chemistry* 86: 514–520. https://doi.org/10.1021/ac403447a.

Shen, Z.J., Xu, D.C., Chen, Y.S., and Zhang, Z. (2017). Heavy metals translocation and accumulation from the rhizosphere soils to the edible parts of the medicinal plant Fengdan (*Paeonia ostii*) grown on a metal mining area, China. *Ecotoxicology and Environmental Safety* 143: 19–27. https://doi.org/10.1016/j.ecoenv.2017.04.042.

Shrivas, K. and Jaiswal, N.K. (2013). Dispersive liquid-liquid microextraction for the determination of copper in cereals and vegetable food samples using flame atomic absorption spectrometry. *Food Chemistry* 141: 2263–2268. https://doi.org/10.1016/j.foodchem.2013.04.067.

Shrivas, K. and Wu, H.F. (2007). Rapid determination of caffeine in one drop of beverages and foods using drop-to-drop solvent microextraction with gas chromatography/mass spectrometry. *Journal of Chromatography. A* 1170: 9–14. https://doi.org/10.1016/j.chroma.2007.09.020.

Shrivas, K., Shankar, R., and Dewangan, K. (2015). Gold nanoparticles as a localized surface plasmon resonance based chemical sensor for on-site colorimetric detection of arsenic in water samples. *Sensors and Actuators B: Chemical* 220: 1376–1383. https://doi.org/10.1016/j.snb.2015.07.058.

Siemon, M., Schiffer, M., Mitra, S., and Walther, G. (2020). Value-based production planning in non-ferrous metal industries: application in the copper industry. *IISE Transactions* 52: 1063–1080. https://doi.org/10.1080/24725854.2020.1711992.

Swain, B., Mishra, C., Hong, H.S., and Cho, S.S. (2015). Treatment of indium-tin-oxide etching wastewater and recovery of In, Mo, Sn and Cu by liquid-liquid extraction and wet chemical reduction: a laboratory scale sustainable commercial green process. *Green Chemistry* 17: 4418–4431. https://doi.org/10.1039/c5gc01244a.

Tan, K.B., Vakili, M., Horri, B.A. et al. (2015). Adsorption of dyes by nanomaterials: recent developments and adsorption mechanisms. *Separation and Purification Technology* 150: 229–242. https://doi.org/10.1016/j.seppur.2015.07.009.

Tang, J., Su, M., Wu, Q. et al. (2019). Highly efficient recovery and clean-up of four heavy metals from MSWI fly ash by integrating leaching, selective extraction and adsorption. *Journal of Cleaner Production* 234: 139–149. https://doi.org/10.1016/j.jclepro.2019.06.198.

Tian, X., Wu, Y., Gong, Y., and Zuo, T. (2015). The lead-acid battery industry in China: outlook for production and recycling. *Waste Management & Research* 33: 986–994. https://doi.org/10.1177/0734242X15602363.

Uikey, P. and Vishwakarma, K. (2016). Review of zinc oxide (ZnO) nanoparticles applications and properties. *International Journal of Emerging Technology in Computer Science and Electronics* 21 (2): 239–242.

ur Rehman, A. and Lee, S.H. (2014). Review of the potential of the Ni/Cu plating technique for crystalline silicon solar cells. *Materials (Basel)* 7: 1318–1341. https://doi.org/10.3390/ma7021318.

Vaiopoulou, E. and Gikas, P. (2020). Regulations for chromium emissions to the aquatic environment in Europe and elsewhere. *Chemosphere* 254: 126876. https://doi.org/10.1016/j.chemosphere.2020.126876.

Varshney, K., Varshney, P.K., Gautam, K. et al. (2019). Current trends and future perspectives in the recycling of spent lead acid batteries in India. *Materials Today: Proceedings* 26: 592–602. https://doi.org/10.1016/j.matpr.2019.12.168.

Vessally, E., Ghorbani-Kalhor, E., Hosseinzadeh-Khanmiri, R. et al. (2018). Application of switchable solvent-based liquid phase microextraction for preconcentration and trace detection of cadmium ions in baby food samples. *Journal of the Iranian Chemical Society* 15: 491–498. https://doi.org/10.1007/s13738-017-1249-z.

Vinod Kumar, V., Anbarasan, S., Christena, L.R. et al. (2014). Bio-functionalized silver nanoparticles for selective colorimetric sensing of toxic metal ions and antimicrobial studies. *Spectrochimica Acta, Part A: Molecular and Biomolecular Spectroscopy* 129: 35–42. https://doi.org/10.1016/j.saa.2014.03.020.

Wahba, O.A.G., Hassan, A.M., Naser, A.M., and Hanafi, A.M. (2017). Preparation and spectroscopic studies of some copper and nickel schiff base complexes and their applications as colouring pigments in protective paints industry. *Egyptian Journal of Chemistry* 60: 25–40. https://doi.org/10.21608/ejchem.2017.517.1000.

Wang, H., Da, L., Yang, L. et al. (2020). Colorimetric fluorescent paper strip with smartphone platform for quantitative detection of cadmium ions in real samples. *Journal of Hazardous Materials* 392: 122506. https://doi.org/10.1016/j.jhazmat.2020.122506.

Wångmar, E. (2022). *Colonial Enterpreneurial Capital in the Industrialization of Southern Sweden*, 161–180. Imp. Underbelly.

Yahaya, Y.A. and Don, M.M. (2014). Pycnoporus sanguineus as potential biosorbent for heavy metal removal from aqueous solution: a review. *Journal of Physical Science* 25: 1–32.

Yan, J., Lu, Y., Chen, G. et al. (2018). Advances in liquid metals for biomedical applications. *Chemical Society Reviews* 47 (8): 2518–2533.

Yin, N.H., Sivry, Y., Benedetti, M.F. et al. (2015). Application of Zn isotopes in environmental impact assessment of Zn-Pb metallurgical industries: a mini review. *Applied Geochemistry* 64: 128–135. https://doi.org/10.1016/j.apgeochem.2015.09.016.

Zamindar, N., Sadat Anari, E., Sadat Bathaei, S. et al. (2020). Application of copper nano particles in antimicrobial packaging: a mini review. *Acta Scientific Nutritional Health* 4: 14–18. https://doi.org/10.31080/asnh.2020.04.0694.

Zhao, Y., Gui, L., and Chen, Z. (2017). Colorimetric detection of Hg2+ based on target-mediated growth of gold nanoparticles. *Sensors and Actuators B: Chemical* 241: 262–267. https://doi.org/10.1016/j.snb.2016.10.084.

Zhou, D., Wang, F., Yang, J., and Fan, L.-z. (2021). Flexible solid-state self-charging supercapacitor based on symmetric electrodes and piezo-electrolyte. *Chemical Engineering Journal* 406: 126825. https://doi.org/10.1016/j.cej.2020.126825.

Zhuang, P., McBride, M.B., Xia, H. et al. (2009). Health risk from heavy metals via consumption of food crops in the vicinity of Dabaoshan mine, South China. *Science of The Total Environment* 407: 1551–1561. https://doi.org/10.1016/j.scitotenv.2008.10.061.

[illegible]

[illegible] (20[illegible]). [illegible] [illegible]

[illegible] (20[illegible]). [illegible] [illegible]

[illegible]

[illegible]

[illegible]

2

Heavy Metals Contamination in Environment

Deeksha Aithani and Jyoti Kushawaha

School of Environmental Sciences, Jawaharlal Nehru University, New Delhi, India

2.1 Introduction

Metals are generally grouped as transition metals, alkaline earth metals, alkali metals, and rare metals in the periodic table. However, the term heavy metals which is used very commonly does not belong to any distinct section in the periodic table. Hence, heavy metals can be defined as per their physical properties, toxicity to living organisms, and chemical reactivity. Heavy metal contamination in the environment is odorless and colorless, hence difficult to be noticed. For shorter time periods, heavy metal contamination does not explicitly damage the environment. However, when heavy metal contamination exceeds the tolerance range of environment or when there is change in environmental conditions, these metals get activated and result in serious ecological damage. Hence, their contamination of the environment is called as chemical time bombs (Pacyna and Pacyna 2001). Figure 2.1 depicts heavy metal transfer in different environmental components.

In the environment, heavy metals occur in different oxidation forms. For example, dominant species of Cr in soil and soil solution are Cr^{3+}, $CrO_4{}^{2-}$ and Cr^{3+}, $CrOH^{2+}$ (at pH 3.5–6.0), $Cr(OH)_4$ (at pH 6–8.5), respectively. Dominant species of Co in soil is Co^{2+}. The dominant species of Cu is Cu^{2+} in soil, and in soil solution, it is Cu^{2+}, Cu-org (at 3.5–6 pH), and Cu-hydroxy species, $CuCO_3{}^0$, Cu-org. The dominant species of Pb in soil is Pb^{2+} and in soil solution it is Pb^{2+}, $PbSO_4{}^0$, Pb-org (at pH 3.5–6.0), and Pb-hydroxy and carbonate, Pb-org (at pH 6–8.5). Dominant species of Ni in soil occur as Ni^{2+} and in soil solution it is Ni^{2+}, $NiSO_4{}^0$, Ni-org (at pH 3.5–6) and Ni^{2+}, $NiHCO_3{}^+$, $NiCO_3$ (at pH 6–8.5). Dominant species of Zn occur as Zn^{2+} in soil and in soil solution it is Zn^{2+}, $ZnSO_4{}^0$, Zn-org (at pH 3.5–6), and Zn^{2+}, Zn-hydroxy and carbonate, Zn-org (at pH 6–8.5) (McLaughlin et al. 2011).

The sources of metal can be classified as geogenic and anthropogenic. The geogenic sources of metals include weathering of parent rock material and volcanic emissions. Substantial amounts of Ni, Cu, Mn, Co, and Zn are released into soil from many igneous rocks including olivine, augite, and hornblende (Acosta et al. 2011). In sedimentary rock system, highest concentration of Co, Ni, Cu, Cr, Mn, Zn, Pb, and Cd are found in shale,

Remediation of Heavy Metals: Sustainable Technologies and Recent Advances, First Edition. Edited by Rangabhashiyam Selvasembian, Binota Thokchom, Pardeep Singh, Ali H. Jawad, and Willis Gwenzi.

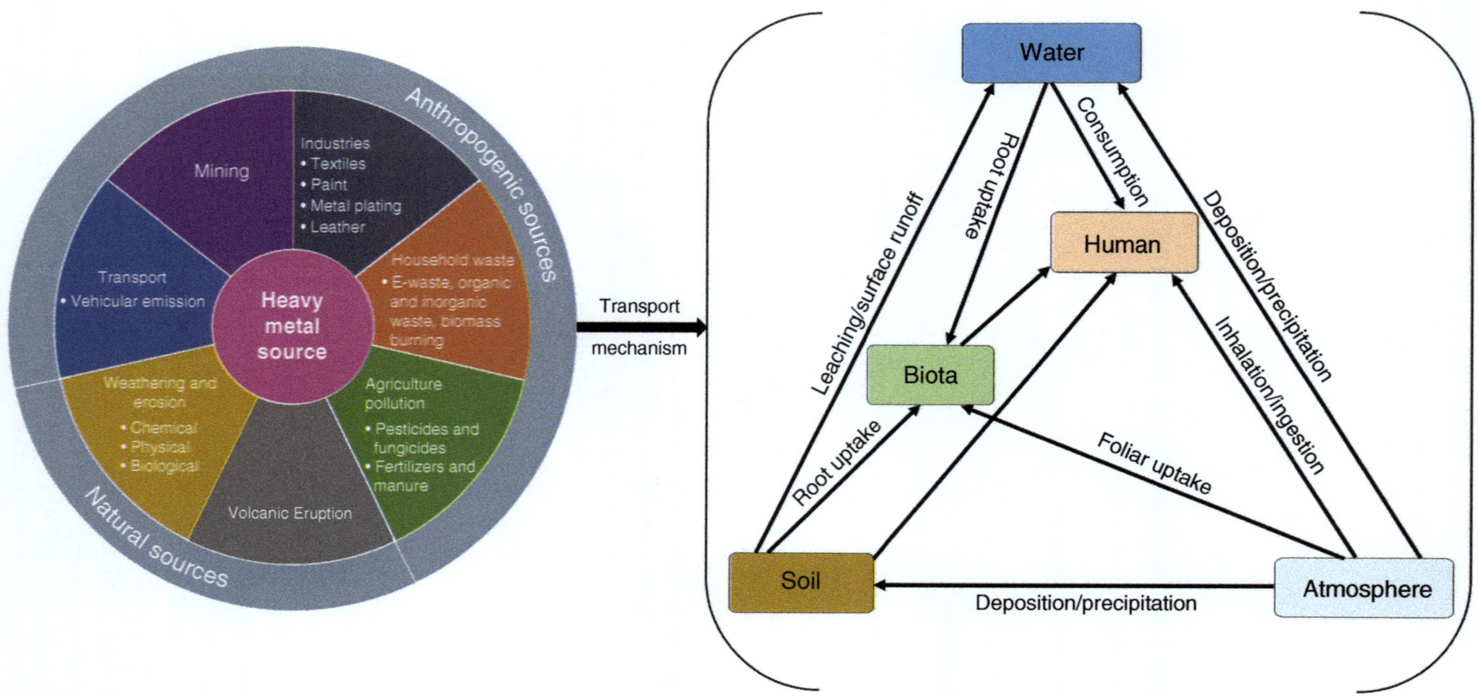

Figure 2.1 Heavy metal sources in the environment.

followed by limestone and sandstone. Metals like Zn, Mn, Pb, Ni, Cu, Cd, and Hg have been emitted by volcanoes in high concentrations (Sandeep et al. 2019). The anthropogenic sources of metals include agricultural activities, industrial emissions, solid waste disposal, and urban activities (Gupta et al. 2019; Sandeep et al. 2019).

One of the most important anthropogenic emission sources of heavy metals is coal combustion (Pacyna et al. 2007). Metals like arsenic (As), antimony (Sb), thorium (Th), beryllium (Be), radon (Rn), cadmium (Cd), selenium (Se), chlorine (Cl), phosphorus (P), chromium (Cr), nickel (Ni), cobalt (Co), mercury (Hg), fluorine (F), manganese (Mn), lead (Pb), and uranium (U) and their compounds are usually found in coal. These are also classified as potentially "hazardous air pollutants" by the US Clean Air Act Amendments of 1990 (Pacyna and Pacyna 2001).

Industrial activities such as mining, electroplating, paints and pigments, polishing and printing operations, burning of coal, battery manufacturing, plumbing, smelting, welding, refineries, porcelain enameling, and metal polishing industries lead to higher metal concentrations in environment (Ayangbenro and Babalola 2017). Agriculture activities such as pesticides, fertilizers, fungicides, use of manure and compost, and irrigation by wastewater are considered to increase metal concentration (Gupta et al. 2019). Disposal of solid waste through open dumps, landfills, sanitary landfills, and wastewater discharge also elevates the metal concentration in the environment (Sandeep et al. 2019). Figure 2.1 represents the heavy metal sources in the environment.

2.2 Heavy Metals in Water

Heavy metal contamination in aquatic environments is a serious problem because of potential toxicity of metals and their accumulation in aquatic habitats. Once heavy metals are released into waterbodies, they are adsorbed into sediments. After metal adsorption on the sediments, the organisms living in these habitats start accumulating the metals at erratic rate. Heavy metals enter the aquatic ecosystem from both sources, i.e. geogenic as well as anthropogenic sources. The sources of heavy metals in aquatic environments are geologic weathering, the discharge of municipal, agricultural, domestic, or industrial waste products, and direct atmospheric precipitation (Figure 2.2) (Khound et al. 2019; Mao et al. 2019).

Heavy metals may be present in groundwater as complexes that vary in their transport via biological barriers. The metal speciation in groundwater is linked to biogeochemical reactivity, which in turn is governed by the microflora, local water, and porous solid properties. A network of connections between aqueous and heterogeneous chemical reactions and the processes of particle coagulation and flocculation control the mobility of metals in groundwater. Processes such as precipitation and adsorption decrease metal availability and transport, whereas processes like hydrolysis and dissolved complexes increase the solubility of metals. The metal transport and availability in groundwater decreased because of precipitation of solids such as metal phosphates, silicates, carbonates, oxides, mixed solids, or sulfides. Adsorption of metals occurs due to heterogenous reaction with solutes in groundwater and soil interstitial water, as soil and aquifer solids provide good surface area. As the water percolates through the soil, it acquires its chemical composition as a result of

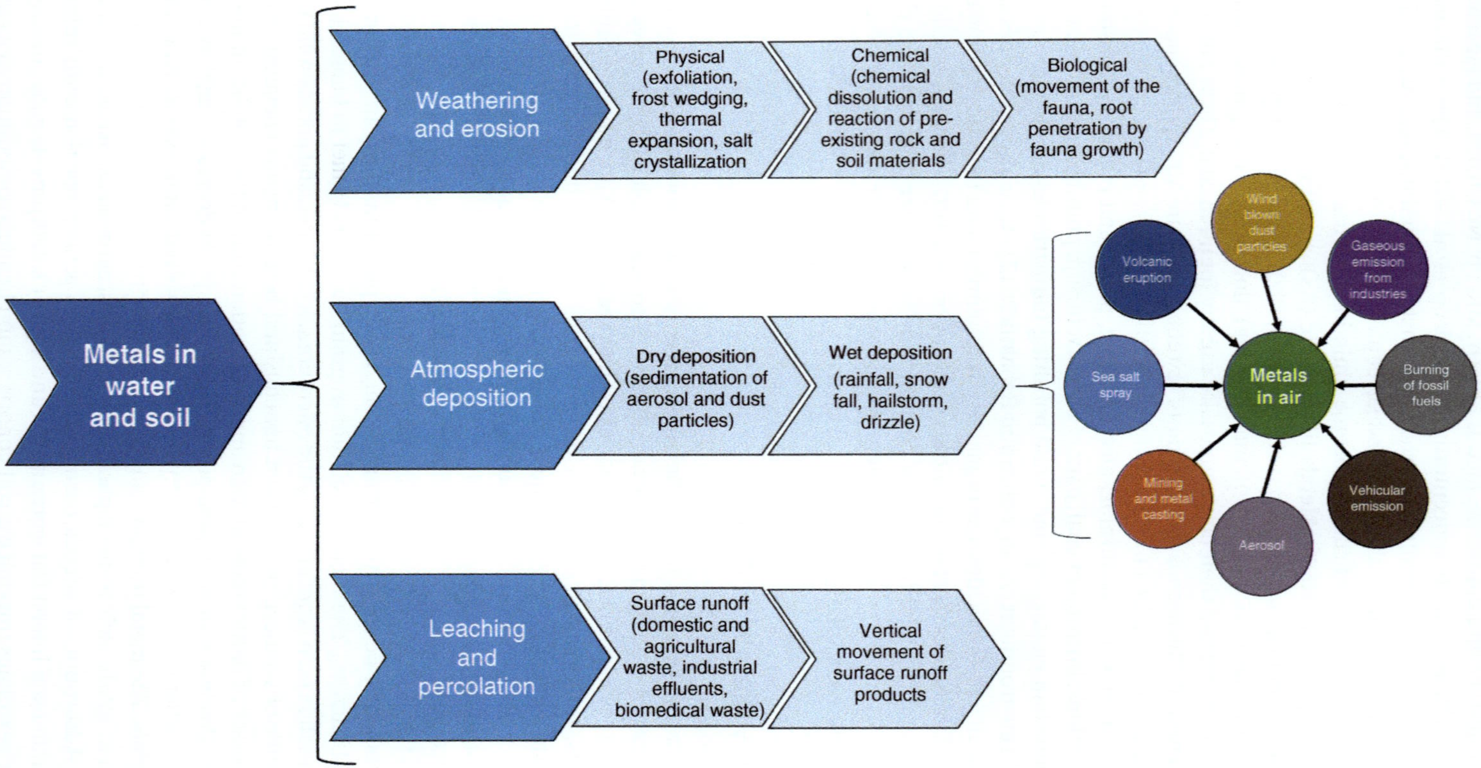

Figure 2.2 Heavy metal sources in water and soil.

water–solid interactions, adjacent biological activity, and physical water mixing. The major factors responsible for increases in the solubility and mobility of solid-bound metals in groundwater are low pH (dissolve and desorb the metals from the solid phase), alteration in redox conditions (moderate to high redox potential), higher concentrations of inorganic salts (competition for dissolved complexation and heterogeneous reactions) and synthetic or natural complexing agents (Wongsasuluk et al. 2014; Kashyap et al. 2018).

Trace metals in surface water may be tied to particulate or colloidal matter and may not be accessible to organisms. In the aquatic system, sediments act as sink source of metals. The changes in environmental conditions such as changes in pH, conductivity, redox potential, and organic matter affect the metal mobility between water and sediment, causing either sorption or desorption of metals (Eliku and Leta 2018; Nyantakyi et al. 2019). The resuspended trace metals in the aquatic system pose toxicity to flora and fauna of aquatic system.

Table 2.1 presents heavy metal contamination in water. Various researchers have studied the heavy metal contamination in river water and groundwater. Research studies suggest that surface and groundwater are contaminated with heavy metals. Heavy metals are present in the river water in different seasons to varying degrees. Figure 2.2 depicts the various pathways of metal contamination in water systems.

2.3 Heavy Metals in Soil

Excessive depositions of heavy metals in soil lead to metal contamination. Heavy metals in the soil comprise some important metals of biological toxicity, such as cadmium (Cd), lead (Pb), chromium (Cr), mercury (Hg), and arsenic (As). They also comprise other heavy metals of certain biological toxicity, such as copper (Cu), nickel (Ni), stannum (Sn), zinc (Zn), and vanadium (V). Heavy metal contamination in soil also affects microbial activity significantly. Metal contamination with Cu, Zn, Pb, and others inhibits soil microbial biomass (Su et al. 2014).

Sources of heavy metals in soil differ as per land use. In urban soil, major sources of heavy metals include emissions from vehicular exhaust and non-exhaust, industrial waste, household waste, and atmospheric depositions. In agriculture soil, major sources are sewage sludge, automobile exhaust, fertilizers, and pesticides. The heavy metal contamination in urban soil directly affects human health as metals are directly transferred to humans via ingestion or skin contact. Heavy metals in soil get adsorbed to soil particles and then get accumulated in crops grown in the contaminated soil. Hence, heavy metals from the agricultural soil affect humans through food chain transfer (Su et al. 2014; Chabukdhara et al. 2016). An estimate has been done by Su et al. (2014) for heavy metal contamination in urban soil and agricultural soil worldwide. They have estimated that the average concentrations of Cr, Cu, Pb, Zn, Ni, and Cd in urban soils are 66.08, 49.60, 1733.94, 289.78, 29.14, and 1.52 $mg\,kg^{-1}$, respectively. For agriculture soil, they reported that the average content of Cr, Cu, Pb, Zn, Ni, Cd, Hg, and As as 46.69, 38.03, 51.19, 117.35, 26.12, 1.50, 0.28, and 21.19 $mg\,kg^{-1}$, respectively (Su et al. 2014). The heavy metal contamination of soil leads to reduced soil fertility, modifies the microbial population and processes, and affects the soil pore spaces and nutrient cycles such as anomalous nitrogen transformation (Srivastava et al. 2017).

Table 2.2 also shows the heavy metal contamination in soil used for different purposes. Soil is used for various purposes such as urban, farmland, park, orchard, and agriculture.

Table 2.1 Heavy metal concentration in water.

Location	Metals (range [mean ± standard deviation])										Reference
	Cu ($\mu g\,l^{-1}$)	Cr ($\mu g\,l^{-1}$)	Co ($\mu g\,l^{-1}$)	Pb ($\mu g\,l^{-1}$)	Ni ($\mu g\,l^{-1}$)	Zn ($\mu g\,l^{-1}$)	Mn ($\mu g\,l^{-1}$)	Fe ($mg\,l^{-1}$)	As ($\mu g\,l^{-1}$)	Cd ($\mu g\,l^{-1}$)	
Ubon Ratchathani province in northeastern Thailand (groundwater)	10.00–323 (60.3)	0.29–2.14 (0.58)	NS	0.95–66.85 (16.7)	0.65–15.60 (6.13)	6.94–302 (63.4)	NS	NS	0.25–6.44 (1.06)	0.13–0.23 (0.15)	Wongsasuluk et al. (2014)
Industrial area, Himachal Pradesh, India (groundwater)	NS	1–322.58 (14.97)	1.75–4225 (92.81)	2.25–53.75 (19.50)	12.50–775 (54.75)	0.02–0.04 (0.03)	450–6375 (1830.27) $mg\,l^{-1}$	0.12–0.60 (0.27)	2.25–117.50 (27.35)	0.92–82.50 (2.85)	Kashyap et al. (2018)
Awash River, Ethiopia, dry season	440–1010 (700 ± 200)	360–1160 (750 ± 290)	NS	310–830 (590 ± 180)	20–90 (50 ± 20)	460–910 (640 ± 150)	NS	1.11–2.73 (2.17 ± 0.61)	NS	30–110 (60 ± 30)	Eliku and Leta (2018)
Awash River, Ethiopia, wet season	820–1690 (1200 ± 340)	300–980 (630 ± 250)	NS	410–1360 (810 ± 290)	30–200 (100 ± 70)	740–1560 (1140 ± 300)	NS	1.82–4.12 (3.36 ± 0.77)	NS	50–240 (130 ± 70)	Eliku and Leta (2018)
Low flow season, Lhasa River basin, Tibet	BDL–1.47 (0.23 ± 0.49)	2.04–5.93 (2.89 ± 1.10)	NS	BDL–0.15 (0.01 ± 0.03)	NS	BDL–10.08 (0.82 ± 2.34)	0.26–70.87 (6.50 ± 16.88)	NS	0.33–144 (20.9 ± 40.13)	BDL	Mao et al. (2019)
High flow season, Lhasa River basin, Tibet	0.44–15.36 (4.65 ± 5.44)	2.11–4.93 (3.61 ± 0.78)	NS	BDL–0.30 (0.07 ± 0.11)	NS	0.61–5.58 (2.11 ± 1.35)	0.59–98.89 (23.97 ± 31.54)	NS	BDL–2.57 (0.14 ± 0.59)	BDL–0.18 (0.02 ± 0.04)	Mao et al. (2019)

Normal flow season, Lhasa River basin, Tibet	0.26–8.98 (1.83 ± 2.25)	1.81–4.90 (3.17 ± 0.8)	NS	BDL–0.18 (0.03 ± 0.05)	NS	BDL–2.64 (0.74 ± 0.71)	0.28–76.10 (17.04 ± 20.80)	NS	BDL–170.45 (19.09 ± 40.93)	BDL–0.04 (0 ± 0.01)	Mao et al. (2019)
Tano River, Ghana	BDL	BDL–687 (266 ± 262)	NS	BDL–929 (381 ± 305)	NS	BDL–1270 (149 ± 266)	NS	NS	BDL–201 (12 ± 33)	BDL–110 (50 ± 31)	Nyantakyi et al. (2019)
Korotoa River, Bangladesh, winter	31–119 (73 ± 33)	43–126 (83 ± 27)	NS	11–64 (35 ± 19)	13–71 (39 ± 23)	NS	NS	NS	15–92 (46 ± 27)	1–22 (11 ± 8)	Islam et al. (2015)
Korotoa River, Bangladesh, Summer	23–96 (61 ± 28)	33–110 (73 ± 27)	NS	8–52 (27 ± 15)	9.3–56 (32 ± 19)	NS	NS	NS	10–77 (37 ± 24)	0.9–18 (8 ± 6)	Islam et al. (2015)
Jia Bharali, north Brahmaputra plain, India Dry season	BDL–180 (100 ± 40)	30–290 (130 ± 60)	BDL–90 (40 ± 30)	BDL–250 (110 ± 60)	BDL–150 (70 ± 40)	BDL–110 (40 ± 30)	BDL–160 (80 ± 40)	0.24–2.90 (0.81 ± 0.56)	BDL–6.18 (2.32 ± 1.42)	BDL–35 (8 ± 9)	Khound et al. (2019)
Jia Bharali, north Brahmaputra plain, India, Wet season	BDL–130 (40 ± 40)	10–140 (70 ± 30)	BDL–60 (20 ± 20)	BDL–140 (60 ± 40)	BDL–100 (40 ± 30)	BDL–70 (20 ± 20)	BDL–130 (40 ± 30)	0.16–1.11 (0.43 ± 0.29)	BDL–2.95 (0.78 ± 0.79)	BDL–12 (3 ± 3)	Khound et al. (2019)
Kali River, monsoon	0.6–3.30 (1.3 ± 1.90)	NS	1–1 (1 ± 0.0)	0.6–3 (1.3 ± 1.14)	NS	0.30–9.40 (6.23 ± 4.12)	2–11.10 (5.66 ± 3.83)	63–104 (81.75 ± 20.84)	NS	0.10–0.10 (0.10 ± 0.0)	Manjunatha et al. (2001)
Kali River, post-monsoon	0.2–1.50 (0.78 ± 0.63)	NS	0.1–0.6 (0.4 ± 0.22)	0.10–0.10 (0.10 ± 0.0)	NS	7.8–26.8 (13.83 ± 9.04)	1.10–3.10 (2.23 ± 0.85)	13–27 (20.25 ± 6.08)	NS	0.10–0.20 (0.15 ± 0.07)	Manjunatha et al. (2001)

Table 2.2 Concentration of heavy metals in soil.

		Metals (range [mean ± standard deviation])									
Location	Type of soil	Cu ($mg\,kg^{-1}$)	Cr ($mg\,kg^{-1}$)	Co ($mg\,kg^{-1}$)	Pb ($mg\,kg^{-1}$)	Ni ($mg\,kg^{-1}$)	Zn ($mg\,kg^{-1}$)	Mn ($mg\,kg^{-1}$)	Fe ($mg\,kg^{-1}$)	Cd ($mg\,kg^{-1}$)	Reference
Beijing, China	Urban soil	13.4–207.9 (31.7 ± 24.9)	44.6–121.5 (61.0 ± 9.82)	NS	4.02–174.4 (23.3 ± 15.5)	17.8–39.0 (24 ± 3.19)	29.4–322.3 (92.9 ± 86.6)	NS	NS	0.003–0.98 (0.13 ± 0.09)	Wang et al. (2012)
Eastern China	Farmland soil	34.55 ± 26.82	14.60 ± 6.34	NS	55.62 ± 13.17	10.52 ± 3.35	103.61 ± 48.01	NS	NS	0.37 ± 0.19	Li et al. (2019)
Eastern China	Orchard soil	17.01 ± 11.57	13.80 ± 6.87	NS	39.73 ± 8.30	10.72 ± 3.91	61.10 ± 31.16	NS	NS	0.28 ± 0.23	Li et al. (2019)
Eastern China	Forestland soil	12.72 ± 8.62	14.89 ± 11.25	NS	50.75 ± 24.00	9.97 ± 8.05	62.26 ± 29.43	NS	NS	0.22 ± 0.14	Li et al. (2019)
Dhaka city, Bangladesh	Playground soil	46–161 (82 ± 53)	2.4–39 (20 ± 17)	NS	13–100 (49 ± 38)	89–143 (117 ± 28)	NS	NS	NS	2–4.2 (3.5 ± 1)	Islam et al. (2015)
Dhaka city, Bangladesh	Park soil	10–92 (46 ± 32)	6.3–58 (37 ± 24)	NS	21–170 (92 ± 56)	8.3–37 (25 ± 11)	NS	NS	NS	2–14 (6.9 ± 4.8)	Islam et al. (2015)
Benin city, Nigeria	Residential soil	9.7 ± 6.7	13.5 ± 15.4	NS	51.3 ± 76.6	2 ± 1.6	83.7 ± 97.5	42.3 ± 25.9	5088.6 ± 2934	0.2 ± 0.1	Iwegbue (2014)
Benin city, Nigeria	Peri-urban farmland soil	31.7 ± 31.3	27.1 ± 12.1	NS	43.2 ± 34.5	7.2 ± 4.9	800 ± 1197.7	100.7 ± 94.3	9697.5 ± 5806.7	0.9 ± 0.7	Iwegbue (2014)
Benin city, Nigeria	Playground soil	12.7 ± 8.3	24.1 ± 8.4	NS	23.3 ± 149	3.8 ± 2.6	94.4 ± 43.2	85 ± 26.4	7839.7 ± 2711.9	0.4 ± 0.2	Iwegbue (2014)
Tarkwa, Ghana	Agriculture soil	2.2–16 (6.2 ± 3.1)	8–77 (21 ± 15)	0.37–9.2 (1.8 ± 1.8)	1.5–27 (7.2 ± 6.2)	1–28 (3.7 ± 5.5)	9.7–118 (39 ± 27)	NS	NS	0.01–0.43 (0.052 ± 0.067)	Bortey-Sam et al. (2015)
Patna, Bihar	Agriculture soil	31.22–119.61 (49.60)			14.53–57.42 (24.45)		76.11–347.46 (136.95)	298.92–642.48 (417.73)	NS		
Ghaziabad, Uttar Pradesh	Agriculture soil	4.05–67.6 (17.8 ± 12.8)	3.6–111.7 (27.3 ± 20.3)	NS	5.6–69.6 (29.8 ± 15.1)	12.8–165 (53.8 ± 36.2)	20.6–282 (87 ± 50)	41.6–677.5 (306.6 ± 137.3)	10865–24873 (16984 ± 3019.3)	0.11–1.1 (0.43 ± 0.22)	Chabukdhara et al. (2016)

NS, heavy metal is not studied.

Research findings suggest that soil used for various purposes is contaminated with different types of heavy metals. The heavy metals contamination of soil poses health hazards to humans and other living organisms. Figure 2.2 shows the various sources of heavy metal in soil system.

2.4 Heavy Metals in Biota

Plants accumulate the metals via two pathways, namely root and foliar uptake. The root absorbs the metals on its surface from the soil solution, followed by their binding to polysaccharides of rhizodermal surface or carboxyl groups of mucilage uronic acid. The metal then penetrates the root passively and diffuses through translocating streams. Once the metal is absorbed by the roots, it is translocated to various parts of vegetable via xylem loading, which occurs through symplastic or apoplastic movement (Farooq et al. 2008; Mandal and Kaur 2019). Transpiration governs the metal transport with different chelates. The major portion of metals is sequestered by root cells, and limited part is translocated to aerial parts except in hyperaccumulators, microbial-assisted, and chelate-assisted translocation. This restricted translocation is because of physical barrier (the presence of casparian strip in root cell), precipitation of metal as insoluble metal salts in intercellular space, and sequestration in vacuole of cortical or rhizodermal cells. In cases of foliar uptake, dust particles encompassing the metals are adsorbed to leaf surface because of wax cover and leaf hairs based on their size. After the adsorption on leaf surface, foliar uptake occurs via two pathways based on phobicity of pollutants. The lipophilic element diffuses through cuticle, while hydrophilic element passes through aqueous pores of stomata and cuticle. The translocation of metals inside the plant from foliar uptake occurs through phloem stream like photosynthates. Translocation process includes penetration through the cuticle layer, followed by desorption in the apoplast and finally binding with subjacent mesophyll cells. Majority of absorbed metals are stored in parenchyma cells of leaf, and only small portion is translocated to root tissues (Farooq et al. 2008; Ramteke et al. 2016).

Factors affecting the accumulation of trace metals in plants are categorized into soil-related factors and plant-related factors. The soil-related factors controlling the bioavailability of metals in soil include soil pH, redox potential, organic matter, soil temperature, soil texture, cation exchange capacity, and interaction between the metals and microorganism of soil. The plant-related factors include physiology (transpiration rate), morphology (leaf and root characteristics), and anatomy of plants (Farooq et al. 2008; Gupta et al. 2019). Moreover, high metal concentration leads to the induced oxidative damage, anomalies in genetic behavior, modification of signaling and biosynthetic pathways, and abnormal reproductive response in the biota (Srivastava et al. 2017).

Vegetables are important part of diet, and they easily take up the heavy metals from the contaminated soil. Heavy metal concentration in vegetables is studied across the globe. Some of the studies are presented in Table 2.3. Research findings suggest that vegetables take up heavy metals and transfer them to humans via food chain.

Table 2.3 Concentration of heavy metals in some of common vegetables.

Location	Vegetable type	Concentration of heavy metals (range [mean ± standard deviation])									Reference
		Cu (mg kg^{-1})	Cr (mg kg^{-1})	Co (mg kg^{-1})	Pb (mg kg^{-1})	Ni (mg kg^{-1})	Zn (mg kg^{-1})	Mn (mg kg^{-1})	Fe (mg kg^{-1})	Cd (mg kg^{-1})	
Katlang, Pakistan	Radish leaf	29.533 ±0.004	32.233 ±0.039	4.133 ±0.006	38.933 ±0.005	11.967 ±0.001	NS	NS	NS	NS	Idrees et al. (2017)
Faisalabad, Pakistan	Radish leaf	0.27 ± 0.02	0.29 ± 0.02	NS	2.03 ± 0.10	NS	1.89 ± 0.05	NS	NS	0.033 ±0.001	Farooq et al. (2008)
Faisalabad, Pakistan	Radish root	0.46 ± 0.03	0.49 ± 0.02	NS	2.25 ± 0.09	NS	0.81 ± 0.04	NS	NS	0.051 ±0.003	Farooq et al. (2008)
Jinxi River areas, China	Radish root	1.8–31.8 (8.1 ± 8.8)	0.3–13.8 (3.4 ± 3.6)	0.1–2.1 (0.9 ± 0.6)	0.6–3.8 (1.5 ± 0.8)	0.7–6.6 (3.7 ± 1.9)	28.6–204 (83.8 ± 58.5)	48.4–189.1 (102.9 ±43.3)	100–500 (200 ± 100)	0.04–0.4 (0.2 ± 0.1)	Abdelhafez and Li (2015)
Urga area of Korba, India	Brinjal	29 ± 11	1.0 ± 0.4		0.7 ± 0.03		24 ± 7	17 ± 9	192 ± 27	0.25 ± 0.07	Ramteke et al. (2016)
Mandi Gobindgarh, India	Brinjal	20.30 ±2.51	NS	NS	NS	Not detected	25.30 ± 4.16	22.00 ±4.35	83.30 ±10.69	NS	Mandal and Kaur (2019)
Allahabad, Uttar Pradesh	Spinach	15.98–202.6 (29.20)	NS	NS	12.21–38.22 (16.20)	56.98–506.2 (66.55)	31.21–92.49 (36.50)	NS	54.6–92.99 (69.98)	11.2–67.66 (15.24)	Yadav et al. (2013)
Mandigobindgarh, India	Spinach	121.50 ±68.73	NS	NS	NS	6.70 ± 4.71	92.20 ±31.42	97.00 ±33.39	452.70 ±116.10	NS	Mandal and Kaur (2019)

NS, heavy metal is not studied.

2.5 Heavy Metals in Air

Major sources of heavy metals in atmosphere are dust and gas, which originate from transport, metallurgy, energy, and manufacture of construction materials. An estimate has been carried out for the global emissions of metals from key anthropogenic sources into the atmosphere in the mid-1990s by Pacyna and Pacyna 2001. Stationary fossil fuel combustion is the major source of Cr (69%), Mn (85%), Sb (47%), Hg (66%), Se (89%), Sn (89%), and Tl (almost 100%) (Pacyna et al. 2007). The major source of Ni (90%) and V (almost 100%) is oil combustion. Major source of atmospheric Pb emissions is the combustion of leaded, low-leaded, and unleaded gasoline, which contributes about 74% to the emission of this metal by anthropogenic sources (Olendrzyński et al. 1996; Pacyna and Pacyna 2001; Pacyna et al. 2007). Non-ferrous metal manufacture is the third major source of metals. It is the largest source of atmospheric As (69%), Cd (73%), Cu (70%), In (100%), and Zn (72%) (Pacyna and Pacyna 2001).

Heavy metals are present in the air in the form of aerosol (Georgescu et al. 2011). These aerosols are then deposited on the soil through natural sedimentation and precipitation processes. Atmospheric deposition is a significant path by which metals enter the environment. The heavy metal's behavior during the atmospheric transport depends on their particle size-distribution and meteorological conditions, such as wind speed, and their physical–chemical properties (Prospero 1999). Atmospheric transport and leaching are the two important pathways for heavy metal transport. The heavy metals emitted into the air are carried to long distance before their final deposition in distant aquatic and terrestrial environments. The leaching process of heavy metals occurs over a shorter distance and depends on the metal's solubility in water and the amount of water available for their transport. Wet deposition (rainfall) and dry deposition (sedimentation) are the two ways in which heavy metals leave the atmosphere. Removal of heavy metals by wet deposition depends on the metal's solubility in the water, except for mercury, as it re-vaporized after it has been deposited on land (Pacyna and Pacyna 2001; Gharaibeh et al. 2010; MohseniBandpi et al. 2018). Table 2.4 represents heavy metal contamination in ambient air, and Figure 2.3 shows the source of metals in the atmosphere. Various researchers have studied the heavy metal contamination in air. Research studies suggest that ambient air is contaminated with heavy metals. Heavy metals are present in the air to varying degrees. There are several health impacts of elevated concentrations of metals in the air including chronic obstructive pulmonary disease (COPD), pneumonia, flu, bronchitis, lung cancer, and heart diseases (Manisalidis et al. 2020).

2.6 Conclusion

Heavy metals are present in different environmental media such as air, water, soil, and biota. Heavy metals are transferred between the media and also reach to humans. Heavy metals are needed in very small concentrations by the living organism as nutrients, but their consequences can be lethal in higher concentrations. The heavy metals toxic to living organisms and their continuous exposure via deferent environmental media lead to detrimental health hazards. In the present study, the various pathways of heavy metal intrusion

Table 2.4 Heavy metal concentration in ambient air.

	Metals (range [mean ± standard deviation])								
Location	**Cu (ng m^{-3})**	**Cr (ng m^{-3})**	**Cd (ng m^{-3})**	**Pb (ng m^{-3})**	**Ni (ng m^{-3})**	**Zn (ng m^{-3})**	**Mn (ng m^{-3})**	**Fe (ng m^{-3})**	**Reference**
Urban area, Irbid city, Jordan	117 (8–316)	NS	NS	111 (15–121)	8 (2–17)	283 (200–400)	23 (8–39)	986 (300–2136)	Gharaibeh et al. (2010)
Rural area, Irbid city, Jordan	55 (4–145)	NS	NS	21 (4–40)	5 (1–11)	213 (118–239)	17 (4–32)	682 (208–1250)	Gharaibeh et al. (2010)
Urban area, Tehran city, Iran	280	70	48	120	30	90	90	1790	MohseniBandpi et al. (2018)
Traffic area, Tehran city, Iran	150	80	50	160	40	170	110	1440	MohseniBandpi et al. (2018)
Suburban area, Tehran city, Iran	39	20	2	130	14	18	120	1060	MohseniBandpi et al. (2018)
Rohtak, India	2–396	NS	ND –15	6–191		3–582	2–198	NS	Dalal et al. (2013)
Dongargarh, Central India	141.8	87.8	81.2	289.0	120.7	320.6	NS	513.9	Ambade (2014)

ND, not detected; NS, not studied.

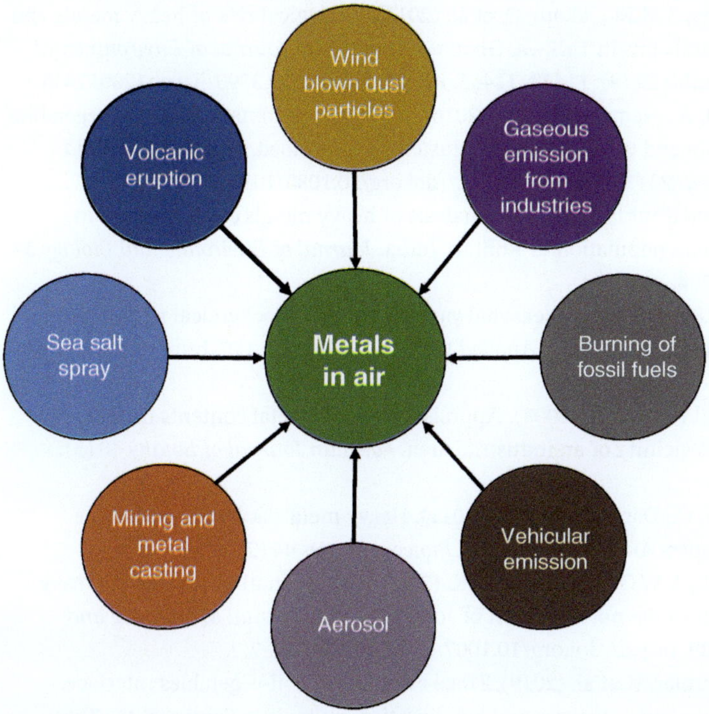

Figure 2.3 Heavy metal sources in the air.

into the various environmental spheres have been discussed. Moreover, an elaborate discussion has been done on the transport mechanisms between different earth spheres. The current study also focuses on heavy metal contamination from various anthropogenic sources such as industrial activities, mining, and agricultural activities that contribute to heavy metal pollution.

References

Abdelhafez, A.A. and Li, J. (2015). Environmental monitoring of heavy metal status and human health risk assessment in the agricultural soils of the Jinxi River Area, China. *Human and Ecological Risk Assessment* 21 (4): 952–971. https://doi.org/10.1080/10807039.2014.947851.

Acosta, J.A., Martínez-Martínez, S., Faz, A., and Arocena, J. (2011). Accumulations of major and trace elements in particle size fractions of soils on eight different parent materials. *Geoderma* 161 (1–2): 30–42.

Ambade, B. (2014). Seasonal variation and sources of heavy metals in hilltop of Dongargarh, Central India. *Urban Climate* 9: 155–165. ISSN 2212-0955, https://doi.org/10.1016/j.uclim.2014.08.001.

Ayangbenro, A.S. and Babalola, O.O. (2017). A new strategy for heavy metal polluted environments: a review of microbial biosorbents. *International Journal of Environmental Research and Public Health* 14 (1): 94.

Bortey-Sam, N., Nakayama, S.M.M., Akoto, O. et al. (2015). Ecological risk of heavy metals and a metalloid in agricultural soils in Tarkwa, Ghana. *International Journal of Environmental Research and Public Health* 12 (9): 11448–11465. https://doi.org/10.3390/ijerph120911448.

Chabukdhara, M., Munjal, A., Nema, A.K. et al. (2016). Heavy metal contamination in vegetables grown around peri-urban and urban-industrial clusters in Ghaziabad, India. *Human and Ecological Risk Assessment* 22 (3): 736–752. https://doi.org/10.1080/10807039.2015.1105723.

Dalal, P., Chaudhry, D., and Shukla, V. (2013). Analysis of heavy metals concentration in ambient air and in human population of Rohtak, India. *Journal of Environmental Biology* 34: 945–949. ISSN: 0254-8704.

Eliku, T. and Leta, S. (2018). Spatial and seasonal variation in physicochemical parameters and heavy metals in Awash River, Ethiopia. *Applied Water Science* 8 (6): 177. https://doi.org/10.1007/s13201-018-0803-x.

Farooq, M., Anwar, F., and Rashid, U. (2008). Appraisal of heavy metal contents in different vegetables grown in the vicinity of an industrial area. *Pakistan Journal of Botany* 40 (5): 2099–2106.

Georgescu, B., Georgescu, C., Dărăban, S. et al. (2011). Heavy metals acting as endocrine disrupters. *Scientific Papers Animal Science and Biotechnologies* 44 (2): 89–93.

Gharaibeh, A.A., El-Rjoob, A.W.O., and Harb, M.K. (2010). Determination of selected heavy metals in air samples from the northern part of Jordan. *Environmental Monitoring and Assessment* 160: 425–429. https://doi.org/10.1007/s10661-008-0706-7.

Gupta, N., Yadav, K.K., Kumar, V. et al. (2019). Trace elements in soil-vegetables interface: translocation, bioaccumulation, toxicity and amelioration—a review. *Science of the Total Environment* 651: 2927–2942. https://doi.org/10.1016/j.scitotenv.2018.10.047.

Idrees, M., Rehman, Z.U., Bibi, S. et al. (2017). Multivariate analysis of the heavy metal concentrations in the vegetable and soil samples-a case study from district Charsadda and district Mardan. *Pakistan Journal of Botany* 49 (3): 1167–1171.

Islam, M.S., Ahmed, M.K., Habibullah-Al-Mamun, M., and Hoque, M.F. (2015). Preliminary assessment of heavy metal contamination in surface sediments from a river in Bangladesh. *Environmental Earth Sciences* 73 (4): 1837–1848. https://doi.org/10.1007/s12665-014-3538-5.

Iwegbue, C.M.A. (2014). Impact of land use types on the concentrations of metals in soils of urban environment in Nigeria. *Environmental Earth Sciences* 72 (11): 4567–4585. https://doi.org/10.1007/s12665-014-3355-x.

Kashyap, R., Verma, K.S., Uniyal, S.K., and Bhardwaj, S.K. (2018). Geospatial distribution of metal(loid)s and human health risk assessment due to intake of contaminated groundwater around an industrial hub of northern India. *Environmental Monitoring and Assessment* 190: 136. https://doi.org/10.1007/s10661-018.6525-6.

Khound, N.J., Phukon, P., and Bhattacharyya, K.G. (2019). Toxic trace metals in the surface water sources of Jia–Bharali river basin, North Brahmaputra Plain, India—a hydrochemical elucidation. *Water Resources* 46 (1): 117–127. https://doi.org/10.1134/S009780781901010X.

Li, X., Chi, W., Tian, H. et al. (2019). Probabilistic ecological risk assessment of heavy metals in western Laizhou Bay, Shandong Province, China. *PLoS One* 14 (3): e0213011. https://doi.org/10.1371/journal.pone.0213011.

Mandal, R. and Kaur, S. (2019). Impact of environmental pollution on trace elements in vegetables and associated potential risk to human health in industrial town Mandigobindgarh (India). *Chemosphere* 219: 574–587. https://doi.org/10.1016/j.chemosphere.2018.12.034.

Manisalidis, I., Stavropoulou, E., Stavropoulos, A., and Bezirtzoglou, E. (2020). Environmental and health impacts of air pollution: a review. *Frontiers in Public Health* 8: 14.

Manjunatha, B., Balakrishna, K., Shankar, R., and Mahalingam, T.R. (2001). Geochemistry and assessment of metal pollution in soils and river components of a monsoon-dominated environment near Karwar, southeast coast of India. *Environmental Geology* 40: 1462–1470. https://doi.org/10.1007/s002540100342.

Mao, G., Zhao, Y., Zhang, F. et al. (2019). Spatiotemporal variability of heavy metals and identification of potential source tracers in the surface water of the Lhasa River basin. *Environmental Science and Pollution Research* 26 (8): 7442–7452. https://doi.org/10.1007/s11356-019-04188-0.

McLaughlin, M.J., Smolders, E., Degryse, F., and Rietra, R. (2011). Chapter 8: Uptake of metals from soil to vegetables. In: *Dealing with Contaminated Sites* (ed. F.A. Swartjes), 325–362. Springer 10.1007/978-90-481-9757-6_8.

MohseniBandpi, A., Eslami, A., Ghaderpoori, M. et al. (2018). Health risk assessment of heavy metals on PM2.5 in Tehran air, Iran. *Data in Brief* 17: 347–355. https://doi.org/10.1016/j.dib.2018.01.018.

Nyantakyi, A.J., Akoto, O., and Fei-Baffoe, B. (2019). Seasonal variations in heavy metals in water and sediment samples from River Tano in the Bono, Bono East, and Ahafo Regions, Ghana. *Environmental Monitoring and Assessment* 191 (9): 570. https://doi.org/10.1007/s10661-019-7744-1.

Olendrzyński, K., Anderberg, S., Stigliani, W. et al. (1996). Atmospheric emissions and depositions of cadmium, lead, and zinc in Europe during the period 1955–1987. *Environmental Reviews* 4 (4): 300–320.

Pacyna, J.M. and Pacyna, E.G. (2001, 2001). An assessment of global and regional emissions of trace metals to the atmosphere from anthropogenic sources worldwide. *Environmental Reviews* 9: 269–298. https://doi.org/10.1139/er-9-4-269.

Pacyna, E.G., Pacyna, J.M., Fudala, J. et al. (2007). Current and future emissions of selected heavy metals to the atmosphere from anthropogenic sources in Europe. *Atmospheric Environment* 41 (38): 8557–8566.

Prospero, J.M. (1999). Long-range transport of mineral dust in the global atmosphere: impact of African dust on the environment of the southeastern United States. *Proceedings of the National Academy of Sciences* 96 (7): 3396–3403.

Ramteke, S., Sahu, B.L., Dahariya, N.S. et al. (2016). Heavy metal contamination of vegetables. *Journal of Environmental Protection* 7 (7): 996–1004. https://doi.org/10.4236/jep.2016.77088.

Sandeep, G.S., Vijayalatha, K.R., and Anitha, T. (2019). Heavy metals and its impact in vegetable crops. *International Journal of Chemical Studies* 7 (1): 1612–1621. https://www.researchgate.net/publication/341508109.

Srivastava, V., Sarkar, A., Singh, S. et al. (2017). Agroecological responses of heavy metal pollution with special emphasis on soil health and plant performances. *Frontiers in Environmental Science* 5: 64.

Su, C., Jiang, L.Q., and Zhang, W.J. (2014). A review on heavy metal contamination in the soil worldwide: situation, impact and remediation techniques. *Environmental Skeptics and Critics* 3 (2): 24–38.

Wang, M., Markert, B., Chen, W. et al. (2012). Identification of heavy metal pollutants using multivariate analysis and effects of land uses on their accumulation in urban soils in Beijing,

China. *Environmental Monitoring and Assessment* 184 (10): 5889–5897. https://doi.org/10.1007/s10661-011-2388-9.

Wongsasuluk, P., Chotpantarat, S., Siriwong, W., and Robson, M. (2014). Heavy metal contamination and human health risk assessment in drinking water from shallow groundwater wells in an agricultural area in Ubon Ratchathani province, Thailand. *Environmental Geochemistry and Health* 36 (1): 169–182. https://doi.org/10.1007/s10653-013-9537-8.

Yadav, A., Yadav, K., and Shukla, D.N. (2013). Investigation of heavy metal status in soil and vegetables grown in urban area of Allahabad, Uttar Pradesh, India. *International Journal of Scientific and Research Publications* 3 (9): 1–7.

3

A Brief Study of the Effects of Heavy Metals and Metalloids on Food Crops

Ngangbam R. Devi

Department of Chemistry, Indian Institute of Technology Guwahati, Guwahati, Assam, India

3.1 Introduction

A heavy metal is defined as metallic or metalloid element that has a comparatively higher atomic density (4–5 times heavier density than water) and is toxic or poisonous even in small quantities (Hawkes 1997). Soil is the primary sustenance of food crops/vegetables. Plants often take up their necessary micronutrients from the surrounding soil system and some plant species thrive in a particular metal-enriched area; for example, the plant community of serpentine prefers soils enriched with Ni, Cr, and Co; calamine on zinc (Zn-) and cadmium (Cd-) rich soils and uraniferous on uranium (U-) rich soils (Raskin et al. 1994). Contamination and consumption of intoxicated food crops by heavy metals via polluted water and soil are the major concerns and crises faced by today's modern world. Heavy metals can be broadly distinguished as essential and non-essential elements, with metals like Fe, Zn, and Cu classified as essential elements and required by food crops/plants in lower concentrations (Singh and Kalamdhad 2011). On the other hand, metals such as Cr, Pb, As, Cd, and Hg are classified under health-risk heavy metals. Consumption even in lower concentration can cause serious concern in plants and can be fatal in higher animals and humans (Kaur et al. 2019).

Over a period of time, these heavy metals can bio-accumulate and bio-magnify in the ecosystem inside plants and animals. In a human body, these metals cause cancer, mental retardation, CNS breakdown, damage to renal and hepatic systems, etc. (Misra and Mani 1991). Thus, soil-food crops/vegetables act as a linkage between abiotic-biotic systems in the environment. Soil polluted with heavy metals has severe effects on soil–plants/food crops. Heavy metal pollution reduces soil fertility and affects the soil-microbial population as well as the nitrogen transformation cycle, which, in turn, causes modification in plant genetics, anomalous biosynthetic pathways, along with modification in reproductive outputs (Srivastava et al. 2017). Heavy metals in water and food can be analyzed by using atomic absorption spectrophotometry. Sources of heavy metals found in soil and plants can be classified broadly as natural sources, industrial sources, agricultural sources, and

Remediation of Heavy Metals: Sustainable Technologies and Recent Advances, First Edition. Edited by Rangabhashiyam Selvasembian, Binota Thokchom, Pardeep Singh, Ali H. Jawad, and Willis Gwenzi.

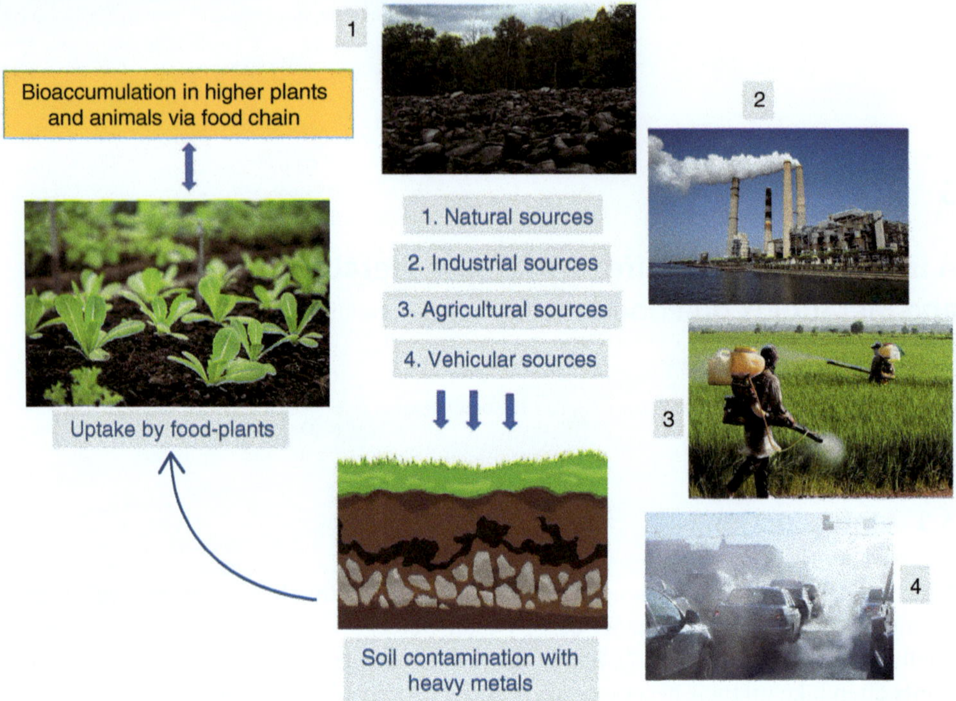

Figure 3.1 An illustration showing different sources of heavy metals that lead to contamination of soil and food crops. *Source:* Eric Dale Creative/Adobe Stock Photos, landbysea/Getty Images, Tan Kian Khoon / Adobe Stock, Tan Kian Khoon / Adobe Stock, Nady/Adobe Stock Photos.

domestic/vehicular sources (Figure 3.1). All such absorbed and bioaccumulated minerals and heavy metals found in the soil directly enter into plants and vegetables through roots which further modify the health benefits of vegetables/food crops, either can upgrade or downgrade their nutritive values. Latif et al. analyzed and reported that out of vegetables such as pumpkin, spinach, gourd, and brinjal, green leafy vegetable spinach has a higher absorption capacity of heavy metals like Mn, Cr, and Fe due to the availability of more absorption surface area. It can be said that different food crops have different capacities to accumulate heavy metals (Latif et al. 2018). Some of the food crops affected by heavy metal pollution are discussed in Table 3.1.

3.2 Sources of Heavy Metals in Soils and Food Crops

Sources of heavy metal pollution in soils and plants/food crops can be broadly classified into five main types, natural, industrial, agricultural, domestic, and vehicular sources.

- Natural sources: Heavy metals are one of the major components of the earth's crust and the natural breakdown of rocks or the earth's crust led to the release of such metals, causing persistent contamination of soil. Volcanic eruptions release various dust particles along with toxic heavy metals such as Pb, Ni, Cu, and Hg, thus polluting air, water, and

Table 3.1 Heavy metal pollution on some food crops from diverse sources (Arora et al. 2008; Chary et al. 2008; Zhuang et al. 2009; Luo et al. 2011; Amin et al. 2013; Khan 2013; Osma et al. 2013; Chang et al. 2014; Chauhan 2014; Parvin et al. 2014; Shaheen et al. 2016; Sharma et al. 2016; Setia et al. 2021).

S. No.	Heavy metals	Remarks	Area: food crops	Reference
1.	Cr, Pb, Zn, and Ni	Vegetables grown in inadequately treated wastewater	India: spinach and amaranthus	Chary et al. (2008)
2.	Cu and Zn	Long-term irrigation with wastewater resulted in the intoxication of the food crops by heavy metals	India: mint, spinach, and carrot	Arora et al. (2008)
3.	Cu, Zn, Cr, Fe, and Pb	Vegetables irrigated with wastewater were prone to contamination with diverse heavy metals	Mardan, Pakistan: onion, garlic, tomato, and eggplant	Amin et al. (2013)
4.	Cd, Ni, and Pb	Vegetables grown in industrial areas and along roadsides are highly polluted	Istanbul, Turkey: capsicum, common bean, and eggplant	Osma et al. (2013)
5.	Cd, Pb, and Zn	Uncontrolled e-waste-processing plants posing serious threats to local soils and vegetables	China: green peas., rapeseed, lettuce, radish, taro, shallot, cabbage, lettuce, and carrot	Luo et al. (2011)
6.	Fe, Zn, Cu, Pb, Cd, Mn, and Cr	Cement factories in Rewa, India affect vegetables grown in adjoining areas	Rewa, India: spinach, cabbage, wild cabbage, eggplant, okra, tomato, and radish	Chauhan (2014)
7.	Cd and Pb	Due to the dumping of untreated industrial waste, high levels of cadmium and copper were reported in various food crops	Amritsar, Punjab, India: coriander, fenugreek, spearmint, spinach, eggplant, bottle gourd, okra, capsicum, radish, mustard, onion, and garlic	Sharma et al. (2016)
8.	Hg, As, Cr, Pb, and Cd	Uses of untreated sewage affluents and agricultural fertilization were the major pollutants in green leafy vegetables	Pearl River Delta, South China: leafy vegetables such as cabbage, Chinese cabbage, lettuce, and pakchoi were mainly contaminated	Chang et al. (2014)
9.	As, Cd, Ni, Pb, and Cr	Higher content of lead in mango was found at the production site	Bangladesh: *Lycopersicum esculentum* Mill.	Shaheen et al. (2016)
10.	Cu, Pb, and Cr	Concentration of Pd found near the industrial area of Chittagong in leafy vegetables was a health risk	Chittagong, Bangladesh: green arum leaves, water spinach, jute leaves, bottle gourd, and sweet gourd	Parvin et al. (2014)
11.	Cr and Pb	Vegetables grown in heavy metal-polluted groundwater were affected	Pakistan: vegetables such as wheat, tomato, radish, spinach, eggplant, and carrot are highly affected	Khan (2013)

(Continued)

Table 3.1 (Continued)

S. No.	Heavy metals	Remarks	Area: food crops	Reference
12.	Cd, Cr, and Ni	Irrigation with sewage water from Manka canal containing effluents from domestic and industrial wastewater affected the adjoining soil system	Dera Ghazi Khan, Pakistan: pumpkin, ridged gourd, spinach, and brinjal	Latif et al. (2018)
13.	Cu, Zn, Pb, and Cd	Health of the local community was affected due to the consumption of rice and vegetables found near the Dabaoshan mine	Dabaoshan mine, South China: mostly rice and green leafy vegetables such as water spinach, celery, lettuce, etc., are contaminated	Zhuang et al. (2009)
14.	Pb, Cd, and Cr	Wheat and paddy grown in the flood plains of Sutlej River possessed carcinogenic Cd in highest concentration posing serious health risks	Sutlej River, India: wheat (*Triticum aestivum*) and rice (*Oryza sativa*)	Setia et al. (2021)
15.	Cu, Ni, and Pb	Semi-treated sewage and industrial effluents caused serious threats to green leafy vegetables	India: leafy and root vegetables along with grains are affected	Rattan et al. (2005)
16.	Cd and Pb	Besides contamination from soil, vegetables grown in and around market and production sites were also contaminated with heavy metals	Varanasi, India: beet *vulgaris*, and wild cabbage	Sharma et al. (2009)

soil (Seaward and Richardson 1989). Similarly, during forest fires, toxic chemicals such as Cd, Pb, and Zn are released, contributing to heavy metal pollution. Wildfire also emits volatile heavy metals like Se (selenium) and Hg, which are part of carbonaceous matter and adds to environmental degradation (Ross 1994; Nagajyoti et al. 2010).

- Industrial sources: Apart from natural sources, industrial source is one of the primary contaminants of soil pollution. Industrial wastes released from thermal and heavy power plants, coal and ore mining industries, chemical factories, tanneries, and oil refinery industries constitute some of the qualitative and quantitative sources of toxic chemicals. Mining of coal and petroleum, and gold are major contributors to As and Cd pollution, which can further contaminate soil and adjoining vegetation (Nagajyoti et al. 2010). Heavy metals like Se, Cd, Ni, and Cs are emitted by coal and petroleum industries, and nuclear power plants (Ahmed and Ahmaruzzaman 2016; Zhu et al. 2016). Another major contributor of toxic chemical pollution to the soil includes processing plants of plastics and textiles, along with electronics and wood preservation industries. Recently, Liu et al. revealed that Pb-acid battery production factories also emit metal particulates and can deposit in soils affecting the surrounding ecosystem (Liu et al. 2014). Coal and ore mining led to the increased bioaccumulation of Cr and Fe into the soil and thus provided stress to plants affecting the yield of crops and germination of seeds (Sethy and Ghosh 2013).

- Agricultural sources: Inorganic fertilizers like phosphate, nitrate, and potash, which are used for agricultural purposes, contain various heavy metals such as Cr, Cd, Zn, Ni, and Pb. Agricultural tools like herbicides, pesticides, and solid waste manure also contain toxic metals and contaminate soil or vegetation resulting in health risks for various animals and mankind. As a result, there are several reports on public health repercussions from using untreated or semi-treated sewage effluents (Sipter et al. 2008; Tóth et al. 2016). Untreated sewage sludge or effluents contain toxic metals like Cr, Pb, Ni, and Cd and food crops irrigated with such water absorbed these metals and showed visible side effects. The utilization of sewage sludge is one of the major contributors to heavy metal pollution in the soil (Kumar et al. 2016; Sharma et al. 2017).
- Vehicular sources: Transportation and/or traffic emissions are also the main global contributors to environmental pollution with heavy metals. Vehicular sources include emissions from diesel, gasoline, and petrol engines, and carbonaceous smoke, along with toxic metals such as Cu, Pb, Mg, and Cd (Nagajyoti et al. 2010). Metals like Cr, Ni, and Zn are released from aerosol emissions; wear and tear of vehicular tires also contribute to Zn, Cd, and microplastic dust particles' pollution in the environment (Chen et al. 2013; Gope et al. 2017).
- Domestic sources: Domestic sources include mainly effluents consisting of wastewater from domestic use and wastewater containing minerals from domestically treated reverse osmosis as well as untreated or semi-treated solid waste that is directly dumped into landfills.

3.3 Impacts on Soil–Plants/Food Crops

BAF, or bio-accumulation factor, is an index used to quantify the amount of metal concentration in soils. The higher the value of BAF, the greater the potential of the plant to accumulate metal. Their concentrations in soils and crops were taken as dry weight. The BAF is calculated as follows:

$$\text{BAF} = [\text{Concentration}]_{\text{plant}} / [\text{Concentration}]_{\text{soil}} \tag{3.1}$$

where $[\text{Concentration}]_{\text{plant}}$, $[\text{Concentration}]_{\text{soil}}$ represent the heavy metal concentration in edible parts of vegetables and soils, respectively (Ghosh and Singh 2005).

Like all other living organisms, plants are affected heavily by both deficiently and abundantly supplemented metal ions. For example, heavy metals like As, Hg, Cd, Pb, or Cr are not essential in plant growth. However, others like Cu, Fe, Mn, Ni, and Zn are essential metal elements for the normal functioning and growth of plants under optimal concentrations but can cause poisoning when exceeded the permissible limit (Table 3.2) (Singh and Kalamdhad 2011; Marschner and Rengel 2012; Gall et al. 2015).

Table 3.2 Permissible limit of some toxic heavy metals in soils and food crops.

Elements	Target value of soil (mg kg^{-1})[a]	Permissible value in food crops
Cr	100.0	1.30
Cd	0.8	0.02
Cu	36.0	10.0
Pb	85.0	02.0
Zn	50.0	0.60
Ni	35.0	10.0
As	20.0[b]	1.40
Hg	10.0–40.0	—

[a] Maximum levels of metals in unpolluted soils: Denneman and Robberse 1990, Ministry of Housing, Netherlands 1994.
[b] *Source:* European Union.

3.3.1 Metal Ions Transportation in Plants

Ionic forms, which are more soluble, are generally taken up by the plants from the soil, but depending upon the type of metals and different vegetables, absorption capacity is different (McLaughlin et al. 2011). Plant roots, stems, and leaves can absorb the metals, but out of these, roots play the most important role as they are directly attached to the soil (Krzesłowska 2011). The mechanism of uptake can be of two types: (i) foliar uptake of heavy metals involving the phloem vascular system as reported by Schreck et al. (2012, 2013) and (ii) transportation of heavy metals from the roots via the xylem, also called the root uptake mechanism (Shahid et al. 2017). Several literatures are reported based on the effects of heavy metals on soils and plants and some of them are discussed below briefly.

Due to the increase in population growth, sewage and industrial effluents are increasingly utilized for irrigation purposes in mostly semi-urban agricultural fields. Rattan et al. reported that soils consuming sewage effluents for irrigation showed a significant increase in the concentration of Fe, Ni, and Pb in comparison to the adjacent tube well-irrigated soil. Sethy et al. concluded that heavy metal contamination in plants overall affected in the germination of seed, resulting in stunted growth of root and shoot, and loss of nutrients, thus decreasing yields (Sethy and Ghosh 2013). Another literature revealed that plants can also acquire heavy metals other than from soil; green vegetables found in production sites were cleaner as compared to those found near the marketplaces, thus concluding that transportation and distribution also played a major role in elevating heavy metal contaminants in vegetables (Sharma et al. 2009). Furthermore, physical factors such as pH, temperature, humidity, as well as surface area available to plants affect the rate of absorption and accumulation of heavy metals in plants. In support, Sharma et al. reported that the uptake of heavy metals like Cd and Cr in Spinach is influenced by the change of season and is found to be higher during hot weather; in contrast, Pb and Ni were accumulated more during the cold season (Sharma et al. 2007). Similarly, Ahmed et al. also reported that the concentration of Cu and Pb in soil and vegetables is seasonal in most part of Bangladesh. Concentrations of these metals are higher during the off-season as compared to the rainy season (Ahmed et al. 2019). The greater the accumulation of Pb in soils, the lower the quality of soil and its fertility but in contrast, there have been reports that plants grown in lower concentrations of Pb tend to inhibit vital metabolic processes of plants, such as photosynthesis and mitosis (Bhattacharyya et al. 2008). In addition to the above efficiencies, another finding reported that heavy metals in plants result in phytotoxicity, causing chlorosis and necrosis and thus unable to contribute to nitrogen fixation (Guala et al. 2010). Akoto et al. concluded that aquatic marine life also often suffered from heavy metal toxicities (Akoto et al. 2008). As a result of the dumping of industrial waste into the Sabata River in central Ethiopia, an abnormally high concentration of Pb is observed in red beet grown in and around the Sabata River (Gemeda et al. 2021). Direct and indirect effects of heavy metal toxicity are observed in certain plants; some of the direct toxic effects include damage to cell walls, oxidative stress, etc. (Van Assche and Clijsters 1990; Jadia and Fulekar 2009) whereas indirect toxic effects result in the substitution of essential metals by non-essential metals (Taiz and Zeiger 2002). Kibra (2008) observed an adverse effect in rice plants grown in mercury-polluted soil, such as inhibition of the growth of tiller and panicle, while Cd toxicity in wheat plants reduces the shoot and root elongation (Ahmad et al. 2012). It is reported that inhibition of photosynthesis, reduction in mineral nutrition, and reduced enzyme activities are some of the factors responsible for stunted growth in plants grown in heavy metal-contaminated soil (Kabata-Pendias and Pendias 2001).

3.4 Heavy Metals and Soil Microbes

Heavy metals affect not only soil, plants, and animals but also useful microorganisms. These microorganisms play a key role in our ecosystem and provide special services in maintaining soil fertility via organic matter decomposition, nutrient cycling, and transformation in the nitrogen cycle. However, their population (expressed in colony-forming units [CFUs]) and functions are affected by external factors like temperature and heavy metal pollution (Schimel et al. 2007; Paz-Ferreiro and Fu 2016). Heavy metal pollution in soil has severe problems for soil microorganisms affecting litter decomposition, carbon mineralization, and nitrogen transformations. Certain insects and invertebrates, like earthworms, are beneficial in farming. However, their efficiency is greatly reduced due to metal contamination (Rai et al. 2019). Especially the contamination reduces the organic matter decomposition capacity resulting in the formation of an undecomposed litter layer in the soil. Singh et al. reported that microbial function in soil was reduced tremendously and saprophagous invertebrates were completely destroyed in the polluted soil (Srivastava et al. 2017). Soil microbial actions of the nitrogen cycle are erupted by heavy metal pollution, causing variation in soil properties and decrease in soil quality and indirectly affecting food crops and vegetables (Dai et al. 2004; Vásquez-Murrieta et al. 2006). It is observed that heavy metals have a serious impact on nitrogen metabolism which results in a reduction in protein synthesis and a low rate of photosynthesis (Lin et al. 2007).

3.5 Effect of Chromium (Cr) on Plants

According to Garnier et al., Cr is commonly present in the earth's crust in higher oxidation states of +3 and +6. These two metal ions have characteristic physical and chemical properties as well as toxicities. Cr^{+6} has a higher toxicity level, whereas Cr^{+3} is a micronutrient and is less toxic (Garnier et al. 2006). Cr^{+6} has been reported to alter the beneficial microbial composition of the soil (Shun-hong et al. 2009). Chromium stress in plants includes a decline in growth, disturbance in the photosynthetic process, and reduction in seed germination (Asati et al. 2016). In various plants, Cr has affected photosynthesis by altering the production of pigments such as chlorophyll and anthocyanin (Boonyapookana et al. 2002).

3.6 Effect of Lead (Pb) on Plants

Absorption of Pb from contaminated soil in plants is mostly via roots. So, retention is highest in roots. It is one of the most hazardous elements found in the soil. It has an adverse effect on plant growth, morphology, and photosynthesis, as well as inhibition of early germination of seeds in certain plants (Mrozek Jr and Funicelli 1982). In maize (*Zea mays*), Pb contamination causes stunted growth resulting in reduced plant biomass (Hussain et al. 2013). Excess concentration of Pb in Chinese cabbage (*Brassica pekinensis*) causes a reduction in nitrogen content and inhibits the functioning of the nitrate reductase enzyme (Xiong et al. 2006).

3.7 Effect of Arsenic (As) on Plants

As with Pb and Hg, it is cancerous in all of its forms and higher-level exposure is fatal. In the environment, As exists in different oxidation states: As(0), As(±3), and As(+5), out of which arsenate (As(+5)) and arsenite (As(+3)) prevail naturally and are highly toxic (Abbas et al. 2018). In plants such as tomatoes (*Lycopersicon esculentum*), arsenic contamination reduces fruit productivity and decreases leaf biomass (Barrachina et al. 1995). Further, in rice plants, As contamination leads to a decrease in seed germination and leaf biomass (Abedin et al. 2002).

3.8 Effect of Cadmium (Cd) on Plants

Cd is one of the most toxic heavy metals with a density around eightfold heavier than water. Das et al. reported that Cd is found to interfere with the transportation of certain micronutrients such as Ca, Mg, P, and K (Das et al. 1997). Another report suggested that roots (carrot and beetroot) and green leafy vegetables (spinach and lettuce) are able to accumulate Cd from soil much faster than other toxic metals (Fytianos et al. 2001). Plants grown in Cd-contaminated soil show chlorosis, inhibition of growth and development, and even untimely death in small concentrations (Di Toppi and Gabbrielli 1999). In the soyabean plant, nitrogen assimilation and metabolism are affected as a result of excess treatment of Cd (Balestrasse et al. 2003). Yourtchi et al. reported that Cd toxicity in certain varieties of wheat reduces shoot elongation along with inhibition of cell division (Yourtchi and Bayat 2013).

3.9 Effect of Mercury (Hg) on Plants

Hg is non-essential in plant growth and metabolism. Higher contamination of mercury in soil is the result of the usage of fertilizers, lime, sludges, and manures in agricultural fields. It exists in various forms, such as Hg^0, Hg^{+2}, HgS, and methyl-Hg. Hg^{+2} can readily accumulate in higher and aquatic plants and is extremely phytotoxic to plant cells (Kamal et al. 2004; Israr et al. 2006). In rice plants, a higher concentration of Hg decreases the biomass of both the shoot and root of seedlings, and reduces tiller and panicle formation as well as productivity (Du et al. 2005; Kibra 2008). Treatment of Hg in tomatoes resulted in late germination, stunted growth, and finally, chlorosis appears on the whole plant; however, the results are significantly contrasting in lower concentrations of Hg (Shekar et al. 2011).

3.10 Effect of Nickel (Ni) on Plants

Ni is required by plants in small quantities. However, in nature, Ni concentration is much higher than that is actually required by plants. The main contributors include mining and burning of coal and oil, and usage of agricultural fertilizers. Ni^{+2} toxicity in monocot and dicot plants mainly causes problems in water uptake (Pandey and Sharma 2002; Gajewska et al. 2006). In wheat, Ni^{+2} toxicity mainly shows effects on growth inhibition, reduced water content and dry matter production (Pandolfini et al. 1992). Whereas, in rice, reduced root growth is mainly observed (Lin and Kao 2005). In the common bean variety (*Phaseolus vulgaris*), it causes toxicity in the roots, a decrease in yield and a reduction in leaf chlorophyll (Piccini and Malavolta 1992) (Table 3.3).

3.11 Future Perspectives

Heavy metal toxicity remains a significant threat to the environment and human health. However, their effects cannot be undermined but can be lessened by converting them to a lesser toxic form. Proper treatment of industrial effluents will somewhat help minimize the global environmental problems caused by heavy metals. Treatment methods such as adsorption, membrane filtration, chemical precipitation, and ion exchange can be utilized. New scientific methods based on the use of microorganisms such as bacteria, fungi, and algae are also implemented to remove toxic heavy metals (Igiri et al. 2018). They transform the heavy metals into a more tolerable form.

Table 3.3 Some detailed effects of chromium, lead, arsenic, cadmium, mercury, and nickel on plants (Mhatre and Chaphekar 1985; Sun and Wu 1998; Boominathan and Doran 2002; Patra et al. 2004; Vassilev et al. 2005; Sundaramoorthy et al. 2010; Garg and Singla 2012; Oliveira 2012; Rizwan et al. 2016; Kumar et al. 2017; Abbas et al. 2018).

S. no.	Heavy metals	Effects on plants	References
1.	Cr	In general, Cr toxicity provides a hindrance to photosynthetic activity and growth in plants (Asati et al. 2016). Reduction in uptake of nutrients, such as Zn, Cu, P, and N, in rice plants has been observed due to Cr overdose. In capsicum, the level of a chemical called carotenoid increases due to the treatment of water containing Cr (Sundaramoorthy et al. 2010; Oliveira 2012)	Asati et al. (2016), Sundaramoorthy et al. (2010), Oliveira (2012)
2.	Pb	Pb in plants are mostly accumulated through roots and hence plants like sweet potatoes and carrots are highly intoxicated (Kumar et al. 2017). In maize, a reduction in germination and growth is observed (Hussain et al. 2013). Excess contamination of Pb in Chinese cabbage results in the reduction of nitrogen content (Xiong et al. 2006)	Kumar et al. (2017), Abedin et al. (2002), Das et al. (1997)
3.	As	As toxicity lowers the fruit yields in tomatoes (Barrachina et al. 1995). As contamination in pea plants results in a decrease in water content (Garg and Singla 2012)	Barrachina et al. (1995) and Garg and Singla (2012)
4.	Cd	Cd stress in wheat plants causes a reduction of water content (Rizwan et al. 2016) and leaf chlorosis (Abbas et al. 2018). In pea plants, Cd toxicity affects photosynthetic functions (Vassilev et al. 2005)	Rizwan et al. (2016), Abbas et al. (2018), Vassilev et al. (2005)
5.	Hg	Accumulation of Hg chloride in corn plants inhibits elongation of primary roots and in higher concentrations of it, mung bean (*Vigna radiata*) seedlings show low respiration rates (Patra et al. 2004). Aquatic plants such as *Hydrilla verticillata* Presl, *Pistia stratiotes* L. are also affected by exposure to Hg (Mhatre and Chaphekar 1985)	Patra et al. (2004) and Mhatre and Chaphekar (1985)
6.	Ni	Ni stress can result in chlorosis and necrosis of leaves in plant species such as barley, water spinach, and wheat (Sun and Wu 1998). In *Nicotiana tabacum* plants, the roots are severely affected; as a result, the overall growth of a plant is inhibited (Boominathan and Doran 2002)	Sun and Wu (1998) and Boominathan and Doran (2002)

3.12 Conclusion

Food crops or plants grown in heavy metal-contaminated soil generally underperformed in many physiological and biochemical functions. Consumption of such contaminated vegetables and food crops often leads to serious health hazards. Therefore, there is a need to understand and minimize the environmental pollution caused by these toxic heavy metals. As a remedial precaution, new scientific methods should be performed and analyzed to study the effects of heavy metal toxification and detoxification in soil, water, and food crops.

References

Abbas, G., Murtaza, B., Bibi, I. et al. (2018). Arsenic uptake, toxicity, detoxification, and speciation in plants: physiological, biochemical, and molecular aspects. *International Journal of Environmental Research and Public Health* 15 (1): 59.

Abbas, T., Rizwan, M., Ali, S. et al. (2018). Effect of biochar on alleviation of cadmium toxicity in wheat grown on cadmium contaminated soil. *Environmental Science and Pollution Research* 25: 25668–25680.

Abedin, M.J., Cotter-Howells, J., and Meharg, A.A. (2002). Arsenic uptake and accumulation in rice (*Oryza sativa* L.) irrigated with contaminated water. *Plant and Soil* 240 (2): 311–319.

Ahmad, I., Akhtar, M.J., Zahir, Z.A., and Jamil, A. (2012). Effect of cadmium on seed germination and seedling growth of four wheat (*Triticum aestivum* L.) cultivars. *Pakistan Journal of Botany* 44 (5): 1569–1574.

Ahmed, M.J.K. and Ahmaruzzaman, M. (2016). A review on potential usage of industrial waste materials for binding heavy metal ions from aqueous solutions. *Journal of Water Process Engineering* 10: 39–47.

Ahmed, M., Matsumoto, M., Ozaki, A. et al. (2019). Heavy metal contamination of irrigation water, soil, and vegetables and the difference between dry and wet seasons near a multi-industry zone in Bangladesh. *Water* 11 (3): 583–585.

Akoto, O., Bruce, T.N., and Darko, D. (2008). Heavy metals pollution profiles in streams serving the Owabi reservoir. *African Journal of Environmental Science and Technology* 2 (11): 354–359.

Amin, N., Hussain, A., Alamzeb, S., and Begum, S. (2013). Accumulation of heavy metals in edible parts of vegetables irrigated with waste water and their daily intake to adults and children, District Mardan, Pakistan. *Food Chemistry* 136: 1515–1523.

Arora, M., Kiran, B., Rani, S. et al. (2008). Heavy metal accumulation in vegetables irrigated with water from different sources. *Food Chemistry* 111: 811–815.

Asati, A., Pichhode, M., and Nikhil, K. (2016). Effect of heavy metals on plants: an overview. *International Journal of Application or Innovation in Engineering & Management* 5: 56–66.

Balestrasse, K.B., Benavides, M.P., Gallego, S.M., and Tomaro, M.L. (2003). Effect on cadmium stress on nitrogen metabolism in nodules and roots of soybean plants. *Functional Plant Biology* 30: 57–64.

Barrachina, A.C., Carbonell, F.B., and Beneyto, J.M. (1995). Arsenic uptake, distribution, and accumulation in tomato plants: effect of arsenite on plant growth and yield. *Journal of Plant Nutrition* 18 (6): 1237–1250.

Bhattacharyya, P., Chakrabarti, K., Chakraborty, A. et al. (2008). Fractionation and bioavailability of Pb in municipal solid waste compost and Pb uptake by rice straw and grain under submerged condition in amended soil. *Geosciences Journal* 12 (1): 41–45.

Boominathan, R. and Doran, P.M. (2002). Ni induced oxidative stress in roots of the Ni hyperaccumulator *Alyssum bertolonii*. *New Phytologist* 156: 205–215.

Boonyapookana, B., Upatham, E.S., Kruatrachue, M. et al. (2002). Phytoaccumulation and phytotoxicity of cadmium and chromium in duckweed *Wolffia globosa*. *International Journal of Phytoremediation* 4 (2): 87–100.

Chang, C.Y., Yu, H.Y., Chen, J.J. et al. (2014). Accumulation of heavy metals in leaf vegetables from agricultural soils and associated potential health risks in the Pearl River Delta, South China. *Environmental Monitoring and Assessment* 186 (3): 1547–1560.

Chary, N.S., Kamala, C.T., and Raj, D.S.S. (2008). Assessing risk of heavy metals from consuming food grown on sewage irrigated soils and food chain transfer. *Ecotoxicology and Environmental Safety* 69 (3): 513–524.

Chauhan, G. (2014). Toxicity study of metals contamination on vegetables grown in the vicinity of cement factory. *International Journal of Scientific and Research Publications* 4: 1–8.

Chen, B., Stein, A.F., Maldonado, P.G. et al. (2013). Size distribution and concentrations of heavy metals in atmospheric aerosols originating from industrial emissions as predicted by the HYSPLIT model. *Atmospheric Environment* 71: 234–244.

Dai, A., Trenberth, K.E., and Qian, T. (2004). A global dataset of Palmer Drought Severity Index for 1870-2002: relationship with soil moisture and effects of surface warming. *Journal of Hydrometeorology* 5 (6): 1117–1130.

Das, P., Samantaray, S., and Rout, G.R. (1997). Studies on cadmium toxicity in plants: a review. *Environmental Pollution* 98: 29–36.

Di Toppi, L.S. and Gabbrielli, R. (1999). Response to cadmium in higher plants. *Environmental and Experimental Botany* 41 (2): 105–130.

Du, X., Zhu, Y.-G., Liu, W.-J., and Zhao, X.-S. (2005). Uptake of mercury (Hg) by seedlings of rice (*Oryza sativa* L.) grown in solution culture and interactions with arsenate uptake. *Environmental and Experimental Botany* 54 (1): 1–7.

Fytianos, K., Katsianis, G., Triantafyllou, P., and Zachariadis, G. (2001). Accumulation of heavy metals in vegetables grown in an industrial area in relation to soil. *Bulletin of Environmental Contamination and Toxicology* 67 (3): 0423–0430.

Gajewska, E., Skłodowska, M., Słaba, M., and Mazur, J. (2006). Effect of nickel on antioxidative enzyme activities, proline and chlorophyll contents in wheat shoots. *Biologia Plantarum* 50 (4): 653–659.

Gall, J.E., Boyd, R.S., and Rajakaruna, N. (2015). Transfer of heavy metals through terrestrial food webs: a review. *Environmental Monitoring and Assessment* 187 (4): 1–21.

Garg, N. and Singla, P. (2012). The role of Glomus mosseae on key physiological and biochemical parameters of pea plants grown in arsenic contaminated soil. *Scientia Horticulturae* 143: 92–101.

Garnier, J., Quantin, C., Martins, E.S., and Becquer, T. (2006). Solid speciation and availability of chromium in ultramafic soils from Niquelandia, Brazil. *Journal of Geochemical Exploration* 88: 206–209.

Gemeda, F.T., Guta, D.D., Wakjira, F.S., and Gebresenbet, G. (2021). Occurrence of heavy metal in water, soil, and plants in fields irrigated with industrial wastewater in Sabata town, Ethiopia. *Environmental Science and Pollution Research* 28 (10): 12382–12396.

Ghosh, M. and Singh, S.P. (2005). A review on phytoremediation of heavy metals and utilization of its by-products. *Applied Ecology and Environmental Research* 3: 1–18.

Gope, M., Masto, R.E., George, J. et al. (2017). Bioavailability and health risk of some potentially toxic elements (Cd, Cu, Pb and Zn) in street dust of Asansol, India. *Ecotoxicology and Environmental Safety* 138: 231–241.

Guala, S.D., Vega, F.A., and Covelo, E.F. (2010). The dynamics of heavy metals in plant-soil interactions. *Ecological Modelling* 221 (8): 1148–1152.

Hawkes, S.J. (1997). What is a “heavy metal”? *Journal of Chemical Education* 74 (11): 1374.

Hussain, A., Abbas, N., and Arshad, F. (2013). Effects of diverse doses of lead (Pb) on different growth attributes of *Zea mays* L. *Agricultural Sciences* 4 (5): 262–265.

Igiri, B.E., Okoduwa, S.I., Idoko, G.O. et al. (2018). Toxicity and bioremediation of heavy metals contaminated ecosystem from tannery wastewater: a review. *Journal of Toxicology* 2018 (9): 2568038–2568054.

Israr, M., Sahi, S., Datta, R., and Sarkar, D. (2006). Bioaccumulation and physiological effects of mercury in *Sesbania drummonii*. *Chemosphere* 65: 591–598.

Jadia, C.D. and Fulekar, M.H. (2009). Phytoremediation of heavy metals: recent techniques. *African Journal of Biotechnology* 8 (6): 921–928.

Kabata-Pendias, A. and Pendias, H. (2001). *Trace Elements in Soils and Plants*. Boca Raton, FL: CRC Press Inc.

Kamal, M., Ghalya, A.E., Mahmouda, N., and Cote, R. (2004). Phytoaccumulation of heavy metals by aquatic plants. *Environment International* 29: 1029–1039.

Kaur, R., Sharma, S., and Kaur, H. (2019). Heavy metals toxicity and the environment. *Journal of Pharmacognosy and Phytochemistry* SP1: 247–249.

Khan, M.U. (2013). Human health risk from heavy metal via food crops consumption with wastewater irrigation practices in Pakistan. *Chemosphere* 93: 2230–2238.

Kibra, M.G. (2008). Effects of mercury on some growth parameters of rice (*Oryza sativa* L.). *Soil & Environment* 27 (1): 23–28.

Krzesłowska, M. (2011). The cell wall in plant cell response to trace metals: polysaccharide remodeling and its role in defense strategy. *Acta Physiologiae Plantarum* 33 (1): 35–51.

Kumar, V., Chopra, A.K., and Srivastava, S. (2016). Assessment of heavy metals in spinach (*Spinacia oleracea* L.) grown in sewage sludge–amended soil. *Communications in Soil Science and Plant Analysis* 47 (2): 221–236.

Kumar, B., Smita, K., and Flores, L.C. (2017). Plant mediated detoxification of mercury and lead. *Arabian Journal of Chemistry* 10: 2335–2342.

Latif, A., Bilal, M., Asghar, W. et al. (2018). Heavy metal accumulation in vegetables and assessment of their potential health risk. *Journal of Environmental Analytical Chemistry* 5 (1): 234.

Lin, Y.-C. and Kao, C.-H. (2005). Nickel toxicity of rice seedlings: cell wall peroxidase, lignin, and $NiSO_4$-inhibited root growth. *Crop, Environment & Bioinformatics* 2: 131–136.

Lin, R.Z., Wang, X.R., Luo, Y. et al. (2007). Effects of soil cadmium on growth, oxidative stress and antioxidant system in wheat seedlings (*Triticum aestivum* L.). *Chemosphere* 69: 89–98.

Liu, G., Yu, Y., Hou, J. et al. (2014). An ecological risk assessment of heavy metal pollution of the agricultural ecosystem near a lead-acid battery factory. *Ecological Indicators* 47: 210–218.

Luo, C., Liu, C., Wang, Y. et al. (2011). Heavy metal contamination in soils and vegetables near an ewaste processing site, South China. *Journal of Hazardous Materials* 186: 481–490.

Marschner, P. and Rengel, Z. (2012). Nutrient availability in soils. In: *Marschner's Mineral Nutrition of Higher Plants*, 315–330. Academic Press.

McLaughlin, M.J., Smolders, E., Degryse, F., and Rietra, R. (2011). Uptake of metals from soil into vegetables. In: *Dealing with Contaminated Sites*, 325–367. Dordrecht: Springer.

Mhatre, G.N. and Chaphekar, S.B. (1985). The effect of mercury on some aquatic plants. *Environmental Pollution Series A, Ecological and Biological* 39 (3): 207–216.

Misra, S.G. and Mani, D. (1991). *Soil Pollution*. Ashish Publishing House.

Mrozek, E. Jr. and Funicelli, N.A. (1982). Effect of zinc and lead on germination of *Spartina alterniflora* Loisel seeds at various salinities. *Environmental and Experimental Botany* 22 (1): 23–32.

Nagajyoti, P.C., Lee, K.D., and Sreekanth, T.V.M. (2010). Heavy metals, occurrence and toxicity for plants: a review. *Environmental Chemistry Letters* 8 (3): 199–216.

Oliveira, H. (2012). Chromium as an environmental pollutant: insights on induced plants toxicity. *Journal of Botany* 2012: 1–8.

Osma, E., Serin, M., Leblebici, Z., and Aksoy, A. (2013). Assessment of heavy metal accumulation (Cd, Cr, Cu, Ni, Pb, and Zn) in vegetable and soils. *Polish Journal of Environmental Studies* 22: 1449–1455.

Pandey, N. and Sharma, C.P. (2002). Effect of heavy metals Co^{+2}, Ni^{+2}, and Cd^{+2} on growth and metabolism of cabbage. *Plant Science* 163: 753–758.

Pandolfini, T., Gabbrielli, R., and Comparini, C. (1992). Nickel toxicity and peroxidase activity in seedlings of *Triticum aestivum* L. *Plant, Cell and Environment* 5 (6): 719–725.

Parvin, R., Sultana, A., and Zahid, M.A. (2014). Detection of heavy metals in vegetables cultivated in different locations in Chittagong, Bangladesh. *IOSR Journal of Environmental Science, Toxicology & Food Technology* 8 (4): 58–63.

Patra, M., Bhowmik, N., Bandhopadhyay, B., and Sharma, A. (2004). Comparation of mercury lead and arsenic with respect to genotoxic effects on plant systems and development of genetic tolerance. *Environmental and Experimental Botany* 52 (3): 199–223.

Paz-Ferreiro, J. and Fu, S. (2016). Biological indices for soil quality evaluation: perspectives and limitations. *Land Degradation & Development* 27 (1): 14–25.

Piccini, D.F. and Malavolta, E. (1992). Effect of nickel on two common bean cultivars. *Journal of Plant Nutrition* 15 (11): 2343–2350.

Rai, P.K., Lee, S.S., Zhang, M. et al. (2019). Heavy metals in food crops: health risks, fate, mechanisms, and management. *Environment International* 125: 365–385.

Raskin, I., Kumar, P.N., Dushenkov, S., and Salt, D.E. (1994). Bioconcentration of heavy metals by plants. *Current Opinion in Biotechnology* 5 (3): 285–290.

Rattan, R.K., Datta, S.P., Chhonkar, P.K. et al. (2005). Long-term impact of irrigation with sewage effluents on heavy metal content in soils, crops and groundwater—a case study. *Agriculture, Ecosystems and Environment* 109: 310–322.

Rizwan, M., Ali, S., Abbas, T. et al. (2016). Cadmium minimization in wheat: a critical review. *Ecotoxicology and Environmental Safety* 130: 43–53.

Ross, S.M. (1994). *Toxic Metals in Soil-Plant Systems*. John Wiley and Sons Ltd.

Schimel, J., Balser, T.C., and Wallenstein, M. (2007). Microbial stress-response physiology and its implications for ecosystem function. *Ecology* 88 (6): 1386–1394.

Schreck, E., Foucault, Y., Sarret, G. et al. (2012). Metal and metalloid foliar uptake by various plant species exposed to atmospheric industrial fallout: mechanisms involved for lead. *Science of the Total Environment* 427: 253–262.

Schreck, E., Laplanche, C., Le Guédard, M. et al. (2013). Influence of fine process particles enriched with metals and metalloids on *Lactuca sativa* L. leaf fatty acid composition following air and/or soil-plant field exposure. *Environmental Pollution* 179: 242–249.

Seaward, M.R.D. and Richardson, D.H.S. (1989). Atmospheric sources of metal pollution and effects on vegetation. In: *Heavy Metal Tolerance in Plants Evolutionary Aspects* (ed. A.J. Shaw), 75–94. Boca Raton, FL: CRC Press.

Sethy, S.K. and Ghosh, S. (2013). Effect of heavy metals on germination of seeds. *Journal of Natural Science, Biology and Medicine* 4 (2): 272–275.

Setia, R., Dhaliwal, S.S., Singh, R. et al. (2021). Phyto availability and human risk assessment of heavy metals in soils and food crops around Sutlej River, India. *Chemosphere* 263: 128321.

Shaheen, N., Irfan, N.M., Khan, I.N. et al. (2016). Presence of heavy metals in fruits and vegetables: health risk implications in Bangladesh. *Chemosphere* 152: 431–438.

Shahid, M., Dumat, C., Khalid, S. et al. (2017). Foliar heavy metal uptake, toxicity and detoxification in plants: a comparison of foliar and root metal uptake. *Journal of Hazardous Materials* 325: 36–58.

Sharma, R.K., Agrawal, M., and Marshall, F. (2007). Heavy metal contamination of soil and vegetables in suburban areas of Varanasi, India. *Ecotoxicology and Environmental Safety* 66 (2): 258–266.

Sharma, R.K., Agrawal, M., and Marshall, F.M. (2009). Heavy metals in vegetables collected from production and market sites of a tropical urban area of India. *Food and Chemical Toxicology* 47: 583–591.

Sharma, A., Katnoria, J.K., and Nagpal, A.K. (2016). Heavy metals in vegetables: screening health risks involved in cultivation along wastewater drain and irrigating with wastewater. *Springer Plus* 5 (1): 1–16.

Sharma, B., Sarkar, A., Singh, P., and Singh, R.P. (2017). Agricultural utilization of biosolids: a review on potential effects on soil and plant grown. *Waste Management* 64: 117–132.

Shekar, C.H.C., Sammaiah, D., Shasthree, T., and Reddy, K.J. (2011). Effect of mercury on tomato growth and yield attributes. *International Journal of Pharma and Bio Sciences* 2 (2): B358–B364.

Shun-hong, H., Bing, P., Zhi-hui, Y. et al. (2009). Chromium accumulation, microorganism population and enzyme activities in soils around chromium-containing slag heap of steel alloy factory. *Transactions of Nonferrous Metals Society of China* 19: 241–248.

Singh, J. and Kalamdhad, A.S. (2011). Effects of heavy metals on soil, plants, human health and aquatic life. *International Journal of Research in Chemistry and Environment* 1 (2): 15–21.

Sipter, E., Rózsa, E., Gruiz, K. et al. (2008). Site-specific risk assessment in contaminated vegetable gardens. *Chemosphere* 71 (7): 1301–1307.

Srivastava, V., Sarkar, A., Singh, S. et al. (2017). Agroecological responses of heavy metal pollution with special emphasis on soil health and plant performances. *Frontiers in Environmental Science* 5: 64.

Sun, E.J. and Wu, F.Y. (1998). Along vein necrosis as indicator symptom on water spinach caused by nickel in water culture. *Botanical Bulletin of Academia Sinica* 39: 255–259.

Sundaramoorthy, P. et al. (2010). Chromium stress in paddy: (i) nutrient status of paddy under chromium stress; (ii) phytoremediation of chromium by aquatic and terrestrial weeds. *Comptes Rendus Biologies* 333 (8): 597–607.

Taiz, L. and Zeiger, E. (2002). Photosynthesis: physiological and ecological considerations. *Plant Physiology* 9: 172–174.

Tóth, G., Hermann, T., Da Silva, M.R., and Montanarella, L. (2016). Heavy metals in agricultural soils of the European Union with implications for food safety. *Environment International* 88: 299–309.

Van Assche, F. and Clijsters, H. (1990). Effects of metals on enzyme activity in plants. *Plant, Cell & Environment* 13 (3): 195–206.

Vásquez-Murrieta, M.S., Migueles-Garduño, I., Franco-Hernández, O. et al. (2006). C and N mineralization and microbial biomass in heavy-metal contaminated soil. *European Journal of Soil Biology* 42 (2): 89–98.

Vassilev, A., Perez-Sanz, A., Semane, B. et al. (2005). Cadmium accumulation and tolerance of two Salix genotypes hydrophonically grown in presence of cadmium. *Journal of Plant Nutrition* 28 (12): 2159–2177.

Xiong, Z.-T., Zhao, F., and Li, M.-J. (2006). Lead toxicity in *Brassica pekinensis* Rupr.: effect on nitrate assimilation and growth. *Environmental Toxicology* 21: 147–153.

Yourtchi, M.S. and Bayat, H.R. (2013). Effect of cadmium toxicity on growth, cadmium accumulation and macronutrient content of durum wheat (Dena CV.). *International Journal of Agriculture and Crop Sciences* 6 (15): 1099–1103.

Zhu, C., Tian, H., Cheng, K. et al. (2016). Potentials of whole process control of heavy metals emissions from coal-fired power plants in China. *Journal of Cleaner Production* 114: 343–351.

Zhuang, P., McBride, M.B., Xia, H. et al. (2009). Health risk from heavy metals via consumption of food crops in the vicinity of Dabaoshan mine, South China. *Science of the Total Environment* 407: 1551–1561.

4

Impact of Heavy Metals on Human Health

Retwik Parui[1], Geetmani S. Nongthombam[1], Maimur Hossain[1], Laxmi R. Adil[1], Rajdikshit Gogoi[1], Soumalya Bhowmik[2], Debika Barman[1], and Parameswar K. Iyer[1,2]

[1] *Department of Chemistry, Indian Institute of Technology Guwahati, Guwahati, Assam, India*
[2] *Centre for Nanotechnology, Indian Institute of Technology Guwahati, Guwahati, Assam, India*

4.1 Introduction

Metals are essential components of our environment that majorly contribute to the healthy flow of our ecosystem and play an exceptional role in the regulation of the metabolic cycle ongoing on human health. Several heavy metals, having greater specific density than water, like zinc (Zn), copper (Cu), and iron (Fe), control enzyme activity and regulate many biological reactions in living organisms (Andreini et al. 2008; Zhang and Zheng 2020). The contribution of these heavy metals to advance industrial processes, evolving healthcare modality, telecommunication, etc., are very notable (Nriagu 1989). However, regular development of industries and modernization are causing excessive exposure of these heavy materials to the environment. The undesired export of these heavy metals toward environmental systems like air, soil, and water enhances excessive accumulation and leads to an unhealthy ecosystem. Those accumulated metals generally transform into various forms depending upon their solubility and stability in different phases of the food chain resulting in biomagnification and rigorously spreading their harmful impact. For example, in Japan, mercury sulfate was released with industrial waste, which converted to methyl-mercury (MeHg) poison with the help of microorganisms and entered into the food chain. Later, the step-by-step migration of this MeHg from tiny creatures to the human body resulted in a life-threatening disease named Minamata (James et al. 2020). This deadly neurological disease has ruled for several years and taken many animals' lives including cats, pigs, dogs, humans, etc. Besides Hg, other heavy metals like Cd, As, Pb, Cu, nickel (Ni), etc., also have many toxic effects on our health. The intake of those heavy metals across their upper limits directly affects different parts of our body, such as the gastrointestinal (GI) tract, heart, brain, kidney, and lungs following various biological reaction mechanisms. Hence, it is noteworthy to control the anthropogenic release of toxic material into the environment. As per the World Health Organization (WHO), the toxicity percentage of heavy metals

Remediation of Heavy Metals: Sustainable Technologies and Recent Advances, First Edition. Edited by Rangabhashiyam Selvasembian, Binota Thokchom, Pardeep Singh, Ali H. Jawad, and Willis Gwenzi.

should be maintained at ppm level. Despite certain awareness, extensive discharge of heavy materials remains high in many developing countries, mostly in India and China (Cheng 2003). Hence, understanding heavy metals' sources and their biological impact is highly desirable. Moreover, the continuous effort of the researchers has developed many heavy metal recognition systems, which can detect specific metals through changes in colorimetric, fluorometric, and conductometric properties (De Acha et al. 2019; Lian et al. 2020). Apart from that, many medicines have been established using metal-chelating agents which bind with heavy metals and follow the urine excretion pathway (Flora and Pachauri 2010). Therefore, in this chapter, we have summarized the complete discussion about the source, the pathway to enter our body, the consumption limit, the mode of action in the biological system, and different detection and remedial techniques of numerous heavy materials that could be highly impactful in the near future.

4.2 Mercury

Hg, a toxic heavy metal, is positioned as a nontransition metal in the periodic table, renowned for its detrimental impact on the human body along with plants and other animals. This noxious element exists in numerous forms owing to different color and texture. Elemental Hg appears as a shiny silver color liquid and forms poisonous vapors due to low vapor pressure and low latent heat. The oxidized vapors have a strong tendency to be soluble in lipids and turn out as health threats through bioaccumulation. The elemental forms are mostly used in light bulbs and thermometers. Besides the elemental form, it also exists as Hg(II) and Hg(I) inorganic salts, which exhibit higher solubility in water, causing easier entry inside the body. In the organic form, MeHg shows better lipophilicity which provides easy biomembrane permeability and affects different organs of our body.

4.2.1 Source and Entry of Mercury Metal into Our Body

Hg ore accumulates in the earth's crust in its elemental form or sulfide via volcanic explosion, outgassing of rocks, and side products of various metal deposition processes. The process of ore deposition also opens up the formation of the secondary phase of Hg as a mine waste which produces greater harm than the primary ore owing to better solubility. Along with the natural sources, anthropogenic activity fastens Hg contamination into the different biochains like soil, water, food, etc. Although the quantifications of natural and human exposures are quite difficult, there are several human exposures that take place regularly in day-to-day life. Usually, people use Hg as dental amalgam in dental treatment, thermometers as a temperature indicator, preservatives in many cosmetics, and various bulbs for efficient lighting, etc. There are some medicines like thimerosal, antiseptics (mercurochrome), and many pesticides that contain a few percentages of Hg. Apart from these, industrial processes such as the pharmaceutical industry, electrical goods manufacturing industry, and soft drinks preparation companies consume Hg at different stages and excrete it as a surplus material which impacts the human body following the bioaccumulation through the food chain.

Generally, the natural elemental ore and the waste generated from human activity are transported and settled in water. In the presence of microorganisms, those settled metals change into the methyl and ethyl forms of Hg consumed by small creatures in the water and biomagnified from small fish to big fish in the water ecosystem. The stored methyl- or ethyl-mercury in the fish tissue further transfers to other animals including humans during its ingestion. This transformed derivative perceives better biopermeability and causes more bioaccumulation along with long-term toxicity in our bodies. As per the statement from WHO, the maximum percentage (~80%) of Hg vapors absorbed in our body is released from dental amalgam (World Health Organization and International Programme on Chemical Safety 1991). Moreover, it is observed that in many cosmetics, Hg concentration is more than 65 ppm which exceeds the upper limit prescribed by WHO and FDA, and directly toxicates our body through skin absorption (Naujokas Marisa et al. 2013).

4.2.2 Biological Impact of Mercury Metal

4.2.2.1 Sulfhydryl Affinity

The higher binding potency of Hg toward the thiol unit causes the complexation between the most abundant thiol-containing amino acids present in our body leading to the major pathway of Hg toxicity (Nordberg et al. 2022). It is reported that the mercury-sulfur (Hg—S) bond poses very high stability inside the cell and the Hg ion follows Rabenstein's ligand exchange reaction process depending upon the availability and reactivity of the thiol unit. The rate of thiol exchange reaction depends on the nucleophilicity of the reactant thiol residue and in the presence of similar nucleophilicity, the reaction follows a rapid first-order reaction. Apart from protein binding, the low molecular weight transporter was also found to interact with the Hg ion following Pearson's acid–base principle. According to this principle, generally, Hg binds with soft nucleophiles like HS^-, S^{2-}, etc., rather than the S or Se present in their positive redox state. After studying the blood serum sample of a Hg-exposed mice model, it was found that Hg binds more predominantly to the albumin residue than the other protein unit. Another recent study established that Hg^{2+} majorly binds to GSH peroxidase 3 (GPX3), hemoglobin, along with the albumin residue, whereas increment in the Hg^{2+} concentration consequence the binding with ApoE and transferrin units (Li et al. 2018). However, the understanding of the priority of binding Hg to the bioresidues with the increment in Hg exposure also helps very much in further detoxification therapy.

4.2.2.2 ROS Generation Ability

The cytotoxicity of Hg^{2+} ion produces oxidative stress inside the cell which results in the oxidation of thiol residue with an increment in the GSSG/GSH ratio and H_2O_2 concentration (Zefferino et al. 2017). The increment of both parameters is generally considered marker Hg for toxification. Later, to understand the mechanism, an examination of intracellular free thiol concentration residue with respect to the divalent Hg^{2+} ion was performed by the Capitanio group, which confirmed no direct participation of Hg^{2+} in this oxidation reaction (Piccoli et al. 2012). Further investigation established that the selenol group poses a negative charge owing to a lower pKa value which exhibits a complexation with Hg^{2+} and reduces the enzyme activity of GSH peroxidase.

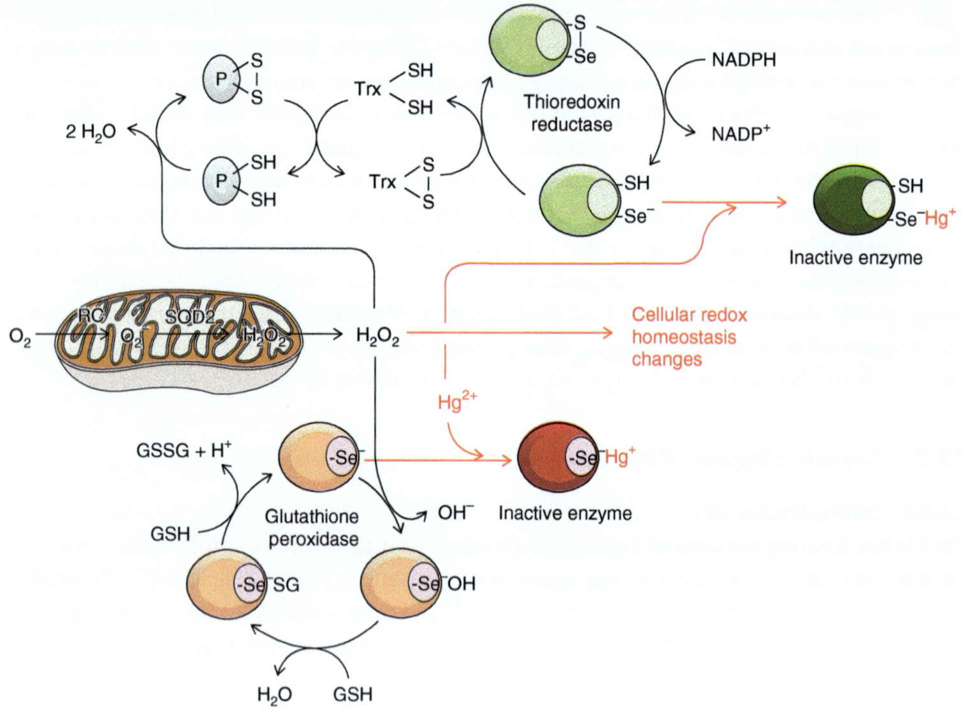

Figure 4.1 The ROS generation mechanism of Hg. *Source:* Zefferino et al. 2017/Hindawi/CC BY 4.0.

As these enzymes play an essential role in producing H_2O from H_2O_2, this enzyme inactivation inhibits water production and increases the level of reactive oxygen species (ROS) (Figure 4.1).

4.2.2.3 Nephrotoxicity

Although maximum absorption of Hg follows the inhalation of its vapors to the lungs, the long-term impacts on the other parts of the body are quite threatening. The lipophilicity of the vapor form helps cross the alveoli to blood serum and further into the central nervous system (CNS), crossing the blood–brain barrier. In different organs, the presence of catalase can convert the elemental Hg to its inorganic salt, which produces greater toxicity compared to the organic form, especially in the renal system. The higher ligating ability of inorganic salts toward sulfhydryl groups facilitates the complex formation between the excess glutathione (GSH) present in the renal system binding in a 2 : 1 ratio (2 GSH molecules, 1 Hg) (Jan et al. 2015). Moreover, the sulfhydryl affinity also disrupts the work function of a sulfur-containing enzyme like *N*-acetyl-β-D-glucosaminidase present in the renal tubule. Further inspection of the Hg uptake process in the proximal tubule confirms the accumulation of a maximum entity of mercury-GSH conjugates transporting through the transporter present at the luminal and basolateral surface. The γ-glutamyltransferase present at the luminal membrane cleaves the γ-glutamylcysteine and facilitates Hg transportation. However, the use of inhibitors like acivicin inhibits the uptake process and helps excrete accumulated Hg through urine.

4.2.2.4 Neurotoxicity

The organic form of Hg accumulates in the CNS and causes neurotoxicity. Mainly MeHg, a major derivative expressed in an aquatic system, is absorbed via the GI tract and crosses the blood–brain barrier by complexation with cysteinyl residue. Due to the structural similarity between the methionine and cysteinyl complex of MeHg, these complexes easily enter CNS through amino acid transporter. The long-term accumulation of MeHg in the brain induces the calcium influx in the neuron, which leads to a disturbance in neurotransmitter formation and hampers balance in brain development as well (Denny and Atchison 1996). Apart from the neuronal disbalance, the high binding tendency toward thiol and selenol units results in the 1 : 1 complexation with protein-containing GSH or thioredoxin residue and interferes with the corresponding enzyme activity.

4.2.2.5 Cardiotoxicity

In addition to neurotoxicity and nephrotoxicity, Hg has a strong impact on the heart, attributed to a detrimental effect on blood pressure and heart rate. According to a report from the Valera group, a strong impact on the blood pressure and heart rate variation was observed for the adults, whereas only abnormality in heart rate was noticed for children, who are indirectly connected to Hg toxicity (Valera et al. 2008, 2012). Besides, myocardial problems and coronary dysfunction are also observed as a result of Hg toxicity. As per the earlier discussion, Hg has a huge role in ROS generation and deactivation of sulfur, a selenium-containing antioxidant enzyme. Along with that, Hg also increases the oxidation of low-density lipoprotein (LDL) and helps disrupt the phospholipid integrity present in the plasma membrane, which directly forms many cardiovascular diseases. Additionally, Hg also inactivates the work function of the paraoxonase enzyme, which controls the function of high-density lipoprotein (HDL) and acts as an antioxidant for LDL (Gonzalvo et al. 1997; Kulka 2016). Hence, deactivation of this enzyme activity enhances the risk of severe cardiovascular diseases and myocardial infarction (MI).

4.2.3 Detection and Remedial Techniques for Mercury Metals

To prevent more consumption of Hg, controlling the emission of Hg from different exposure like fuel plants, precritical industry, etc., under the WHO limit is the most favorable pathway that can be followed rather than detoxification. The intake of foods that are possibly contaminated by direct Hg emission should be minimal or properly detoxified before its consumption. An aquatic plant named *Chlorella* poses excellent behavior on Hg elimination from different parts of our body naturally (Volesky 1990). Some nano filters like carbon nanotubes are very efficacious for the purification of water containing Hg (Pandy et al. 2011). However, continuous research on the detection process also established some chemosensors which display an effective change in their photophysical nature through selective binding to Hg ions. For example, Lee group established an aggregation-induced emission-based peptidyl probe that selectively chelates the Hg metals. Due to the chelation-induced aggregation, the probe exhibits excellent fluorescence "*turn-on*" features for the detection of Hg^{2+} ions (Neupane et al. 2016).

Now in terms of medical treatment, there are numerous precautions that can be followed for a human exposed to or inhalation of Hg. If a patient's skin or eye came in contact with

Hg, washing with normal water or saline would be the best option to follow. A patient who has inhaled Hg vapor and faced congestion in breathing needs to perform his chest X-ray and observe his blood oxygen saturation level. If the oxygen level becomes too low, the patient should be kept under an artificial ventilation chamber. For liquid Hg, the intubation process could be useful for its removal from the body. However, in the case of inorganic Hg salt, no instant excretion can harm the GI tract, and intravenous therapy need to adopt for the betterment of health. There are some chelating agents like British Anti-Lewisite (BAL), D-penicillamine (DPCN), dimercaptosuccinic acid (DMSA) commonly used for Hg-infected patients (Ye et al. 2016). Generally, these chelating agents form a water-soluble complex with the target metal and follow the urine excretion pathway. Moreover, combination therapy where the use of a combined dose of DMSA and MiADMSA produces a better result than the administration of a single chelating agent. If the patient's condition is very severe and no improvement is found with chelating agents, hemodialysis could be the best procedure to make the patient recover from toxic effects.

4.3 Arsenic

4.3.1 Source and Entry of Arsenic Metal into Our Body

As is a naturally occurring metalloid that is ranked 53rd in terms of abundance and accounts for about 1.5 ppm (0.00015%) of the earth's crust. It is estimated that the atmospheric concentration of As is within $3\,ng\,m^{-3}$, whereas in the soil, it is about $100\,mg\,kg^{-1}$. Its prevalence in freshwater, vegetation, and seawater is around $10\,\mu g\,l^{-1}$, $400\,\mu g\,kg^{-1}$, and $1.5\,\mu g\,L^{-1}$, respectively (Rieuwerts 2017). Many industrial processes involving wood preservatives such as copper chrome arsenate (CCA) contain high amounts of As contaminants. As is also employed in the pharmaceutical and glass industries. Their presence has been found in microelectronics, pesticides, insecticides, etc. The widespread use of sodium arsenite as a weed killer is also responsible for vegetation contamination (Chung et al. 2014).

The longevity of As as particulates in the air can be a contributing factor to major air pollution. A substantial amount of methylated As components comprising trivalent and pentavalent compounds have been found in the air of urban and industrial areas. Even the minute presence of As ranging up to $30\,ng\,m^{-3}$ in the air can have an adverse effect on the human body on continuous exposure. Although mining and industrial activities are partly responsible for the increase in the concentration of As in the environment, the primary liability is attributed to natural causes, viz. erosion, landslides, volcanoes, etc.

4.3.2 Biological Impact of Arsenic Metal

For years, As has been romanticized by many fictional writers as a secret murder accessory, but lately, it has been found to be a vital micronutrient in animals. It plays a significant role in gene silencing and also in the metabolism of the amino acid, methionine (Uthus 2003). The recommended dose of As for human beings, as extrapolated from mammalian studies, confines between 12.5 and 25 μg per day. Human beings have been exposed to As since time immemorial via contaminated groundwater, which has been the ultimate source of drinking water and agricultural activities. As is found in its dissolved forms in water sources, and it tends to exist as arsenite, arsenate, dimethylarsinic acid (DMA), and methylarsonic acid.

In many countries, the natural presence of As in groundwater is high and poses an extreme health hazard, especially in its inorganic form. Crops irrigated from contaminated sources coupled with food prepared from these sources lead to the increased accumulation of As in the human body, which goes way beyond the permissible limits. Even the intake of meat, fish, and other dairy foodstuffs can lead to the unwanted build-up of As in the body. Smoking also increases the chances of As poisoning exponentially, thereby heightening the risk of lung, kidney, and bladder cancer, along with many other heart ailments. Tobacco plants have been found to take up the As naturally present in the soil and the blatant use of lead arsenate insecticides has made the matter worse. Inadvertently, humans get directly exposed to As through the use of cosmetics. The billion-dollar cosmetic industry has very low priority when it comes to consumer safety assessments due to the wrong perception that it does not have any severe effect on our physical well-being. Cosmetics come in contact with the bare skin directly and a number of beauty products have As impurities. The As content may be minute; however, on regular usage, it becomes detrimental, and it has been a leading cause for its direct exposure (Borowska and Brzóska 2015). Currently, the lack of an international standard for heavy metal impurities in cosmetics is also a major concern.

In continuation of our prior discussions, we now know that long-term exposure to As through water and food leads to cancer, skin lesions (hyperkeratosis), diabetes, cardiovascular disease, etc. Early childhood exposure to As has been associated with adverse cognitive development and increased mortality in young adults (Quansah et al. 2015). In Taiwan, long-term As exposure has been linked to a blood vessel-affecting disease called "*Blackfoot disease*" which ultimately leads to gangrene. Short-term As exposure would lead to red or swollen skin, skin changes (viz. new warts or lesions), nausea and vomiting, abdominal pain, diarrhea, muscle cramps, tingling of fingers and toes, abnormal heart rhythm, etc. Currently, no credible method is known that can help in distinguishing cancers induced by As or by other factors. Therefore, the effect of As poisoning and its exact impact on human biological metabolism is still undergoing immense studies so as to correlate its association with other diseases worldwide. As content in potable water is deemed to be one of the most significant contributors to cancer caused by environmental factors in the world. Therefore, the WHO has recommended several precautionary measures to avoid As exposure and it has listed it to be one of the top 10 chemicals which pose major public health concerns.

4.3.3 Detection and Remedial Techniques for Arsenic Metals

Consumption of microbiologically safe rainwater or treated surface water is being encouraged worldwide. Attempts are being made to make provisions for distinguishing between low and high As-containing water by painting tube wells and hand pumps with different colors after conducting regular tests accompanied by imparting effective education to the masses. The high As-containing water is used for bathing and washing, while the low As-containing water is used for irrigation and consumption. Several technologies for As removal have also been installed in centralized or domestic scales and they employ principles based on coagulation-flocculation, oxidation, ion exchange, absorption, membrane techniques, etc. The WHO also publishes guidelines for limiting As poisoning and is actively involved in educating the masses with the involvement of local governments. Currently, the recommended limit of As in drinking water is $10\,\mu g\,l^{-1}$; however, millions of people globally are exposed to concentrations of $100\,\mu g\,l^{-1}$ or greater.

Further research on the detection and remediation of As is ongoing and work toward a cleaner and healthier water for consumption is an imperative challenge for mankind.

4.4 Iron

4.4.1 Source and Entry of Iron Metal into Our Body

The magnanimous presence of Fe in the earth's crust is indisputable. In terms of natural abundance by mass, no element can surpass Fe, and therefore studies regarding its impact on human health are of crucial importance. Natural water sources and groundwater contain a substantial amount of Fe, and in certain regions around the world, the Fe content in potable water is abnormally high. Mining and metallurgical industries are also responsible for Fe hazards.

Pediatric ingestion of potent adult medicines, viz. prenatal vitamins, can lead to life-threatening complications. The widespread presence of Fe in numerous over-the-counter (OTC) multivitamins is a leading factor in Fe ingestion poisoning. Accidental ingestions are more likely in children below the age of six, while intentional ingestion by adults is often associated with suicide attempts. Lack of proper diagnosis and treatment of Fe poisoning results in multi-organ failure, which ultimately ends in a loss of life. Medical procedures involving multiple blood transfusions associated with acute disorders, viz. anemia, blood cancers, thalassemia, etc., build up Fe toxicity in the body. The nontoxic ingestion limit of elemental Fe for adults is around $20\,mg\,kg^{-1}$, and as we increase it to around $60\,mg\,kg^{-1}$, moderate symptoms will start appearing. Ingestion beyond $60\,mg\,kg^{-1}$ is precarious and results in severe morbidity.

4.4.2 Biological Impact of Iron Metal

Iron oxides impact human bodies negatively and these are enlisted as hazardous materials by Occupational Safety and Health Administration (OSHA). Iron oxide is detrimental to human beings when inhaled, and its fumes are responsible for metal fume fever which exhibits symptoms such as fever and chills, chest tightness, aches, cough, and metallic taste sensation. It was also found that prolonged or repeated exposure to these oxides can cause eye discoloration, leading to permanent Fe staining. In addition, people who are repeatedly exposed to iron oxide fumes suffer from pneumoconiosis (siderosis). As per OSHA, the permissible exposure limit (PEL) of airborne iron oxide is $10\,mg\,m^{-3}$ over an eight-hour work shift (Madiwale and Liebelt 2006).

Ingested Fe damages the GI mucosa resulting in perforation, hematemesis, and peritonitis. Hemorrhagic necrosis of the GI mucosa causes abdominal pain, diarrhea, nausea, and vomiting. At the cellular level, free Fe deposits in the mitochondria and hinders the cellular metabolism in the CNS, heart, and liver. It also discriminately impairs oxidative phosphorylation and causes unwanted catalysis of lipid peroxidation, which results in cell death (Baranwal and Singhi 2003).

Post-ingestion of Fe, the serum Fe level takes two to four hours to attain its peak however, sustained-release formulations warrant serial levels due to their erratic nature. Studies have shown that around 10% of the ingested Fe binds to transferrin via intestinal absorption (Singhi and Baranwal 2003). The serum Fe levels under normal circumstances fall in the range of 50–150 $\mu g\,dl^{-1}$; whereas the total iron-binding capacity (TIBC) varies from 300 to 400 $\mu g\,dl^{-1}$. As the Fe concentration rises in the body, it leads to transferrin saturation, which in turn causes the build-up of free Fe circulation in the blood, ultimately impairing the target organs.

The five successive stages of Fe poisoning:

- Stage I (30 minutes to 6 hours): The affected person exhibits the common GI symptoms, viz. vomiting, stomachache, hematemesis, loose motion, etc.
- Stage II (six hours to one day): This is the healing period where the GI symptoms subside, although toxic amounts of Fe still persist inside the body.
- Stage III (one to three days): Recurrence of GI symptoms coupled with metabolic acidosis. In this stage, renal failure, Fe-induced coagulopathy, cardiomyopathy, and hepatic dysfunction are normally observed.
- Stage IV (one to four days): Spike in aminotransferase levels with the possibility of liver dysfunction.
- Stage V (0.5–2 months): This is basically the recovery period for GI mucosa injuries as well as proximal and pyloric bowel scarring.

4.4.3 Detection and Remedial Techniques for Arsenic Metals

Treatment and remediation of patients with Fe poisoning are usually conducted by a group comprising physicians, internists, intensivists, toxicologists, pharmacists, and hematologists. Acute Fe toxicity can be addressed by the reversal of hypovolemia and hypotension. In severe circumstances of metabolic acidosis, chelating agents such as deferoxamine can be employed. Sometimes if the patients exhibit hemodynamic instability, they usually get themselves admitted to the ICU. Awareness regarding the potential toxicity of Fe should be well communicated to health professionals. Extra care should be taken for the well-being of the kidney, and a nephrologist should be consulted for the possibility of dialysis.

4.5 Manganese

Manganese (Mn) is one of the most crucial trace elements for human health, which is usually found in various foodstuffs and dietary supplements. Mn is vital for the normal functioning of pyruvate carboxylase, Mn superoxide dismutase, arginase, etc.

4.5.1 Source and Entry of Manganese Metal into Our Body

Mn is present in different kinds of foods, including grains, nuts, soybeans and peas, clams, oysters, mussels, tea, coffee, and black pepper (Nielsen 2012). Little amounts of Mn (1–100 $\mu g\,l^{-1}$) are also present in drinking water. A significant amount of Mn is available in all forms of milk. Breast milk contains Mn concentrations of 3–10 μg l^{-1} and 30–100 μg l^{-1} in cow's milk-based infant formulas. Soy-based infant foods contain high Mn (200–300 μg l^{-1}). Recent research indicates that Mn gets significantly absorbed from human milk as compared to soy source and cow milk-based products (3.1%). Infants and children are likely to absorb more Mn than 18 years or older (Chen et al. 2018).

Although adequate levels of Mn intake are important for human health, exposure to high Mn levels can have a serious detrimental impact on human health. There are several reports which highlight the adverse impact resulting from Mn exposure in humans. Occupational exposure like mining or industrial jobs related to Mn is associated with most of the reported cases of Mn toxicity. Mn metal is utilized in the processing of aluminum (Al) cans, steel, fertilizers, fungicides, and electronics. Pesticides like manganese ethylene bis-dithiocarbamate (MANEB) can also threaten human health upon excessive exposure. Mn is also present in an addictive drug called "Bazooka", which is made of cocaine and contains manganese carbonate. Another major concern for human health is Mn aerial particulates from the additives like methylcyclopentadienyl manganese tricarbonyl (MMT) used in fuel. Combustion of MMT-containing fuel produces airborne Mn sulfate and phosphates-based products. The intake of Mn-tainted water has also been responsible for severe adverse intoxication in our bodies (Crossgrove and Zheng 2004).

4.5.2 Biological Impact of Manganese Metal

Mn participates in cholesterol, carbohydrate, glucose, and amino acid metabolism; scavenging reactive oxygen species; the formation of bone; and immunity system reproduction (Li and Yang 2018; Institute of Medicine (US) Panel on Micronutrients 2001). Along with vitamin K, Mn also regulates the hemostasis process. It is necessary for the healthy growth of the human brain, nervous system response, and many enzymatic processes. A significant amount of Mn gets bonded to metalloproteins (e.g. glutamine synthetase) (Takeda 2003). The presence of Mn in the synaptic cleft can influence synaptic neurotransmission. Mn is also essential for bone quality, including its development, functioning, and maintenance. The combination of Mn with nutrients and other elements like calcium, Zn, and Cu, controls the bone mineral density, which can play a critical role for adults (Pisani et al. 2016).

The excess exposure or deficiency of Mn can be harmful to our health. In Table 4.1, the recommended intake of Mn is listed according to the following age group variants

Table 4.1 Adequate Mn intake requirements for different age groups.

Stage of life	Age (years)	Males (mg day^{-1})	Females (mg day^{-1})
Infants	0–1	0.003–0.6	
Children	1–13	1.2–1.9	1.2–1.6
Adolescents	14–18	2.2	1.6
Adults and older	18+	2.0	

Source: Adapted from Food and Nutrition Board 2001 Report.

(Source: Food and Nutrition Board 2001 Report) (Institute of Medicine (US) Panel on Micronutrients 2001).

Although Mn deficiency has minimal evidence in humans, it can cause weakness of bones, skin disease, hampered growth in children, depigmentation of hair, lower serum cholesterol, etc. (Nielsen 2012). Mn deficiency might also adversely impact carbohydrate and lipid metabolism (Li and Yang 2018). On the other hand, excess Mn exposure can cause serious toxic effects on the nervous system. Mn neurotoxicity can cause muscle contraction involuntarily with a tremble in which the patient walks on his toes leaning forward (called "cock-walk"). The detection of Mn toxicity at an early stage is difficult without patient history (occupation, food habit, daily routine, etc.). When industrial workers' health was analyzed for epidemiological studies, it showed a direct correlation between the dysfunction neurological system and cumulative Mn exposure (Lucchini et al. 1999). Chronic Mn excess exposure can cause Mn aggregation in the brain. The excess deposition of Mn in the basal ganglia usually leads to Parkinsonism-like characteristics. The dopaminergic systems can also be impacted due to Mn accumulation. There are also a few reports which highlight the toxic impact of excess Mn on cardiac muscle cells and tissues. The detailed mechanism of Mn-induced neurotoxicity is not yet adequately understood. As per the previous reports, Mn-induced neurotoxicity may be due to the participation of trace elements (Cu, Al, Fe, and Zn) coupled with Mn (Verity 1999; Crossgrove and Zheng 2004). Excess Mn can modify the extent of Fe concentrations in both cerebrospinal fluid (CSF) and blood. Moreover, the participation of Mn in Fenton reactions potentially enhances the ROS level, which further promotes oxidative cell ablation (Crossgrove and Zheng 2004).

4.5.3 Detection and Remedial Techniques for Manganese Metals

Long-term high intake or exposure can lead to excess Mn content in the blood and urine of patients. Therefore, the presence of Mn can be evaluated from body fluids and hair samples. Along with this, the food and occupational history of patients and symptoms such as cock walk bradykinesia, impaired speech, etc., are equally crucial for diagnosis. The recommended ranges of Mn extents in body fluids are 0.4–0.85 $\mu g l^{-1}$ in serum, 4–15 $\mu g l^{-1}$ in blood, and 1–8 $\mu g l^{-1}$ in urine.

The treatment of Mn delivers different outcomes among patients. Treatment with a drug called levodopa (l-3,4-dihydroxyphenylalanine) has a partial impact on the improvement of symptoms. Chelation therapy is also another recommended method to remove the excess Mn in the body and mitigate the symptoms. The EDTA excretion therapy was applied in some cases. Calcium disodium edetate ($CaNa_2EDTA$) was utilized intravascularly as a chelating agent (Crossgrove and Zheng 2004).

4.6 Zinc

Zn is another useful element essential for the normal growth and development of plants and humans. Zn is crucial for the regulation of enzyme activity involved in carbohydrate metabolism, protein synthesis, gene expression, etc. Usually, the optimum amount of Zn found in an adult human body is 3 g. It can almost be found in the whole human body (tissues, organs, body fluids, bones, and cells). It regulates the functioning of 300 enzymes in the human body which are important for activating overall growth, immune system, cell division, fertility, taste, smell, appetite, etc. (Sangeetha et al. 2022).

4.6.1 Source and Entry of Zinc Metal into Our Body

The most important sources of Zn are dairy products and nonvegetarian food like red meat, poultry, and seafood. The main source of Zn for humans is meat consumption. The relatively high Zn content is available in oysters ($25\,mg\,100\,g^{-1}$), nuts ($3\,mg\,100\,g^{-1}$), red meat ($5.2\,mg\,100\,g^{-1}$), poultry ($1.5\,mg\,100\,g^{-1}$), eggs ($1.3\,mg\,100\,g^{-1}$), dairy products ($1.2\,mg\,100\,g^{-1}$), etc. (Sangeetha et al. 2022).

There are three main pathways by which Zn can enter the human body: (1) breathing, (2) skin, and (3) ingestion. Based on the type of exposure, different parts of the body can be affected. Industrial workers can get exposed to Zn-containing smoke and inhaling excess Zn-containing smoke can harm human health. Moreover, the military smoke from bombs also contains ZnO or $ZnCl_2$, making soldiers very vulnerable to inhalation. The most usual effect of Zn-containing smoke inhalation is known as metal fume fever. The absorption of Zn can also occur through the skin. However, the mechanism is not adequately established, and limited studies were focused on this. Although the Zn oral uptake of small amounts is very crucial for the human body, excess consumption or exposure can have a serious impact on health. The recommended dietary allowance (RDA) of Zn for the different age groups is listed in Table 4.2 (International Zinc Nutrition Consultative Group (IZiNCG) et al. 2004).

4.6.2 Biological Impact of Zinc Metal

Zn plays an important role in the functioning of ~300 enzymes to carry out vital chemical reactions. It is also a key player in the formation of DNA, cell growth, protein-building blocks, healing injured tissues, and supporting the immune system. As it also regulates the smooth functioning of the immunity system, even a moderate deficiency can diminish the activity of neutrophils, macrophages, and lymphocytes. Therefore, Zn is required during

Table 4.2 RDA limits for different age groups.

Life stages	Reference body wt. (kg)	RDA ($mg\,day^{-1}$)
6–12 months	9	0.84
1–3 years	13	0.74
4–8 years	22	1.20
8–13 years	40	2.12
14–18 years, male	64	3.37
14–18 years, female	57	3.02
Pregnancy		4.1–5.0
Lactation		3.8–4.5

Source: Adapted from Food and Nutrition Board 2004 Report.

the rapid growth of infants, teenagers, and pregnancy. Zn also modulates the senses of smell and taste (Sangeetha et al. 2022). It also controls or influences the functioning of ~3000 proteins like transcription factors, signaling enzymes, etc. (Gammoh and Rink 2019).

The deficiency of Zn can also affect the physical growth and fast-developing stages of life (infants, adolescence, and pregnancy). Abnormal weight gain, hair fall, lack of appetite, weak immunity, diarrhea, delayed wound healing, susceptible infections, fertility issues, etc., are a few well-studied detrimental effects of Zn deficiency (Sangeetha et al. 2022). Moreover, excess exposure can also impact human health adversely. The focal neuronal deficit, lethargy, respiratory disorder, metal fume fever (MFF), diarrhea, and high risk of prostate cancer are the few categorical impacts on human health of Zn excess exposure. There are multiple mechanisms through which Zn excess can impact because the role of Zn is diverse in the human body. Zn and Cu have a competitive absorption in the small intestine. Therefore, the excess Zn intake can directly impact Cu-related metabolism in the human body. Zn intake above the RDA for a long time can interfere with the effective absorption of Cu in the intestine, which can cause Cu deficiency. As Cu regulates Fe-based metabolism in the body, which is important for red blood cell (RBC) and white blood cell (WBC) production. Therefore, Cu deficiency can cause diminished Fe mobilization, anemia, leukopenia, hypocupremia, neutropenia, etc. (Lazarchick 2012).

4.6.3 Detection and Remedial Techniques for Zinc Metals

There is a lack of consensus about the best method for determining the Zn-level analysis of a particular population. However, some indirect signs, such as Fe deficiency, have been used prevalently. This is an inaccurate approach and can produce flawed results of Zn detection because of various interfering factors. Food intake data and occupational history may provide a rough idea about Zn exposure or its level. Currently, plasma and urinary Zn detection are relatively more direct and accurate indicators. Surgical removal may be needed if any patient consumes significantly high amounts of Zn. Otherwise, $CaNa_2EDTA$ or copper sulfate-based medications can be prescribed by doctors to treat Zn poisoning. $CaNa_2EDTA$ assists the body in getting rid of excess Zn by binding to it in the blood (Agnew and Slesinger 2022).

4.7 Lead

Pb has numerous significant customs in the present day, from roofing sheets to screens for radioactive emissions and X-rays (Watson 2001). Mostly, Pb is used in the manufacturing of Pb-based batteries like solar cells (Law et al. 2014; Hossain et al. 2021), metal products, ammunition, and paints. Despite all these significant utilities, the industrial use of Pb has been drastically minimized from caulking, paints, and ceramic products as the potential health risk of Pb remains a topic of concern and debate. Severe and chronic human health impacts can be realized upon sufficient exposure to Pb. The Institute for Health Metrics and Evaluation (IHME) assessed the loss of 13.9 million healthy life including 540,000 deaths worldwide, partial or severe disability-adjusted life years due to Pb toxicity. IHME also prophesied that in almost 63.8% of Pb-poisoning cases, there was potential accountability to meet intellectual disability (Caravanos et al. 2016).

4.7.1 Sources and Exposure of Lead Metal

Pb is one of the most pervasive heavy metals, which have been mostly perceived in the common environmental elements like soil, air, ground and surface water, sediment, etc.; also detected in the biological systems. In surface water, Pb exists in the form of some insoluble compounds. It undergoes some complex formation with the other particles present in the soil and sediment system, thereby shrinking its biological availability to the living organisms in those environments. Vegetation absorbs a minor quantity of Pb from the soil via root absorption or via atmospheric deposition. The main sources of Pb from where a human body gets exposure to it are gasoline and house paint, storage batteries, pewter pitchers, plumbing pipes, toys, faucets, etc. (Thürmer et al. 2002). Medically and scientifically, it has been realized that the prime route of Pb exposure to children is oral ingestion of Pb-comprising dust and Pb-containing paint. Likewise, for adults, the key route of exposure is the inhalation of Pb-contaminated fumes and dust in occupational and industrial settings, mostly during refining, smelting, and mining procedures or during the Pb-based battery-manufacturing process. The blood-lead level of workers in the battery reclamation service was found to be significantly high due to the frequent Pb exposure, as reported by Gittleman and group in 1994 (Gittleman et al. 1994). The exposure may also take place while smoking or eating in a Pb-contaminated environment (Baxter et al. 1985). Notably, drinking water can also contribute as a Pb exposure source to human beings through uptaking from Pb-containing plumbing pipes. However, reports of toxicity due to lead contaminated water is rare in case of humans (Houk et al. 1989). An additional direction of Pb exposure is found to be food ingestion. It may be present in food naturally as well as through atmospheric deposition. Pb is subconsciously familiarized with food, during packaging, processing, or transportation through Pb-contaminated dust and water used for cooking. The global daily average consumption of Pb, assessed by the United Nations Environment Program (UNEP 1991), is $40\,\mu g\,day^{-1}$ from daily drinking water and $80\,\mu g\,day^{-1}$ from daily consumed food (Juberg et al. 1997).

4.7.2 Health and Biological Impact of Lead

The teeth and bones of an adult cover the major portion of the total body mass (>95%) and thus carry the burden of Pb. Children are largely impacted by Pb because of their susceptibility to faster growth and rapid metabolism (Castro-González and Méndez-Armenta 2008). Pb toxicity or poisoning can be either chronic or acute. Acute toxicity can result in abdominal pain, hypertension, loss of appetite, headache, renal dysfunction, arthritis, fatigue, vertigo, hallucinations, and sleeplessness. Chronic exposure also results in some concerning health issues including allergies, weight loss, birth defects, paralysis, psychosis, autism, hyperactivity, dyslexia, muscular weakness, kidney damage, and brain damage and may even cause death. The adverse effect on the neuro-development in children is one of the critical health issues caused by Pb. Even a very low accumulation of Pb in children's bodies (aged between 0 and 5 years) can create a burden on the normal growth of the children with subsequent lowering of IQ. Pb exposure can affect children in utero because of Pb carriage via the placental barrier, which results in a major concern of Pb exposure to pregnant women.

Pb is a silvery bright-colored metal that looks a little bluish in a dry environment. On contact with air, it begins to form several complex compounds depending on the atmospheric composition. Despite the low geochemical mobility of Pb, it has been dispersed worldwide by man in extreme quantities. Pb creates toxic effects in living cells by creating oxidative stress (Debnath et al. 2019). In living cells, oxidative stress is activated by the imbalance between the antioxidant generation and the free radical production to detoxify and repair the reactive intermediates and resulting damages. Antioxidants are very important for living cells as they prevent the cells from some specific free radicals like H_2O_2. For example, GSH is one of the most significant antioxidant-producing cells. GSH stabilizes ROS, which in turn transforms to glutathione disulfide (GSSG) and reverts to GSH by glutathione-disulfide reductase enzyme.

Pb deactivates GSH by binding to the sulfhydryl group of GSH and increases the amount of ROS with a concomitant decrease in antioxidant level, resulting in an insufficient amount of antioxidant enzymes to prevent the cells from such radicals. Lipid peroxidation is an additional biomarker for oxidative stress, as electrons from the lipid molecules of the cell membrane are collected by the free radicals, eventually resulting in lipid peroxidation (Flora et al. 2008). As a result, very high concentrations of ROS may introduce structurally damaged cells, nucleic acids, proteins, and membrane lipids ensuing in a state of stress at the cellular level.

Another way of Pb toxicity involves an ionic mechanism that involves the replacement of some other biologically active bivalent or monovalent cations like Mg^{2+}, Ca^{2+}, Fe^{2+}, Na^{+}, etc., by Pb ions, which eventually interrupts the biological cell metabolism. This ion exchange phenomena of Pb creates noteworthy changes in several biological processes such as protein folding, enzyme regulation, apoptosis, release of neurotransmitters, and ionic transportation. Pb toxicity can severely affect memory storage and neuronal excitation since Pb is capable of substituting calcium even at very low concentrations, which affects the protein kinase C (Flora et al. 2008).

In 1994, Waterman and co-workers (Waterman et al. 1994) reported that Pb exposure leads to generate autoantibodies in contradiction of neural proteins, including glial fibrillary acidic protein and myelin basic protein, thereby assisting neurological defects by elevating the immunogenicity of the proteins of the nervous system. Min group (Min et al. 2008) reported that blood Pb level has a noteworthy connotation with elevated bronchial responsiveness that came from the increment of the Ig-E level with Pb contamination. It may trigger to cause asthma problem and other allergic situations. Notably, Li and co-workers (Li et al. 2005) found that Pb exposure to children leads to a reduction of $CD4^{+}$ T lymphocytes that might cause immune function alteration in children.

4.7.3 Detection and Control of Lead Exposure

Pb is regarded as a major toxic heavy metal, which may cause severe health effects even at very low exposure levels. The level of Pb consumption is usually verified through the blood samples of an individual. However, this tactic does not always exactly imitate the intoxication level due to the susceptibility of Pb to move from the vascular system, which deposits in the human bones. Thus, blood sample testing does not assess the overall level of exposure to Pb as it cannot detect the Pb stored in the bones (UNEP 2009). The changes that occur in the bones can be spotted by X-ray.

An efficient tactic to decline Pb toxicity is to make awareness among people about Pb exposure and poisoning. The best way to prevent Pb toxicity is precaution. Staying away from the exposure sources as much as possible is the best way to minimize Pb toxicity. Children have a frequent tendency to put their hands into their mouth. Thus, parents should prevent them from habitually doing this and need to encourage their children to wash their hands frequently. The possibility of Pb exposure from hot water is more as it contains a higher amount of Pb compared to cold water. It is recommended to use cold water as much as possible (Wani et al. 2016). Taking vitamin C-containing foods is suggested to minimize Pb toxicity and oxidative stress, as vitamin C exhibits antioxidant characteristics, which is accomplished to eliminate free radicals. An efficient therapeutic approach to remove Pb from the body is chelation therapy. If the blood Pb levels are found to be $45\,\mu g\,dl^{-1}$ or larger, chelation therapy can be a novel tactic. In this approach, the advantage of higher chelation affinity of the chelating agents toward Pb over other metal ions like calcium is used to bind with Pb (Ferreirós-Martínez et al. 2011), which is further excreted in urine.

4.8 Chromium

Cr is a silvery-bright metal that has an inherent brittle nature. Due to the brittleness of Cr, its content in most of the Cr alloys is not more than 30–40% by weight. Cr is a commonly used alloying element in dental medications, bridges, implants, and some other important metallic applications. The most prolific advantage of Cr-based alloys is the generation of inert protective oxides (Cr_2O_3) on the exterior of the alloy in a suitable oxidizing environment that provides better surface stability to the alloy, e.g. stainless steels. Cr can exhibit with a number of oxidation states [Cr(II)—Cr(VI)], where most of the stable Cr-based compounds are found at the oxidation state of Cr(III) in nature, e.g. ferrochromite. Another stable state of Cr is the oxidation state of Cr(VI), whereas Cr(0) is naturally unavailable. However, the increment in the concentration of Cr in the environment has been associated with wastewater and air discharge of Cr, largely from chemical as well as metallurgical industries, which has become a topic of concern. The main industrial Cr-based pollutant is coming out in the hexavalent [Cr(VI)] form and toxicity wise classified as a human carcinogen.

4.8.1 Sources and Exposure of Chromium

Naturally, though the abundance of comparatively lesser toxic Cr(III) is more, however, recently presence of Cr(VI) has been detected in both surface and groundwater at a level that exceeds $50\,\mu g\,l^{-1}$, which is the limit set by the WHO for drinking water (Velma et al. 2009). It has been estimated that annually more than three lakh workers are exposed to Cr-containing complexes in their workplaces. Being an essential nutrient in the human body, Cr(III) is important for fat and glucose metabolism, and in the potentiation of the effect of insulin that plays an important role in protein metabolism. Despite this, occupational exposure has become a major threat due to the potential risk of Cr-assisted diseases in workers placed occupationally got exposed to Cr(VI). In most of the cases, nonoccupation Cr exposure takes place through Cr-contaminated water and food, whereas inhalation is the main route to undergo occupational exposure. It is estimated that the Cr contents in various elements of the environment are 5–800 μg l^{-1} in seawater, 1–3000 mg kg^{-1} in soil, and 5.2 mg l^{-1} to 26 μg l^{-1} in lakes and rivers, respectively (Kanakoudis and Keramaris 2020). The content of Cr in foods (<10–1300 μg kg^{-1}) varies significantly, which mainly depends on the preparation method and processing. Nevertheless, the primary way of Cr exposure is via inhalation, resulting in the lung as the prime target organ of infection. However, noteworthy Cr exposure in humans also has been reported via the skin. For example, the occurrence of dermatitis observed among the workers in construction sites is ascribed to their exposure to Cr-containing cement (Shelnutt et al. 2007).

4.8.2 Health and Biological Impact of Chromium

The primary influences governing the toxicity of Cr complexes are their oxidation states and corresponding solubilities. Hexavalent Cr complexes are very influential oxidizing agents and therefore tend to be corrosive and possess higher toxicity than Cr(III) complexes (Connett and Wetterhahn 1983). However, a proper mechanism of Cr-interaction in biological systems is still indeterminate. The degree of toxicity may be allied to the ease with which Cr can penetrate via cell membranes and its successive intracellular oxidation/reduction to reactive intermediates. Cr(III) is hardly passed through any route. Thus, the toxicity is primarily associated with the Cr(VI) form. It may be absorbed by the GI tract, lungs, and intact skin to a certain extent. Under physiological circumstances of cells, Cr(VI) can be reduced by ascorbic acid, GSH reductase, and hydrogen peroxide (H_2O_2), resulting in forming Cr(V), Cr(VI), hydroxyl radicals, thiyl radicals, and finally, Cr(III), which are quite reactive intermediates. These intermediates have the ability to attack DNA, membrane lipids, and proteins, thereby troubling the standard cellular functions (Mattia et al. 2004).

The harmful effect of Cr(VI) in mammals was reported after the experiments with animal models. Cr(VI) can cause severe progressive proteinuria and hepatic lipid peroxide in rats (Kim and Na 1991). Gumbleton and Nicholls (1988) reported Cr(VI)-tempted renal damage in rats induced by single subcutaneous injections. In the case of humans, respiratory cancers have been reported in workers exposed to Cr(VI)-containing compound (Costa 1997). Lipid peroxidation content in urine and breaking of DNA strand in peripheral lymphocytes was also observed in case of Cr exposure to workers (Gambelunghe et al. 2003). Oxidative damage is regarded as the reason behind some genotoxic impacts of Cr including breaking of DNA strands (Xie et al. 2005) and chromosomal abnormalities (Wise et al. 2002). High Cr contamination on breathing can lead to the nose irritation as well as ulcers. The major health issues observed in humans and animals arising from Cr(VI) exposure via ingestion include ulcers in the small intestine, male reproductive system damage, sperm damage, and anemia. Although some strong carcinogenicity of Cr in mammals and humans are reported so far, the mechanism associated to causing cancer is not thoroughly understood.

4.8.3 Safety Limits and Control

It has been estimated that annually 33 tons of Cr are released into the atmosphere (Wilbur et al. 2012). The United States OSHA stand a "safe" limit of 5 μg m^{-3}, for an eight-hour time-weighted average. Nevertheless this revised safe level may still stance a carcinogenic hazard. In 2003, the WHO recommended a limit for Cr in drinking water at 50 μg l^{-1}, whereas Office of Environmental Health Hazard Assessment (OEHHA) recommended a safe limit of 0.02 μg l^{-1} for Cr(VI) in 2011. For harmless everyday dose, the WHO set a safety limit of 2–8 μg Cr(III) per day, that is equivalent to 0.03–0.13 μg Cr(III) per day per kg of body weight for a human adult of 60 kg (Zhang et al. 2016).

4.9 Copper

Cu is one of the most important elements for the global economy due to its utilization in several consumer goods. It was one of the first metals that were used by humans. Right from ancient times, it was employed as a restorative material for tooth replacement, cooking utensils, and creating valuable art objects. Cu also exhibits broad importance in the field of technology because of its good electrical and thermal conductivity. It is utilized for manufacturing microelectronic devices, electrical wire, in photography and electroplating, etc. Cu is naturally present in the earth's crust in its metallic forms, ores, and minerals. Cu is poisoned in third transition series of the periodic table and usually exists in three oxidation states Cu^0, Cu^1, and Cu^2. Most of the Cu compounds and complexes are generally blue or green in color and are highly soluble in water. Cu is an essential trace element found in almost every cell of an organism. The human body contains approximately 100 mg of Cu. However, its content varies over time, age, and gender (Bost et al. 2016). Liver and brain are found to have the highest concentration of Cu, i.e. 8–15%. However, 50% of Cu is stored in the muscles and bones (Sass-Kortsak 1966). Cu is crucial for the survival of animals and humans due to its ability to cycle between Cu^{1+} and Cu^{2+}, which permits it to behave as an electron donor or acceptor and is involved in the functioning of several enzymes in the biological system. It is an interesting element as its deficiency and overdose both lead to serious health issues. Therefore, a balanced and precise regulation of Cu is essential in a body to prevent undesired harmful effects or dysbiosis.

4.9.1 Source and Entry of Copper Metal into Our Body

The natural processes like windblown, volcanoes, forest fires, sea spray, decaying vegetation, etc., can emit Cu into the atmosphere. However, anthropogenic emission including coal combustion, waste incineration, metal production, nonferrous metal mining, and fuel combustion also participated in a huge percentage of Cu contamination. Since the global use and demand for Cu increases with time, it becomes an easily available metal that elevates the Cu level drastically in the environment. Although some Cu is naturally found in its elemental form, however, most of it is produced commercially from oxide and sulfide minerals like chalcopyrite, covellite, and chalcocite (Georgopoulos et al. 2001). The extensive use of Cu in industry, mining, smelting, fossil fuel combustion, and the use of pesticides for agriculture upraised the concentration of Cu which could lead to water, soil, and air contamination. Artisanal mining and small-scale mining are a form of exploration and processing of minerals from ore. These involves around 30 million people worldwide (Covre et al. 2022). Waste management is one of the biggest problems in mining industries. Mishandling those waste from extraction and processing can impose severe health risk for the people living in the area due to their harmful impacts into the ecosystem. Around 2.2–3 tons of Cu waste were estimated to be generated per ton of Cu produced, which results in the production of 24.6 million tons of waste per year. Such Cu waste will transfer to water, soil, and plants and are not biodegradable. Thus, these can be accumulated in living organisms and lead to several human health hazards through inhalation, ingestion, and dermal contact.

The primary source of Cu exposure in humans is via drinking water and food. However, the relative intake of Cu via food and water depends on the geographical location. Cu mining, smelting, and industries dump their Cu wastes into the nearest water bodies which further accumulate in the body of fish and other aquatic organisms, thus leading to dietary intake of Cu via the food chain. Also, pipes used for the water distribution system and bath fixtures are often made of Cu or Cu alloys. These elevate the Cu concentration in drinking water by leaching process from the pipe and bath fixtures. Drinking water becomes the major source of Cu intake for humans, and blue or green stains in bath fixtures signify the presence of Cu in water. However, soft acidic groundwater could increase the corrosion process of Cu pipe, thus increasing the Cu concentration in water. Another source of Cu exposure to humans is via agricultural use as pesticides and water treatment to eliminate algae. Cu content in food items varies depending upon the location. Variation in the Cu concentration in soil, use of Cu-based fungicides or bactericides, and slurry or manure spreading could affect the Cu present in fruits, vegetables, and cereals and, to some extent, in meat and animal products. Even with these variations, many food items are found to be rich in Cu content, such as nuts and offal, and to some extent, fruits and cereals are also regarded as good source of Cu.

4.9.2 Utility and Biological Impact of Copper

Cu is an essential and third most abundant metal present in a human body, necessary for the functioning of several biological mechanisms in the body of a living organism. In animals and humans, Cu has two primary functions: first, it acts as an enzyme in oxidation–reduction reactions with examples including lysyl oxidase, cytochrome c oxidase, ferroxidase, 2-furoate-CoA dehydrogenase, superoxide dismutase, hemocyanin, etc. Secondly, it acts as a component in several structural and functional proteins due to its electron transfer capabilities such as neurotransmitter function, connective tissue biosynthesis, and Fe metabolism (Georgopoulos et al. 2001; Taylor et al. 2020). Cu-containing enzymes are responsible for multiple systems like the nervous, respiratory, and immune system, brain function and energy generation in the brain. Many studies suggest that Cu is essential for several physiological processes like child or infant growth, bone strength, host defense mechanisms, cholesterol metabolism, myocardial contractility brain development and function, immune function, and pigment formation (Uriu-Adams and Keen 2005). Cu is also necessary in human for the formation of red and white blood cells, glucose metabolism and for normal Fe metabolism. Maintaining the proper balance of Cu is mandatory to avoid health issues occurring due to deficiency or toxicity of Cu. It was found that Cu is required for fetal development, however a neurodegenerative disorder with progressive ataxia and myelopathy could develop due to Cu deficiency during pregnancy (Araya et al. 2007). Some other effects due to deficiency of Cu include low numbers of white blood cells, anemia, defects in connective tissue leading to skeletal problems, and osteoporosis in infants and children.

After the intake of Cu, it was assumed to be absorbed primarily in the proximal portion of the small intestine which is further transferred to the liver via the portal vein (Stern et al. 2007). In 1989, Turnlund et al. reported that an average of 56% Cu absorption occur in adult humans when the Cu diet was 0.78 mg Cu/d, 36% for 1.68 mg Cu/d, and 12% for 7.53 mg Cu/d. Several parameters that influence the absorption rate of dietary Cu include competition by other metals (like Zn, Fe, Cd, molybdenum (Mo), or Pb), gender, age, and food type. Cu released from the intestine cells, 75% of portal Cu is taken up by the liver and the rest flows to the peripheral circulation. Once in the liver, 80% of Cu either gets incorporated into the Cu-requiring proteins or transported to extrahepatic tissues by amino acids, albumin or mostly by ceruloplasmin and the rest 20% is excreted into the bile (Taylor et al. 2020). Bile is the crucial pathway for the excretion of extra Cu and a suitable way to control the liver's Cu level. Despite low toxicity, failure to balance Cu homeostasis in the body leads to Cu excess or deficiency. However, both result in toxicity and diseases. Cu dysregulation in humans results in two hereditary diseases: Wilson's disease, which is caused due to excess disorder, and Menkes disease, which is caused due to deficiency disorder (Rehman et al. 2019). Both diseases are caused due to genetic defects in Cu transport genes for Cu-ATPases.

Hepatolenticular degeneration or Wilson's disease is a rare autosomal recessive hereditary condition of Cu transport. It involves high Cu concentration into ceruloplasmin because of impaired ability to excrete biliary Cu. Individuals with Wilson's disease are susceptible to nonfunctional ATPase, ATP7B, which produces Cu toxicosis because of its excess accumulation in the liver, brain, to some extent in the eyes, kidneys, and other organs, thus causing hepatocyte death, liver failure, and oxidative damage. Initial symptoms of Wilson's disease include hepatic, psychiatric, or neurologic manifestations and can be caused in the age group of 3 to 50 years.

Menkes disease, first reported in 1962 by physician John Menkes, is an X-linked Cu deficiency syndrome that occurs at a frequency of 1 out of 200,000 live births (Stern et al. 2007; Taylor et al. 2020). This disease is caused due to impaired Cu delivery by protein ATP7A out of intestinal cells and is accompanied by neurological or connective tissue abnormalities due to insufficiency of various Cu-based enzymes necessary for brain development. Menkes patient shows significant symptoms including kinky and brittle hair due to lack of melanin, or loose skin, fragile bones, and aortic aneurysms due to lysyl oxidase activity owing to connective-tissue abnormalities. Menkes disease normally kills its victims by the age of 10, while a handful have lived up to their early 20s.

Acute Cu poisoning or toxicity represents the case when the individuals directly intake Cu-contaminated beverages or water with high quantity of Cu salts. In many acute Cu toxicity cases, this ingestion represents suicide attempts and the dose often exceeds 20 g on a single oral intake. Symptoms reported in these cases include nausea, abdominal pain, headache, vomiting, diarrhea, respiratory difficulties, lethargy, GI bleeding, tachycardia, liver and kidney failure, and death. The mechanisms behind the acute Cu toxicity in humans are still unclear but it is logical to speculate that they possibly represent a significant oxidative stress at different points in the human body. However, direct inhalation of Cu fumes can induce irritation in the upper respiratory tract and cause nausea, vomiting, anorexia, hepatomegaly, and malaise (Ae et al. 2018). Abnormalities in Cu metabolism is responsible for some other diseases like idiopathic copper toxicosis (ICT), and Indian childhood cirrhosis (ICC), or non-ICC (Georgopoulos et al. 2006). These infancy syndromes appear to have a contribution from elevated Cu concentration and genetic component and are similar in their presentation and apparent etiology.

4.9.3 Detection and Remedial Techniques of Copper

The data or facts analyzed in previous paragraphs suggest a clear need for more information about Cu status within the body due to its significant function in several biological mechanisms. Urine and serum sample are often used in clinical laboratory to detect the status of Cu content in a body. The analytical methods that are often applied traditionally for Cu determination are flame and electrothermal atomic absorption spectroscopy (FAAS and ETAAS), and atomic emission spectroscopy (AES), and some other photometric kinetic methods, which have various limitations like bulky, needed costly instruments and professional operational skills. Later square wave anodic stripping voltammetry (SWASV) became the most powerful and sensitive electroanalytical technique for heavy metal detection in a trace level. However, SWASV method also suffers from some disadvantages like poor reproducibility, high reagent use, and instability. Later, lab-on-a-chip (LOC), also known as microfluidics, has gained immense interest due to its numerous advantages like reduced reagent consumption, higher sensitivity, precise control, lowered detection time, detection automation, and manipulation of low-volume fluids (Sharma et al. 2015). Nevertheless, besides having numerous advantages, LOC system still lacks in some aspects. The flow detection system with SWASV has been reported, which increases the accumulation efficiency by assuring that fresh solution flows through the electrode but due to use of external corollary equipment for sample pre-processing, it increases the complexity in the detection process. Thus, integration of microfluidics and flow detection system can overcome all the shortcomings and develops an ideal next-generation system known as point-of-care testing (POCT) system (Jiang et al. 2022).

4.10 Cadmium

4.10.1 Source and Entry of Cadmium Metal into Our Body

Cd is a naturally occurring element which is somewhat rarer as compared to other heavy elements. It majorly finds its use in paints, pigments, and battery industries. About three-quarters of the Cd produced gets used in the electrode production of alkaline cells. This element occurs in nature in mineral forms such as carbonate, sulfide, and selenides. Mostly industrial processes lead to the admixture of Cd in water streams, sewage sludge, groundwater, and soil. Once inside soil, Cd can reside for decades in sediments until it gets absorbed by the plants. Vegetation from the contaminated soil can build up threatening amounts of the element and ingestion of such vegetables and foods cooked out of it can act as a direct exposure pathway to humans. However, contaminated water can act as a more direct route to humans. Aerosolized Cd from air can possess a similar airborne threat which may be generated from burning industrial and municipal wastes already contaminated with Cd. It has also been found that significant amounts of Cd can enter through cigarette smoke, eventually elevating Cd amounts to critical levels in blood.

Ingestion of Cd-contaminated water can disrupt essential mechanisms in body, which can result in several short- and long-term disorders. The International Agency for Research on Cancer (IARC) has identified Cd as a human carcinogen, categorizing it in Group 1 (Kim et al. 2020). Cd exposure to humans can also be possible as a part of an occupational hazard for people working in battery, alloy, paint, and electroplating industries. Vegetation polluted with Cd is a severe threat to humans as it can provide a route toward mass exposure. Seafood, grains and several vegetables were found to contain substantial amounts of Cd contaminants. Upon ingestion of such foodstuff, a significant amount of Cd can get absorbed in the body. It is to be noted carefully that the mass outbreak of Cd poisoning in Japan, named as the Itai-itai disease, was mainly through mass Cd toxication via foodstuff and water supply. The Itai-itai patients suffered from a severe bone disease, which was very painful and degenerative, severe gastrointestinal and lung diseases and kidney failure in acute exposure (Nishijo et al. 2017). It has been observed that Cd is absorbed more in lungs than through gastrointestinal route, which makes Cd a concerning airborne toxicant. Inhalation of Cd-polluted dust in both acute and chronic scenarios can lead to severe lung damage and renal tubular dysfunction. Researchers have found that the blood of smokers (mean blood concentration = $1.3\,\mu g\,l^{-1}$) contained almost thrice the amount of Cd as that of nonsmokers (mean blood concentration = $0.4\,\mu g\,L^{-1}$) (Batáriová et al. 2006). This finding can be attributed to the fact that tobacco plants naturally accumulate higher levels of Cd in their leaves as compared to other parts (Proshad et al. 2020).

4.10.2 Toxicology of Cadmium Poisoning

Human studies have unfolded that Cd binds with metallothionein protein, inducing a dispositional tolerance. This type of induction is important biologically as it harnesses the biodistribution of Cd to crucial cellular macromolecules, thus protecting organs like the liver. In cases of heavy amounts of Cd, however, metallothionein binding is not enough and free unbounded Cd can result in toxic effects (Lech and Sadlik 2017). Studies on non-poisoned population in a 10-year-long examination reveal the details of Cd level in human tissues. Autopsy was performed on several organs, such as heart, liver, lungs, kidneys, small intestines, stomach, and brain. The brain was analyzed to have the lowest level of Cd ($0.02\,\mu g\,g^{-1}$ digested samples), whereas in the case of kidneys and liver, concentrations are alarmingly high, $16\,\mu g\,g^{-1}$ and $1.5\,\mu g\,g^{-1}$, respectively. These figures show that the biodistribution of Cd is mainly toward these two internal organs (Lech and Sadlik 2017). Metallothionein-bound Cd is absorbed by kidneys immediately and the complex thus formed is extremely toxic. Mechanistically, this toxic complex is a part of the chronic nephrotoxicity of Cd poisoning. On the other hand, Cd is also able to replicate the functionality of essential metals in the body, such as Zn, for example. Similar to Zn, Cd also shows albumin binding in plasma. As a consequence of this binding, dysregulation of Fe, Ca, and Zn homeostasis has been observed to occur (Schaefer et al. 2020). Specifically, dysfunction of calcium homeostasis can be attributed to liver injuries caused by Cd poisoning. Research shows that mitochondrial fragmentation can be caused by an increased level of dynamin-related protein 1 (Drp 1), which in turn, can be induced by disruption of calcium homeostasis (Xu et al. 2013).

Apart from these, bone-related issues such as fractures and osteoporosis can also be caused by Cd toxicity. Studies on people living in Cd-polluted areas vs. a nonpolluted control region show results through which high Cd intake levels can be associated directly with osteoporosis for people living in polluted areas. The cause of this osteotoxicity may be due to the reduction of parathyroid hormone (PTH) in serum due to high concentrations of Cd, which can induce calcium release from bone tissues (Schutte et al. 2008).

Cd can be associated with many cardiovascular diseases caused by smoking, among them peripheral artery disease and coronary heart disease being the most common. A study on 4304 middle-aged women and men shows that there can be a deep association of Cd exposure with the pathway for cardiovascular diseases (Li et al. 2019). Carcinogenicity of Cd is quite well known and many forms of kidney, pancreas, lung, prostate, breast, and GI cancer can be attributed to it (Djordjevic et al. 2019). Researchers propose a mechanism leading toward carcinoma through a series of events in the order of ROS generation, increase in levels of tumor necrosis factor-alpha (TNF-α), upregulation of nuclear factor erythroid 2-related factor 2 (Nrf2) and, finally, digressive gene expression leading to mal regulation of cell proliferation and apoptosis resistance (Wang et al. 2018). Nephrocarcinogenicity can also be caused after prolonged and substantial exposure to Cd (Pesch et al. 2000). Researchers hypothesize that the β-catenin signaling pathway imparts a significant role behind renal and nephrocarcinogenesis induced by Cd. This signaling can cause β-catenin protein to accumulate in the cytoplasm. It must be understood that the β-catenin protein is a vital transcriptional factor. The β-catenin overexpression can induce circumvention of the ROS-reliant/β-catenin signaling, promoting kidney cancer (Son et al. 2012).

4.10.3 Detection and Remedial Techniques of Cadmium

For the assessment of Cd exposure and its effects, urinary Cd is recurrently used as a technique. Several studies have established a connection between urinary Cd levels with deficient lung function and renal disorders. It is speculated in this regard that there may be a relation which is dose-responsive. In cases of Cd-induced poisoning assessment, urinary beta 2-microglobulin (B2M) level can be reckoned as a biomarker of tubular proteinuria. The low concentrations of B2M in urine ($360\,\mu g\,l^{-1}$) can impart a favorable role in this regard (Argyropoulos et al. 2017).

Remediation of Cd poisoning from the human body is not very effective through chelation therapy like in the case of some other heavy metals. Thus, prevention from additional exposure must be considered the first line of defense. Several chelating ligands like ethylenediaminetetraacetic acid (EDTA), dimercaptosuccinic acid (DMSA), 2,3-dimercapto-1-propanesulfonic acid (DMPS), and 2,3-dimercaptopropanol (known commonly as British anti-lewisite, BAL) are available for clinical use. DMPS, EDTA, and DMSA have been found to be increasing renal excretion of Cd. Various studies also show that EDTA is more effective than the others mentioned in the list to mobilize intracellular Cd, and it is approved for use by FDA (Bernhoft 2013).

4.11 Nickel

4.11.1 Source and Entry of Nickel Metal into Our Body

Ni is a silvery white shiny metal and is the 28th element in the periodic table. It exhibits variable valency from −1 to +4. However, +2 is the most common in biologically relevant systems. Ni is extremely biologically relevant and is often found in many metalloenzymes. It can be found widely distributed in the environment, mostly in soil, water, and sometimes, air. Ni gets solubilized in water through various industrial processes and natural weathering from soil deposits. The solubilization process mostly occurs at a lower pH than 6.5 as above a pH of 6.7 Ni mostly remains as insoluble hydroxides. Chloride, nitrate, sulfate, etc., salts of Ni are freely soluble in water, whereas salts like phosphates are insoluble in water. Acidic effluents from Ni refineries are one of the prime sources of soluble Ni contamination both directly and indirectly through solubilization from sludges or deposits. Major uses of Ni involve alkaline battery, stainless steel, mordant, and coating industries, to name a few. Soluble Ni salts are used mostly in electroplating and as a mordant. Other Ni-based compounds, complexes, and the metal itself in powder or nanoparticle form are used thoroughly in various catalytic processes. As a result of all the industrial processing, Ni gets released into the environment and eventually affects humans directly or indirectly (Denkhaus and Salnikow 2002).

Humans can be exposed to Ni in many ways including inhalation, ingestion, and skin contact. Inhalation of airborne particles during grinding and processing of metallic Ni and its alloys, mists containing soluble Ni salts during electroplating and refining can be a prime way for Ni exposure to humans. An occupational health hazard is considered the most common way of Ni toxicity in humans. Notably, a higher magnitude of occupational exposure to Ni has been found in comparison to that nonoccupational exposure, which can enter through food or water. Elevated levels of Ni have been observed in blood, urine, or tissues for people who have been exposed to Ni sources, mostly through inhalation. For the general mass population, food and water are the media through which Ni exposure mostly occurs. Many household items like water heaters, storage tanks, and kettles are made with stainless steel containing Ni or Ni-plating as a corrosion inhibitor. Tests suggest that boiling water in such kettles can significantly leach soluble Ni in the water, which can have fatal consequences for human health.

4.11.2 Toxicology of Nickel Poisoning

Human exposure to heavily Ni-polluted air may cause varying degrees of pathological effects. Incidences were observed for lung and nasal cancer in people working in Ni-based industries for prolonged periods of time (Sunderman Jr et al. 1988). Apart from lung and nasal carcinoma, Ni is often associated with contact dermatitis. Like other toxicants, the toxicology of Ni is dependent on the pathway it takes to get inside an organism. Inhalation, ingestion, and absorption via the dermis are different contact mechanisms for exposure. However, the cellular entry of Ni in the body is dependent on its chemical form. Nickel Carbonyl ($Ni(CO)_4$) can enter the human body through inhalation due to its volatile nature. Moreover, it is very fat soluble. Thus, it can go across cell membranes via diffusion or with the aid of calcium channels (Cangul et al. 2002). In contrast to carbonyls, insoluble particles can enter cells by phagocytosis (Heck and Costa 1982).

There are various transport proteins which can bind to Ni in the body. Albumin is by far the most common, but binding to histidine and α2-macroglobulin is also known (Glennon and Sarkar 1982). The bound state of Ni with transport proteins is capable of biodistribution throughout the tissues. Research revealed a number of proteins which can bind with Ni, namely α1-lipoprotein, prealbumin, and α1-antitrypsin (Nielsen et al. 1994). Kidneys, lungs, liver, brain, bones, and endocrine glands were found to have the highest concentrations of Ni in the body. However, it was also found in some samples of saliva, nails, hair, and breast milk. Experimental results unveil that Ni is not accumulated in the body and it gets excreted mostly via urinary or fecal routes, and in some cases, via bile and sweat (Valko et al. 2005). Skin contact with Ni in its soluble form can cause serious health issues including allergic dermatitis (Szczepaniak and Prokop 2004; Vahter et al. 2002). Acute toxicity of Ni via the ingestion route includes nausea, vomiting, headache, abdominal cramps, and visual disparity. In larger dosages (>0.5 g), ingestion of Ni can cause death (Daldrup et al. 1983).

As discussed earlier, inhalation is the fatal pathway of exposure to Ni. However, the fatality is dependent on the form of the toxicant. Soluble salts of Ni can be easily absorbed by lungs but also can get excreted through the bloodstream and finally in a urinary pathway. However, insoluble forms of Ni can remain in lungs for a much longer time and keep building up lethal concentrations. Soluble Ni salts upon inhalation can trigger irritation in sinuses and may lead to loss of smell or, in a worse case, perforation in the nasal septum. Serious illnesses like bronchitis, asthma, and breathing shortness can be caused by prolonged exposure. $Ni(CO)_4$ is considered the most toxic and acute form of Ni toxicity. Not only of Ni but it also comes with the toxic effects of CO, thus causing two-way damage. Short exposure with the carbonyl can cause nausea, chest pain, and breathing trouble, but longer exposures can easily trigger pneumonia or even death. IARC 1997 has classified soluble forms of Ni in Group 1 as there are sufficient evidences of carcinogenicity caused by it in humans and other animals. However, in the case of metallic Ni, there is inadequate evidence of human carcinoma. Thus, it was placed in Group 2B as a suspected carcinogen. Researchers have found out that Ni is capable of blocking calcium channels and can cause carcinogenic effects due to altered calcium metabolism in the body (Zamponi et al. 1996). It is also reported in literature outcomes that in a high Ni concentration, it can significantly impair Fe utilization.

Ni has been identified as a cause of inflammatory response by regulating the expression of transcription factors which are involved in inflammatory processes. NF-κB activation induced by Ni can cause changes in tissue and cellular responses. This result also explains the observation of allergic effects and dermal hypersensitivity induced by Ni (Viemann et al. 2007). NF-κB is a known transcription factor which is very important for biological processes like apoptosis, inflammation response, and expressions of adhesion molecules. Experiments have unveiled that soluble Ni(II) is capable of upregulating endothelial leukocyte adhesion molecule-1, intercellular adhesion molecule-1 and vascular cell adhesion molecule-1 when cultured in human endothelial cells (Goebeler et al. 1993). Ni-induced oxidative stress has been researched widely by numerous research groups all across the globe. ROS generation from Ni compounds has been studied in human plasma.

Lipid peroxidation in human blood plasma can be caused by soluble Ni salts in a time- and concentration-dependent fashion. Ni compounds are notoriously known to produce hydroxyl radicals which can destroy α-tocopherol in plasma. Antioxidants such as GSH, catechin, and mannitol were found to be effective in reducing lipid peroxidation and the formation of hydroxyl radicals (Chen et al. 2002).

Further progress in experimental research has aided in disentangling the complexity of molecular pathways involving Ni-induced carcinogenicity. It has been found that Ni compounds can assist hypermethylation in DNA while also aiding histone deacetylation. All of these changes can lead to inactivation of the genetic expression. The mechanisms of Ni-induced DNA hypermethylation is unknown. However, a possible model can be proposed keeping in mind the ability of Ni to replace magnesium, increase condensation of chromatin, and trigger de novo methylation in DNA (Lee et al. 1995).

4.11.3 Remedial Techniques

Chelation therapy is by far the most effective method for the remediation of Ni-based toxicity. For acute poisoning of $Ni(CO)_4$ and any other soluble Ni salts, sodium diethyldithiocarbamate (dithiocarb) has been tested as a very effective and specific antidote. For milder and chronic toxicities, tetraethylthiuram (antabuse, oxidized form of diethyldithiocarbamate) can be used as an antidote but to a lesser efficacy. BAL has very limited therapeutic efficacy in treating Ni poisoning. For Ni-based dermal toxicities, placing the patient on a low Ni diet is found to be effective along with oral administration of dithiocarb or antabuse (Sunderman 1981).

4.12 Radioactive Heavy Metals

There are currently 118 elements reported in our modern periodic table. The nucleus of some elements is not stable, and they transform into new elements by releasing energy through alpha, beta, and gamma decay, and spontaneous fission. Such elements are known as radioactive elements, and the property of these elements is termed radioactivity (Figure 4.2). In the human body, crucial radioactive elements are ^{226}Ra, ^{228}Ra, ^{232}Th, ^{234}U, and ^{238}U (Lerebours et al. 2018; Thomas et al. 2022).

4.12.1 Source of Radioactive Heavy Metals

Uranium (U) element, discovered in 1789, has an atomic number of 92 and is the heaviest element found in nature. There are three naturally occurring isotopes of uranium, and out of the three, ^{238}U constitutes 99%, followed by ^{235}U (0.72%) and ^{234}U (0.0055). Additionally, three more isotopes ^{232}U, ^{233}U, and ^{236}U are also produced by nuclear reactors (Lee et al. 2020). Essentially, U is found in very low concentrations in soil, rocks, and water. It is also significantly found in some substances such as lignite, phosphate rock deposits, etc. In nature, uranium is not found in its elemental form but as minerals, viz. autunite, uranophane, carnotite, brannerite, and uraninite (Pandelus et al. 2021).

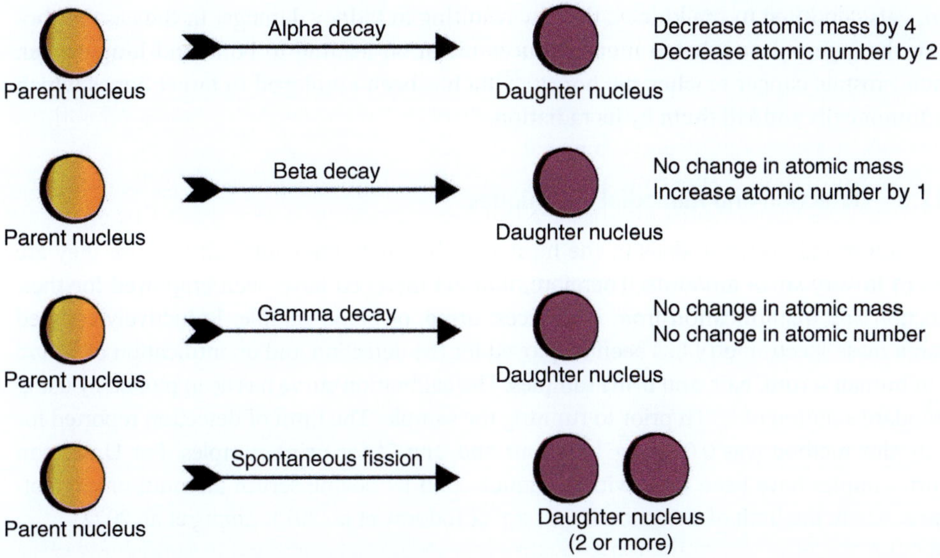

Figure 4.2 Radioactive material decay types.

Thorium (Th), atomic number 90, was discovered in 1828 and is formed from U via radioactive decay. It is found in monazite, thorite, and thorianite minerals and basically exits as ^{228}Th, ^{230}Th, and ^{232}Th isotopes. The primary isotope, ^{232}Th has 99.98% abundance, whereas ^{230}Th isomer is found in 0.0002% in nature.

Radium (Ra), atomic number 88, was discovered in 1898 and is found in uranium and thorium ores. It has 33 isotopes (^{202}Ra to ^{234}Ra), out of which ^{226}Ra has the longest lifetime. In nature, these radioactive heavy metals occur in trace amounts in soil, water, plants, rocks, animals, etc., and humans get inadvertently exposed to them by inhalation, ingestion, and absorption via the skin.

4.12.2 Utility and Biological Impact of Radioactive Metal on Health

Radioactive metals are mostly toxic to human health; however, they were formerly used in medical radiology and cancer therapy. Thorium dioxide was used as a contrast agent in medical radiology until 1950 and radium bromide, and radium chloride was used as radon sources in the earlier 1930 in cancer treatment but was discontinued due to their adverse toxic effects. Everybody is introduced to low levels of uranium because it is found in nature, but higher levels of radioactive materials can be harmful to health. According to EPA, the maximum contamination level is $30\,\mu g\,l^{-1}$ for uranium elements in drinking water and $5\,pCi\,g^{-1}$ for 15 cm of soil. Thorium content present in the human body is 80 mBq (average), out of which 60% is localized in the skeleton (National Research Council (US) Committee on the Biological Effects of Ionizing Radiations 1988).

Although 99% of the ingested radioactive metals leave the body via feces and urine, a small amount will remain for several years in the person's bone. In the case of U, the retained metal releases alpha particles which can kill or damage live cells. If uranium enters the body by wound, it is stacked in the wound or is absorbed in the blood, which in turn, gets deposited in the kidney, thereby resulting in kidney damage. In the case of thorium, the ingested or inhaled metal induces radiation leading to bone and lung cancer. When prostate cancer reaches the bones, ^{223}Ra has been employed to target bones which has tumor cells and kill them by its radiation.

4.12.3 Detection and Remedial Techniques

Detection of radioactive metals in the human body is extremely difficult because they are present in very small amounts. Therefore, indirect methods have been employed for their detection and quantification from hair, feces, urine, or blood sample. Inductively coupled plasma mass spectrometry has been employed for the detection and quantification of U and Th in human serum, hair, and urine samples. The calibration curve has been plotted by using a standard solution of U/Th prior to running the sample. The limit of detection reported for Th by this method was $0.024\,ng\,g^{-1}$ for hair and $2\,ng\,l^{-1}$ for urine samples. For U, human serum samples have been used without liquefaction by bovine serum albumin or chymotrypsin, where the limit of detection is $0.8\,pg\,g^{-1}$ (Todorov et al. 2013; Zhang et al. 2022).

4.13 Conclusion

Heavy metals influence the generation of ROS to a large extent via Fenton, photo-Fenton and similar pathways, thereby impacting the biochemical processes and physiological well-being. Most of these heavy metals cross the blood–brain barrier and impose a lasting detrimental effect on the psychological and neurological aspects of the human psyche. Several anthropogenic activities, viz. mining, paints, and ceramic industry, fertilizers, pesticides, electronics, etc., are actively responsible for the contamination of the environment for healthy human sustenance. However, the primary sources of heavy metal contaminants are natural calamities like volcanoes, erosions, and landslides. Haphazard release of heavy metal wastes, particularly Pb, Fe, Hg, As, etc., as effluents from factories and industrial plants should be strictly regulated. Most of these toxic metals are present in substantial amounts in the drinking water, and so the government and social charitable organizations are actively involved in the remediation of the water sources. Minimization of heavy metals in the food, water, and atmosphere should be globally addressed. In spite of heavy metals posing a threat to the healthy sustenance of mankind, they cannot be blatantly avoided. As the father of toxicology, Paracelsus (1493–1541) famously said, "All things are poison and nothing is without poison; only the dose permits something not to be poisonous." It might appear to be an oxymoron statement. However, it has been found and proven that humans and many other living organisms need small quantities of these "bad" or "toxic" metals to function normally.

References

Ae, S., Ws, A., Dm, E.H., and Ha, S. (2018). Health impact of exposure to copper and zinc among workers in A metal die casting foundry in 10th Ramadan city. *Egyptian Journal of Occupational Medicine* 42 (3): 383.

Agnew, U.M. and Slesinger, T.L. (2022). *Zinc Toxicity*. Treasure Island, FL: StatPearls Publishing.

Andreini, C., Bertini, I., Cavallaro, G. et al. (2008). Metal ions in biological catalysis: from enzyme databases to general principles. *Journal of Biological Inorganic Chemistry* 13 (8): 1205.

Araya, M., Olivares, M., and Pizarro, F. (2007). Copper in human health. *International Journal of Environmental Health* 1 (4): 608.

Argyropoulos, C.P., Chen, S.S., Ng, Y.-H. et al. (2017). Rediscovering beta-2 microglobulin as a biomarker across the spectrum of kidney diseases. *Frontiers in Medicine* 4: 73.

Baranwal, A.K. and Singhi, S.C. (2003). Acute iron poisoning: management guidelines. *Indian Pediatrics* 40 (6): 534.

Batáriová, A., Spěváčková, V., Beneš, B. et al. (2006). Blood and urine levels of Pb, Cd and Hg in the general population of the Czech Republic and proposed reference values. *International Journal of Hygiene and Environmental Health* 209 (4): 359.

Baxter, P.J., Samuel, A.M., and Holkham, M.P. (1985). Lead hazard in British stained glass workers. *British Medical Journal (Clinical Research Ed.)* 291 (6492): 383.

Bernhoft, R.A. (2013). Cadmium toxicity and treatment. *The Scientific World Journal* 2013: 394652.

Borowska, S. and Brzóska, M.M. (2015). Metals in cosmetics: implications for human health. *Journal of Applied Toxicology* 35 (6): 551.

Bost, M., Houdart, S., Oberli, M. et al. (2016). Dietary copper and human health: current evidence and unresolved issues. *Journal of Trace Elements in Medicine and Biology* 35: 107.

Cangul, H., Broday, L., Salnikow, K. et al. (2002). Molecular mechanisms of nickel carcinogenesis. *Toxicology Letters* 127 (1): 69.

Caravanos, J., Carrelli, J., Dowling, R. et al. (2016). Burden of disease resulting from lead exposure at toxic waste sites in Argentina, Mexico and Uruguay. *Environmental Health* 15 (1): 72.

Castro-González, M.I. and Méndez-Armenta, M. (2008). Heavy metals: implications associated to fish consumption. *Environmental Toxicology and Pharmacology* 26 (3): 263.

Chen, C.-Y., Su, Y.-J., Wu, P.-F., and Shyu, M.-M. (2002). Nickel-induced plasma lipid peroxidation and effect of antioxidants in human blood: involvement hydroxyl radical formation and depletion of α-tocopherol. *Journal of Toxicology and Environmental Health. Part A* 65 (12): 843.

Chen, P., Bornhorst, J.B., and Aschner, M. (2018). Manganese metabolism in humans. *Frontiers in Bioscience (Landmark Edition)* 23 (9): 1655.

Cheng, S. (2003). Heavy metal pollution in China: origin, pattern and control. *Environmental Science and Pollution Research* 10 (3): 192.

Chung, J.-Y., Yu, S.-D., and Hong, Y.-S. (2014). Environmental source of arsenic exposure. *Journal of Preventive Medicine and Public Health* 47 (5): 253.

Connett, P.H. and Wetterhahn, K.E. (1983). *Inorganic Elements in Biochemistry*, 93. Berlin, Heidelberg: Springer-Verlag.

Costa, M. (1997). Toxicity and carcinogenicity of Cr(VI) in animal models and humans. *Critical Reviews in Toxicology* 27 (5): 431.

Covre, W.P., Ramos, S.J., da Silveira Pereira, W.V. et al. (2022). Impact of copper mining wastes in the Amazon: properties and risks to environment and human health. *Journal of Hazardous Materials* 421: 126688.

Crossgrove, J. and Zheng, W. (2004). Manganese toxicity upon overexposure. *NMR in Biomedicine* 17 (8): 544.

Daldrup, T., Haarhoff, K., and Szathmary, S.C. (1983). Toedliche nickel sulfate-intoxikation. *Berichte zur Gerichtlichen Medizin* 41: 141.

De Acha, N., Elosúa, C., Corres, J.M., and Arregui, F.J. (2019). Fluorescent sensors for the detection of heavy metal ions in aqueous media. *Sensors* 19 (3): 599.

Debnath, B., Singh, W.S., and Manna, K. (2019). Sources and toxicological effects of lead on human health. *Indian Journal of Medical Specialities* 10: 66.

Denkhaus, E. and Salnikow, K. (2002). Nickel essentiality, toxicity, and carcinogenicity. *Critical Reviews in Oncology/Hematology* 42 (1): 35.

Denny, M.F. and Atchison, W.D. (1996). Mercurial-induced alterations in neuronal divalent cation homeostasis. *Neurotoxicology* 17 (1): 47.

Djordjevic, V.R., Wallace, D.R., Schweitzer, A. et al. (2019). Environmental cadmium exposure and pancreatic cancer: evidence from case control, animal and in vitro studies. *Environment International* 128: 353.

Ferreirós-Martínez, R., Esteban-Gómez, D., Tóth, É. et al. (2011). Macrocyclic receptor showing extremely high Sr(II)/Ca(II) and Pb(II)/Ca(II) selectivities with potential application in chelation treatment of metal intoxication. *Inorganic Chemistry* 50 (8): 3772.

Flora, S.J.S. and Pachauri, V. (2010). Chelation in metal intoxication. *International Journal of Environmental Research and Public Health* 7 (7): 2745.

Flora, S.J.S., Mittal, M., and Mehta, A. (2008). Heavy metal induced oxidative stress & its possible reversal by chelation therapy. *The Indian Journal of Medical Research* 128 (4): 501.

Gambelunghe, A., Piccinini, R., Ambrogi, M. et al. (2003). Primary DNA damage in chrome-plating workers. *Toxicology* 188 (2): 187.

Gammoh, N.Z. and Rink, L. (2019). Zinc and the immune system. In: *Nutrition and Immunity* (ed. M. Mahmoudi and N. Rezaei). Cham: Springer International Publishing doi:10.1007/978-3-030-16073-9_8.

Georgopoulos, G., Roy, A., Yonone-Lioy, M.J. et al. (2001). Environmental copper: its dynamics and human exposure issues. *Journal of Toxicology and Environmental Health, Part B* 4 (4): 341.

Georgopoulos, P.G., Wang, S.W., Georgopoulos, I.G. et al. (2006). Assessment of human exposure to copper: a case study using the NHEXAS database. *Journal of Exposure Science & Environmental Epidemiology* 16 (5): 397.

Gittleman, J.L., Engelgau, M.M., Shaw, J. et al. (1994). Lead poisoning among battery reclamation workers in Alabama. *Journal of Occupational Medicine* 36 (5): 526.

Glennon, J.D. and Sarkar, B. (1982). Nickel(II) transport in human blood serum. Studies of nickel(II) binding to human albumin and to native-sequence peptide, and ternary-complex formation with l-histidine. *The Biochemical Journal* 203 (1): 15.

Goebeler, M., Meinardus-Hager, G., Roth, J. et al. (1993). Nickel chloride and cobalt chloride, two common contact sensitizers, directly induce expression of intercellular adhesion molecule-1 (ICAM-1), vascular cell adhesion molecule-1 (VCAM-1), and endothelial leukocyte adhesion molecule (ELAM-1) by endothelial cells. *The Journal of Investigative Dermatology* 100 (6): 759.

Gonzalvo, M.C., Gil, F., Hernández, A.F. et al. (1997). Inhibition of paraoxonase activity in human liver microsomes by exposure to EDTA, metals and mercurials. *Chemico-Biological Interactions* 105 (3): 169.

Gumbleton, M. and Nicholls, P.J. (1988). Dose-response and time-response biochemical and histological study of potassium dichromate-induced nephrotoxicity in the rat. *Food and Chemical Toxicology* 26 (1): 37.

Heck, J.D. and Costa, M. (1982). Surface reduction of amorphous NiS particles potentiates their phagocytosis and subsequent induction of morphological transformation in Syrian hamster embryo cells. *Cancer Letters* 15 (1): 19.

Hossain, M., Garai, R., Gupta, R.K. et al. (2021). Fluoroarene derivative based passivation of perovskite solar cells exhibiting excellent ambient and thermo-stability achieving efficiency >20%. *Journal of Materials Chemistry C* 9 (32): 10406.

Houk, V.N., Ing, R.T., and Falk, H. (1989). Assessing lead exposure from drinking water. *American Journal of Public Health* 79 (7): 823.

Institute of Medicine (US) Panel on Micronutrients (2001). *Dietary Reference Intakes for Vitamin A, Vitamin K, Arsenic, Boron, Chromium, Copper, Iodine, Iron, Manganese, Molybdenum, Nickel, Silicon, Vanadium, and Zinc*. Washington, DC: The National Academies Press.

International Zinc Nutrition Consultative Group (IZiNCG), Brown, K.H., Rivera, J.A. et al. (2004). International Zinc Nutrition Consultative Group (IZiNCG) technical document 1: assessment of the risk of zinc deficiency in populations and options for its control. *Food and Nutrition Bulletin* 25 (1 Suppl 2): S99–S203.

James, A.K., Nehzati, S., Dolgova, N.V. et al. (2020). Reply to comments on "Rethinking the minamata tragedy: what mercury species was really responsible?". *Environmental Science & Technology* 54 (13): 8488.

Jan, A.T., Azam, M., Siddiqui, K. et al. (2015). Heavy metals and human health: mechanistic insight into toxicity and counter defense system of antioxidants. *International Journal of Molecular Sciences* 16 (12): 29592.

Jiang, S., Zhao, J., Zhao, S. et al. (2022). Hybrid 3D printed integrated microdevice for the determination of copper ions in human body fluids. *Analytical and Bioanalytical Chemistry* 414 (14): 4047.

Juberg, D.R., Kleiman, C.F., and Kwon, S.C. (1997). Position paper of the American Council on Science and Health: lead and human health. *Ecotoxicology and Environmental Safety* 38 (3): 162.

Kanakoudis, V. and Keramaris, E. (2020). *Environmental Sciences Proceedings*, vol. Vol. 2.

Kim, E. and Na, K.J. (1991). Nephrotoxicity of sodium dichromate depending on the route of administration. *Archives of Toxicology* 65 (7): 537.

Kim, T.-H., Kim, J.H., Le Kim, M.D. et al. (2020). Exposure assessment and safe intake guidelines for heavy metals in consumed fishery products in the Republic of Korea. *Environmental Science and Pollution Research* 27 (26): 33042.

Kulka, M. (2016). A review of paraoxonase 1 properties and diagnostic applications. *Polish Journal of Veterinary Sciences* 19 (1): 225–232.

Law, C., Miseikis, L., Dimitrov, S. et al. (2014). Performance and stability of lead perovskite/TiO2, polymer/PCBM, and dye sensitized solar cells at light intensities up to 70 suns. *Advanced Materials* 26 (36): 6268.

Lazarchick, J. (2012). Update on anemia and neutropenia in copper deficiency. *Current Opinion in Hematology* 19 (1): 58–60.

Lech, T. and Sadlik, J.K. (2017). Cadmium concentration in human autopsy tissues. *Biological Trace Element Research* 179 (2): 172.

Lee, Y.W., Klein, C.B., Kargacin, B. et al. (1995). Carcinogenic nickel silences gene expression by chromatin condensation and DNA methylation: a new model for epigenetic carcinogens. *Molecular and Cellular Biology* 15 (5): 2547.

Lee, C.-G., Park, R., Park, J., and Lim, S.H. (2020). A comparative study of ultra-trace-level uranium by thermal ionization mass spectrometry with continuous heating: static and peak-jumping modes. *Nuclear Engineering and Technology* 52 (7): 1532.

Lerebours, A., Gudkov, D., Nagorskaya, L. et al. (2018). Impact of environmental radiation on the health and reproductive status of fish from chernobyl. *Environmental Science & Technology* 52 (16): 9442.

Li, L. and Yang, X. (2018). The essential element manganese, oxidative stress, and metabolic diseases: links and interactions. *Oxidative Medicine and Cellular Longevity* 2018: 7580707.

Li, S., Zhengyan, Z., Rong, L., and Hanyun, C. (2005). Decrease of CD4+ T-lymphocytes in children exposed to environmental lead. *Biological Trace Element Research* 105 (1): 19.

Li, Y., He, B., Hu, L. et al. (2018). Characterization of mercury-binding proteins in human neuroblastoma SK-N-SH cells with immobilized metal affinity chromatography. *Talanta* 178: 811.

Li, H., Fagerberg, B., Sallsten, G. et al. (2019). Smoking-induced risk of future cardiovascular disease is partly mediated by cadmium in tobacco: Malmö Diet and Cancer Cohort Study. *Environmental Health* 18 (1): 56.

Lian, J., Xu, Q., Wang, Y., and Meng, F. (2020). Recent developments in fluorescent materials for heavy metal ions analysis from the perspective of forensic chemistry. *Frontiers in Chemistry* 8: 593291.

Lucchini, R., Apostoli, P., Perrone, C. et al. (1999). Long-term exposure to "low levels" of manganese oxides and neurofunctional changes in ferroalloy workers. *Neurotoxicology* 20 (2–3): 287.

Madiwale, T. and Liebelt, E. (2006). Iron: not a benign therapeutic drug. *Current Opinion in Pediatrics* 18 (2): 174.

Mattia, G.D., Bravi, M.C., Laurenti, O. et al. (2004). Impairment of cell and plasma redox state in subjects professionally exposed to chromium. *American Journal of Industrial Medicine* 46 (2): 120.

Min, J.-Y., Min, K.-B., Kim, R. et al. (2008). Blood lead levels and increased bronchial responsiveness. *Biological Trace Element Research* 123 (1): 41.

National Research Council (US) Committee on the Biological Effects of Ionizing Radiations (1988). *Health Risks of Radon and Other Internally Deposited Alpha-Emitters: BEIR IV*. National Academies Press (US).

Naujokas Marisa, F., Anderson, B., Ahsan, H. et al. (2013). The broad scope of health effects from chronic arsenic exposure: update on a worldwide public health problem. *Environmental Health Perspectives* 121 (3): 295.

Neupane, L.N., Oh, E.-T., Park, H.J., and Lee, K.-H. (2016). Selective and sensitive detection of heavy metal ions in 100% aqueous solution and cells with a fluorescence chemosensor based on peptide using aggregation-induced emission. *Analytical Chemistry* 88 (6): 3333.

Nielsen, F.H. (2012). *Present Knowledge in Nutrition*. Wiley-Blackwell doi:10.1002/9781119946045.ch38.

Nielsen, J.L., Poulsen, O.M., and Abildtrup, A. (1994). Studies of serum protein complexes with nickel using crossed immunoelectrophoresis. *Electrophoresis* 15 (1): 666.

Nishijo, M., Nakagawa, H., Suwazono, Y. et al. (2017). Causes of death in patients with Itai-itai disease suffering from severe chronic cadmium poisoning: a nested case–control analysis of a follow-up study in Japan. *BMJ Open* 7 (7): e015694.

Nordberg, G.F., Nordberg, M., and Costa, M. (2022). Chapter 1 – Toxicology of metals: overview, definitions, concepts, and trends*. In: *Handbook on the Toxicology of Metals: Volume I: General Consideration* (ed. G.F. Nordberg and M. Costa), 1–14. London: Academic Press.

Nriagu, J.O. (1989). A global assessment of natural sources of atmospheric trace metals. *Nature* 338 (6210): 47.

Pandelus, S.B., Kennedy, B.J., Murphy, G. et al. (2021). Phase analysis of Australian uranium ore concentrates determined by variable temperature synchrotron powder X-ray diffraction. *Inorganic Chemistry* 60 (15): 11569.

Pandy, J., Khare, R., Kamboj, M. et al. (2011). Potential of nanotechnology for the treatment of waste water. *Asian Journal of Pharmaceutical Research* 1: 272.

Pesch, B., Haerting, J., Ranft, U. et al. (2000). Occupational risk factors for renal cell carcinoma: agent-specific results from a case-control study in Germany. *International Journal of Epidemiology* 29 (6): 1014.

Piccoli, C., D'Aprile, A., Scrima, R. et al. (2012). Subcytotoxic mercury chloride inhibits gap junction intercellular communication by a redox- and phosphorylation-mediated mechanism. *Free Radical Biology & Medicine* 52 (5): 916.

Pisani, P., Renna, M.D., Conversano, F. et al. (2016). Major osteoporotic fragility fractures: risk factor updates and societal impact. *World Journal of Orthopedics* 7 (3): 171.

Proshad, R., Zhang, D., Uddin, M., and Wu, Y. (2020). Presence of cadmium and lead in tobacco and soil with ecological and human health risks in Sichuan province, China. *Environmental Science and Pollution Research* 27 (15): 18355.

Quansah, R., Armah, F.A., Essumang, D.K. et al. (2015). Association of arsenic with adverse pregnancy outcomes/infant mortality: a systematic review and meta-analysis. *Environmental Health Perspectives* 123 (5): 412.

Rehman, M., Liu, L., Wang, Q. et al. (2019). Copper environmental toxicology, recent advances, and future outlook: a review. *Environmental Science and Pollution Research* 26 (18): 18003.

Rieuwerts, J. (2017). *The Elements of Environmental Pollution*. Routledge.

Sangeetha, V.J., Dutta, S., Moses, J.A., and Anandharamakrishnan, C. (2022). Zinc nutrition and human health: overview and implications. *eFood* 3 (5): e17.

Sass-Kortsak, A. (1966). *Advances in Clinical Chemistry*, vol. Vol. 8 (ed. H. Sobotka and C.P. Stewart). Elsevier.

Schaefer, H.R., Dennis, S., and Fitzpatrick, S. (2020). Cadmium: mitigation strategies to reduce dietary exposure. *Journal of Food Science* 85 (2): 260.

Schutte, R., Nawrot Tim, S., Richart, T. et al. (2008). Bone resorption and environmental exposure to cadmium in women: a population study. *Environmental Health Perspectives* 116 (6): 777.

Sharma, S., Zapatero-Rodríguez, J., Estrela, P., and Kennedy, R. (2015). Point-of-care diagnostics in low resource settings: present status and future role of microfluidics. *Biosensors* 5 (3): 577.

Shelnutt, S.R., Goad, P., and Belsito, D.V. (2007). Dermatological toxicity of hexavalent chromium. *Critical Reviews in Toxicology* 37 (5): 375.

Singhi, S.C. and Baranwal, A.K. (2003). Acute iron poisoning: clinical picture, intensive care needs and outcome. *Indian Pediatrics* 40 (12): 1177.

Son, Y.-O., Wang, L., Poyil, P. et al. (2012). Cadmium induces carcinogenesis in BEAS-2B cells through ROS-dependent activation of PI3K/AKT/GSK-3β/β-catenin signaling. *Toxicology and Applied Pharmacology* 264 (2): 153.

Stern, B.R., Solioz, M., Krewski, D. et al. (2007). Copper and human health: biochemistry, genetics, and strategies for modeling dose-response relationships. *Journal of Toxicology and Environmental Health, Part B* 10 (3): 157.

Sunderman, F.W. (1981). Chelation therapy in nickel poisoning. *Annals of Clinical and Laboratory Science* 11 (1): 1.

Sunderman, F.W. Jr., Dingle, B., Hopfer, S.M., and Swift, T. (1988). Acute nickel toxicity in electroplating workers who accidently ingested a solution of nickel sulfate and nickel chloride. *American Journal of Industrial Medicine* 14 (3): 257.

Szczepaniak, E. and Prokop, J. (2004). Wyprysk kontaktowy. *Postepy Dermatol Alergol* 21 (4): 205–210.

Takeda, A. (2003). Manganese action in brain function. *Brain Research Reviews* 41 (1): 79.

Taylor, A.A., Tsuji, J.S., Garry, M.R. et al. (2020). Critical review of exposure and effects: implications for setting regulatory health criteria for ingested copper. *Environmental Management* 65 (1): 131.

Thomas, J.R., Sreejith, M.V., Aravind, U.K. et al. (2022). Outdoor and indoor natural background gamma radiation across Kerala, India. *Environmental Science: Atmospheres* 2 (1): 65.

Thürmer, K., Williams, E., and Reutt-Robey, J. (2002). Autocatalytic oxidation of lead crystallite surfaces. *Science* 297 (5589): 2033.

Todorov, T.I., Ejnik, J.W., Guandalini, G. et al. (2013). Uranium quantification in semen by inductively coupled plasma mass spectrometry. *Journal of Trace Elements in Medicine and Biology* 27 (1): 2.

United Nations Environment Program (1991). *UNEP Environmental Data Report (1991).* London, UK: GEMS Monitoring and Assessment Research Center.

United Nations Environment Programme (2009). *UNEP Annual Report 2008*. Nairobi, Kenya: UNEP.

Uriu-Adams, J.Y. and Keen, C.L. (2005). Copper, oxidative stress, and human health. *Molecular Aspects of Medicine* 26 (4): 268.

Uthus, E.O. (2003). Arsenic essentiality: a role affecting methionine metabolism. *Journal of Trace Elements in Experimental Medicine* 16 (4): 345.

Vahter, M., Berglund, M., Åkesson, A., and Lidén, C. (2002). Metals and women's health. *Environmental Research* 88: 145–155.

Valera, B., Dewailly, E., and Poirier, P. (2008). Cardiac autonomic activity and blood pressure among Nunavik Inuit adults exposed to environmental mercury: a cross-sectional study. *Environmental Health* 7 (1): 29.

Valera, B., Muckle, G., Poirier, P. et al. (2012). Cardiac autonomic activity and blood pressure among Inuit children exposed to mercury. *Neurotoxicology* 33 (5): 1067.

Valko, M., Morris, H., and Cronin, T.D.M. (2005). Metals, toxicity and oxidative stress. *Current Medicinal Chemistry* 12 (10): 1161.

Velma, V., Vutukuru, S.S., and Tchounwou, P.B. (2009). Ecotoxicology of hexavalent chromium in freshwater fish: a critical review. *Reviews on Environmental Health* 24 (2): 129.

Verity, M.A. (1999). Manganese neurotoxicity: a mechanistic hypothesis. *Neurotoxicology* 20 (2–3): 489.

Viemann, D., Schmidt, M., Tenbrock, K. et al. (2007). The contact allergen nickel triggers a unique inflammatory and proangiogenic gene expression pattern via activation of NF-κB and hypoxia-inducible factor-1α. *Journal of Immunology* 178 (5): 3198.

Volesky, B. (1990). Biosorption and biosorbents. In: *Biosorption of Heavy Meatals* (ed. B. Volesky). Boca Raton, FL: CRC Press.

Wang, Y., Mandal, A.K., Son, Y.-O.K. et al. (2018). Roles of ROS, Nrf2, and autophagy in cadmium-carcinogenesis and its prevention by sulforaphane. *Toxicology and Applied Pharmacology* 353: 23.

Wani, A.L., Ara, A., and Usmani, J.A. (2016). Lead toxicity: a review. *Interdisciplinary Toxicology* 8 (2): 55.

Waterman, S.J., el-Fawal, H.A., and Snyder, C.A. (1994). Lead alters the immunogenicity of two neural proteins: a potential mechanism for the progression of lead-induced neurotoxicity. *Environmental Health Perspectives* 102 (12): 1052.

Watson, D. (2001). *Food Chemical Safety: Volume 1: Contaminants*. Woodhead Publishing.

Wilbur, S., Abadin, H., Fay, M. et al. (2012). *Toxicological Profile for Chromium*. Atlanta (GA): Agency for Toxic Substances and Disease Registry (US). PMID: 24049864.

Wise, J.P., Wise, S.S., and Little, J.E. (2002). The cytotoxicity and genotoxicity of particulate and soluble hexavalent chromium in human lung cells. *Mutation Research* 517 (1): 221.

World Health Organization and International Programme on Chemical Safety (1991). *Inorganic Mercury/Published Under the Joint Sponsorship of the United Nations Environment Programme, the International Labour Organisation, and the World Health Organization; First Draft Prepared by L. Friberg*. World Health Organization.

Xie, H., Wise, S.S., Holmes, A.L. et al. (2005). Carcinogenic lead chromate induces DNA double-strand breaks in human lung cells. *Mutation Research* 586 (2): 160.

Xu, S., Pi, H., Chen, Y. et al. (2013). Cadmium induced Drp1-dependent mitochondrial fragmentation by disturbing calcium homeostasis in its hepatotoxicity. *Cell Death & Disease* 4 (3): e540.

Ye, B.-J., Kim, B.-G., Jeon, M.-J. et al. (2016). Evaluation of mercury exposure level, clinical diagnosis and treatment for mercury intoxication. *Ann Occup Environ Med.* 28 (1): 5.

Zamponi, G.W., Bourinet, E., and Snutch, T.P. (1996). Nickel block of a family of neuronal calcium channels: subtype- and subunit-dependent action at multiple sites. *The Journal of Membrane Biology* 151 (1): 77.

Zefferino, R., Piccoli, C., Ricciardi, N. et al. (2017). Possible mechanisms of mercury toxicity and cancer promotion: involvement of gap junction intercellular communications and inflammatory cytokines. *Oxidative Medicine and Cellular Longevity* 2017: 7028583.

Zhang, Y. and Zheng, J. (2020). Bioinformatics of metalloproteins and metalloproteomes. *Molecules* 25 (15): –3366.

Zhang, G., Chen, D., Zhao, W. et al. (2016). A novel D2EHPA-based synergistic extraction system for the recovery of chromium (III). *Chemical Engineering Journal* 302: 233.

Zhang, Y., Shao, X., Kong, X. et al. (2022). Determination of thorium in the hair and urine of workers and the public in a rare earth mining area. *Radiation Medicine and Protection* 3 (2): 91.

5

Different Approaches for Detecting Heavy Metal Ions

Ziaul Hasan[1,2], Arif Jamal[3], and Tauseef Hassan[4]

[1] Department of Biosciences, Jamia Millia Islamia, New Delhi, Delhi, India
[2] Centre for Interdisciplinary Research in Basic Sciences, Jamia Millia Islamia, New Delhi, Delhi, India
[3] Department of Mechanical Engineering, Jamia Millia Islamia, New Delhi, Delhi, India
[4] Center for Nanoscience and Nanotechnology, Jamia Millia Islamia, New Delhi, Delhi, India

5.1 Introduction

A growing number of human activities have resulted in the discharge of numerous harmful compounds into water resources, endangering aquatic ecosystems and surroundings. Heavy metal ions happen to be harmful to aquatic ecosystems because they are so poisonous, fail to degrade, and tend to biomagnify and bioaccumulate as they go up the food chain. Their widespread presence in water means they may have an influence on biological systems in a variety of ways. Ions composed of heavy metals have been discovered to be very harmful to plants and animals, even when present in the soil beneath our feet. This is because plants absorb them and then release them into the environment, where they might be consumed by animals and humans. It is believed that the sorptive capacity of soil components plays a significant role in mitigating the pollution of water, plant, and agricultural systems, as well as the harm they bring to humans and other animals.

Water pollution has real-world repercussions; therefore, scientists have figured out how to detect and evaluate contaminants made up of heavy metals. Various methods have been developed to detect heavy metal ions. There are three broad categories into which these techniques fall: *spectroscopic, electrochemical* and *optical methods.* It is possible to identify the heavy metal ion using spectroscopic methods such as atomic absorption spectroscopy, graphite furnace atomic absorption spectrometry (AAS), atomic fluorescence spectrometry (AFS), as well as X-ray fluorescence spectrometry (XRFS). Although extremely sensitive, these methods may be costly and time-consuming to implement due to the need for preliminary processing. On the other hand, electrochemical methods are superior in several respects, including reliability, ease of use, cost-effectiveness, and suitability for use in the field. These methods include coulometry for detecting heavy metal ions, potentiometry, amperometry, voltammetry, cyclic voltammetry, pulse voltammetry, stripping voltammetry, and galvanostatic techniques.

Remediation of Heavy Metals: Sustainable Technologies and Recent Advances, First Edition. Edited by Rangabhashiyam Selvasembian, Binota Thokchom, Pardeep Singh, Ali H. Jawad, and Willis Gwenzi.

Due to their straightforward nature and relatively brief analysis times, electrochemical methods may compete well against different spectroscopic approaches in terms of speed. Because of its superior selectivity, mobility, cheap costs, quick analytical pace, and remarkable responsiveness, anodic stripping voltammetry (ASV) has replaced several previous electrochemical methods for the identification of harmful metal ions. Standard luminescence, reflection, or absorption spectrometric methods may be used to identify metal ions. Indicator dyes and chromogenic reagents that react specifically with certain metal ions are often used in optical detection methods. In addition, apparatus such as instruments of the capillary type, integrated optics, optical fibers, and other similar technologies may be helpful in the detection of metal ions.

Here, we provide an in-depth analysis of the current state of metal ion pollution in aquatic bodies and address its possible effects on the well-being of all organisms. The advantages of the many methods that may be used for detection have been stressed. Metals serve crucial biological roles, but the proportions at which they are found may have profound effects on human health. In low concentrations, it has few adverse effects, but in high concentrations, it becomes dangerous and has a number of physiological and cytological consequences. According to assessments by the World Health Organization (WHO) and the Bureau of Indian Standards (BIS), the impacts and the potentially harmful range are shown in Table 5.1.

Table 5.1 summarizes the many sources and consequences of certain most hazardous heavy metal ions, which include mercury (Hg), cadmium (Cd), arsenic (As), lead (Pb), and chromium (Cr), and their exposure has been linked to a number of health problems (Flora et al. 2008; Flora 2009; Bashir et al. 2016). The "existence" of heavy metal ions has nothing to do with the problem; instead, it is their "excess" that is problematic. According to studies published in 2005 (Valko et al. 2005), even minute amounts of metals play important roles in a wide variety of biological activities, including cell signaling and transportation. These metal ions have hazardous effects on living things because, whenever too abundant, they veer off their normal metabolic route and attach to protein sites other than their usual binding locations. The binding capacity of metal ions is responsible for this cascade disruption in cells (Falcó et al. 2006; Singh et al. 2010; Sevcikova et al. 2011).

Since heavy metal ions can cause serious problems if they accumulate to unsafe levels, it is crucial first to locate them, quantify their concentrations, and then remove them using appropriate scientific methods.

5.2 Detection

People and animals alike are very vulnerable to the devastating effects that heavy metal ions may have on their bodies. Heavy metal ions are a major environmental concern, making it crucial to develop methods for detecting them in water, soil, and food samples. Various techniques exist for detecting heavy metal ions, and every method has its own set of pluses and minuses. The term "metal ion detector" refers to any instrument or machine that is purpose-built for identifying the existence of metal ions in the environment. There may be cases when knowing the exact concentration of these metal ions may be useful. Given the importance of having access to water that is free of metal ions, it is crucial to

Table 5.1 Limitations, causes, and outcomes of contamination by various heavy metal ions.

Metal	BIS limits[a] (mg l^{-1})	WHO limits[b] (mg l^{-1})	Potential origins	Impacts
Mercury (Hg)	0.001	0.001	Coal burning, urban trash incineration, and volcanic gases are all major contributors	Abnormalities in the brain's development; gastrointestinal, respiratory, cutaneous, and ocular abnormalities; abnormally high blood pressure, hypotonia, acrodynia, and increased salivation
Cadmium (Cd)	0.01	0.005	Inks, dyes, batteries, electroplated parts, polymers, artificial rubber, photoconductors, photovoltaic cells, engraving techniques, and more	Potential side effects include kidney toxicity, high blood pressure, decreased appetite, fatigue, microcytic hypochromic anemia, lung cancer, lymphocytosis, osteomalacia, lung cancer, pulmonary fibrosis, atherosclerosis, osteoporosis, peripheral neuropathy, and hyperuricemia
Arsenic (As)	0.05	0.05	Discharging untreated effluents that include pyrite (FeS) and arsenopyrite (FeAsS) due to the oxidation of treated wooden power poles containing fertilizers, pesticides, and arsenic-based chemical preservatives	Anorexia, hyperpigmentation, dark pigmentation, localized edema, and skin cancer are only a few of the many adverse effects on the developing fetus and mother's internal organs, as well as the digestive tract, the urinary tract, the blood, and the skin
Lead (Pb)	0.05	0.05	There are many applications for PVC[c] pipes outside of plumbing and farming, including recycled PVC lead paints, jewellery, lunch boxes, and more	Exposure has been associated with an elevated likelihood of developing Alzheimer's along with several kinds of dementia in the elderly because it passes the blood–brain barrier (BBB), which typically keeps dangerous compounds out of the brain. Because of this, it may cause neurological illnesses, decreased intelligence, renal damage, stunted development, behavioral difficulties, extreme irritability, stupor, and ataxia
Chromium (Cr)	0.05	0.05	Industries related to leather, tanning, and chrome plating	Irritating contact dermatitis, lung cancer, skin inflammation, skin ulcers, septal perforation, teratogenicity, carcinogenicity, embryotoxicity, and mutagenicity

(Continued)

Table 5.1 (Continued)

Metal	BIS limits[a] ($mg\,l^{-1}$)	WHO limits[b] ($mg\,l^{-1}$)	Potential origins	Impacts
Silver (Ag)	0.1	0.1	Industries related to jewellery, electroplating, copper, gold, nickel, and other metals refine	Cytopathological effects on fibroblasts, mast cells, and keratinocytes; rheumatism; knotting of cartilage; argyria; gastroenteritis; neurological issues; mental fatigue
Copper (Cu)	1.3	1	Tanneries, fertilizers, and solar cells all fall under this category	Anemia, baldness, autism, rheumatoid arthritis, diabetes, cystic fibrosis, excessive bleeding, and kidney disease are among conditions that may be exacerbated by an inadequate diet
Zinc (Zn)	5	5	Colorants, cosmetics, and soldering are all covered in this	Conditions affecting the respiratory system, including metal fume fever and bronchiolar leucocytes; chances of prostate cancer, neurological issues, erectile dysfunction, and macular degeneration

[a] Permitted levels for heavy metal ions in accordance with the Bureau of Indian Standards.
[b] Allowable limits for heavy metal ions in water as established by the World Health Organization.
[c] Polyvinyl chloride (PVC), which is a lightweight and durable polymer.

devise a strategy to ensure that water sources are always safe for human and animal use. The presence of these dangerous metal ions within the water sample must be identified prior to removal so that an appropriate removal strategy may be selected and a numerical estimation of the amount of contamination can be provided. To this purpose, eco-friendly, low-cost, and quick detection methods need to be developed.

Furthermore, an approach to identification needs to be sufficiently sensitive to identify low concentrations of metal ions. In spite of the fact that there are a few different techniques for detecting ions of heavy metals, there is not a single one that can be used for the detection of all ions.

5.3 Methods of Detection

Heavy metal ion detection techniques may be loosely grouped into three distinct groups.

5.3.1 Spectroscopic Detection

Neutron activation evaluation, inductively coupled plasma mass spectrometry (ICC MS), and inductively coupled plasma-optical emission spectrometry are three of the finest sensitive methods used in spectroscopy for detecting heavy metal ions. They have very low limits of detection (LOD) and can simultaneously measure the concentration of ions made of metals that are toxic for a multitude of elements. However, these techniques are extremely expensive because they require sophisticated machinery and a highly trained workforce to operate them. Flame atomic absorption spectroscopy has been used to detect Pb^{2+}, Cu^{2+}, and Cd^{2+} ions in water, with permissible detecting capabilities of 2, 3, and $0.2\,\mu g\,dm^{-3}$. Low-concentration ions of lead, copper, and mercury have been detected by means of ASV and high-resolution surface-plasmon-resonant spectroscopy.

Below is a discussion of a few major spectroscopic techniques that may be used to detect heavy metal ions.

5.3.1.1 Atomic Absorption Spectrometry

To determine how many atoms are present in a sample, atomic absorption spectroscopy uses light of a predetermined wavelength to excite them from their ground state. The amount of energy absorbed during this process is recorded and correlated with the concentration of atoms in the sample. Figure 5.1 depicts the block schematic for an ordinary spectrometer used for atomic absorption. To produce atoms/ions in the gaseous state for assessment, this apparatus has a primary light source, detector, monochromator, electronic "readout" device, and atomizer.

Only in gaseous media, when ions or atoms are very widely separated from one another, can the determination by atomic absorption spectroscopy be made. Electrothermal atomizers, plasmas, and flames are used to create atoms and ions of analyte in the gas phase. By evaporating and breaking down the sample, atomization releases atoms and ions into the gas phase. Specialized light sources are often employed to excite ions or atoms of specific elements, such as hollowed-out cathode lamps as well as electrodeless discharging lamps constructed of the material under investigation. There are a variety of lights that may be employed to decide a variety of components without the need to change the bulb. Atomic absorption spectroscopy often uses photomultiplier tubes or solid-state detectors as its

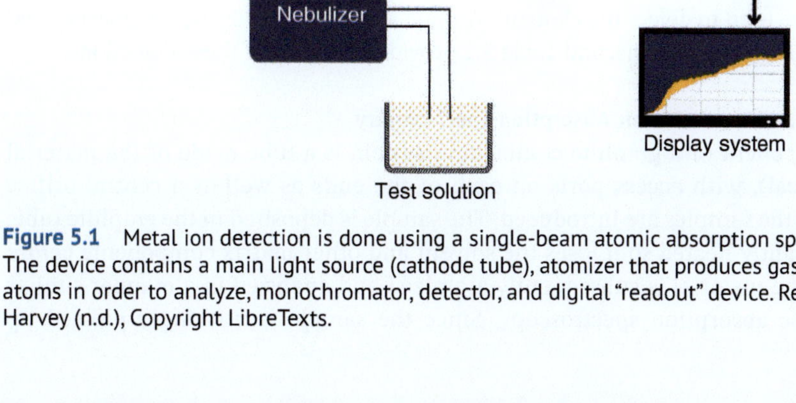

Figure 5.1 Metal ion detection is done using a single-beam atomic absorption spectrophotometer. The device contains a main light source (cathode tube), atomizer that produces gas-phase ions or atoms in order to analyze, monochromator, detector, and digital "readout" device. Reproduced from Harvey (n.d.), Copyright LibreTexts.

Table 5.2 Both AAS and its graphitic furnace counterpart, GF-AAS, are used to detect and identify potentially harmful heavy metal ions released by a wide variety of sources.

Metal ion(s)	Origin	Analytical method	Reference
Mn, Cr, Cd, Pb	Kulfo River, Arba Minch, Gamo Gofa (Ethiopia)	AAS[a]	Tsade (2016)
Pb, Sb, Al, As	Balochistan, Pakistan cased boreholes in District Pishin	AAS	Tareen et al. (2014)
Cd, Cr, Fe, Mn, Ni, Pb, Zn	Therapeutic mud and water	AAS	Radulescu et al. (2014)
Cd, Cu, Fe, Ni, Pb, Tl, Zn	Seawater	AAS	Kojuncu et al. (2004)
Cd, Cr, Cu, Ni, Pb	Chinese tea	GF-AAS[b]	Zhong et al. (2016)
Ca, Cu, Fe, Mg, Zn	Chinese tablets	AAS	Dong and Zhu (2002)
Ca, Cd, Cu, Fe, Mg, Mn	Formula for Chinese taponin tablets	AAS	Dong and Zhu (2003)
Co, Ni	Water specimens	GF-AAS	Minami et al. (2003)
Pb	Water specimen	GF-AAS	Chen et al. (2005)
Fe	Pure, mineral, and organic waters	AAS	Tautkus et al. (2004)
V	Wine	GF-AAS	Wierzbicki and Pyrzyñska (2002)
Cd	Wine	GF-AAS	Cvetković et al. (2006)
Pt	Rocks	GF-AAS	Odonchimeg et al. (2016)
Cr	Grapes and wine originating from France	GF-AAS	Cabrera-Vique et al. (1997)
Hg	Specimens of fish muscle	GF-AAS	Moraes et al. (2013)

[a] Examining the absorption spectrum of individual atoms.
[b] Studying graphite in a graphite furnace using AAS.

detection mechanism. Flow injection mercury systems, a variant of atomic absorption spectroscopy, are used to detect the element. Atomic absorption spectroscopy may be used to detect a number of metal ions, and Table 5.2 provides a review of these detections.

5.3.1.2 Graphite Furnace Atomic Absorption Spectrometry

The main component of a graphite combustion system is a tube made of the material (often cylindrical), with access ports on each of the ends as well as a central orifice through which the samples are introduced. The sample is deposited in the graphite tube, where it is instantly heated so it loses the solvent and other matrix components before atomization. The material is subsequently atomized and analyzed in a manner similar to flame atomic absorption spectroscopy. Since the sample is no longer pressed up against the furnace wall, atomization may take place in a more stable environment, potentially leading to more reliable results.

The drawbacks of using flame sampling instead of GF-AAS as the instrument for heavy metal ion detection are that this approach can only identify a smaller number of elements and that analysis durations are longer. Graphitic furnace atomic absorption spectroscopy has been used to assess the levels of heavy metals, including cadmium, mercury, copper, lead, and arsenic, with permissible detection boundaries of 0.014, 0.061, 0.19, 0.28, and 0.49 ($mg\,l^{-1}$), correspondingly. Lead (Pb) amount of "Yin Qiao Jie Du" (YQJD) herbal formula (medicinal tablet) was also analyzed using a graphite furnace method, and the LOD for Pb resulted to be in a range of 0.1 parts per billion. Table 5.2 reports the use of atomic absorption spectroscopy for successfully identifying particular heavy metal ions in graphitic furnaces.

5.3.1.3 Atomic Fluorescence Spectrometry

The methods and operations of atomic fluorescence spectroscopy are distinct from those of AAS. When conducting AFS, a specimen mixture is atomized and then subjected to light, much like when conducting atomic absorption spectroscopy. The detector picks up the radiation emitted through such excited atoms as they pass through radiative deactivation, allowing for the measurement of atomic fluorescence. By employing a carrier gas like argon to transfer free mercury atoms to the cell and activating them with a collimated UV light source, the cold vapor AF method may be utilized to test mercurial levels. In this case, the UV photodiode or photomultiplier tube is used to detect the fluorescence generated by the excited atoms after they have absorbed the energy. Metal detection by AFS has seen several significant advancements, which are summarized in Table 5.3.

5.3.1.4 X-Ray Fluorescence Spectrometry

It results in the ionization of the substance when it is exposed to X-rays or gamma rays. As demonstrated in Figure 5.2, such a high-intensity radiation bombardment of the material can even cause electrons to be ejected from inner K or L shell orbitals. XRFS is used to detect the photons or fluorescence emitted by electrons in higher energy shells as they fill the vacant shells.

Because each element possesses its own distinct set of energy levels, the resulting fluorescence spectrum will also feature a wide range of peak energies and locations. In reality, XRFS is a very efficient technique for determining an object's elemental composition in a short amount of time. As shown in Figure 5.3, an X-ray source, specimen chamber, fluorescence detection instrument, processing of information, and a display system make up an X-ray fluorescence spectrometer.

To learn more about how X-ray fluorescence is being applied for the detection and identification of dangerous compounds in FDA-regulated products, check out Palmer et al. (2009) and the other references cited in Table 5.3.

5.3.2 Electrochemical Methods of Detection

These kinds of approaches are low-cost, easy to use, and trustworthy, and they have uncomplicated protocols for monitoring contaminated samples. Electrochemical technologies also have very short analysis durations compared to other spectroscopic methods. However,

Table 5.3 AFS and XRFS are helpful methods for the identification of potentially hazardous heavy metal ions.

Metal ion(s)	Specimen	Technique of analysis	Reference
Cd	Brown rice	AFS[a]	Hafuka et al. (2017)
Pb	Water samples	AFS	Beltrán et al. (2015)
As	Chicken meat	AFS	Sánchez-Rodas et al. (2006)
Pb, Cu	Water-based solution	XRFS[b]	Hutton et al. (2014)
As	Mining contaminated soils	AFS	Ruiz-Chancho et al. (2005)
Fe, Ni, Mn, Cu, Zn, Pn	Coastal seawaters	XRFS	Yuan-Zhen et al. (2012)
As, As	Acid mine drainage, Beltran	AFS	Oliveira et al. (2006)
As, As	Seafood	AFS	Schaeffer et al. (2005)
Cr, Cu, As, Ni, V, Zn, Pb	Soil	XRFS	Ene et al. (2010)
Se	Cow milk	AFS	Muniz-Naveiro et al. (2007)
As, U, Pb	Aerial parts of *Origanum sipyleum* L.	XRFS	Durmuşkahya et al. (2016)
Sc	Sesame seeds	AFS	Kápolna et al. (2007)
Sb, Sb	Mollusks, marine algae	AFS	De Gregori et al. (2007)
Cd, Cr, As, Cu, Ni, Pb, Zn	Soils	XRFS	Taha (2017)
Cr, Ni, Cu, Ba, Pb, Sr, Ga, V, Nb Rb, Y, Zn, Zr	Preliminary surface soil analysis	XRFS	El-Bahi et al. (2013)
Hg	Water	AFS	Logar et al. (2002)
As, Cu, Pb, Zn	Soil samples	XRFS	Radu and Diamond (2009)
Cr, Cu, Hg, Ni, Pb, Zn	Fish tissues	XRFS	Zarazúa et al. (2014)
Hg	Samples of polluted water	XRFS	Marguí et al. (2010)

[a] AFS.
[b] XRFS.

among the primary drawbacks associated with these electrochemical methods are their lower sensitivity and higher LOD in comparison to spectroscopic and optical approaches. Sometimes these techniques have their designs tweaked so that they are better able to detect heavy metal ions. For instance, many electrochemical techniques used for various biosensing electrodes are required to modify the electrode material and expand the electrodes' detection range and responsiveness. Water with high concentrations of metal ions may alter the electrochemical system's current, voltage, charge, luminescence, and electrochemical impedance, as reported by Cui et al. (2015). Methods like electrochemiluminescence, amperometry, potentiometry, voltammetry, impedance analysis, and coulometry may be used to determine how the presence of metal ions has altered the electrical signal. The following section contains a comprehensive list of key electrochemical methods.

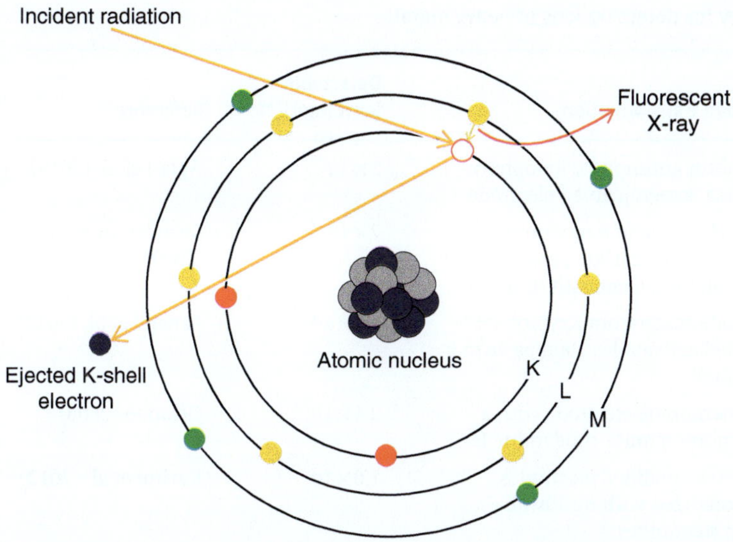

Figure 5.2 Here we see a schematic representation involving the K-capture process used in XRFS, in which an atom's K-shell electrons are ejected by incredibly energetic X-rays or gamma rays, creating an energy gap that is replenished by an electron from the subsequent energy band up. Reproduced from Rebollo (n.d.).

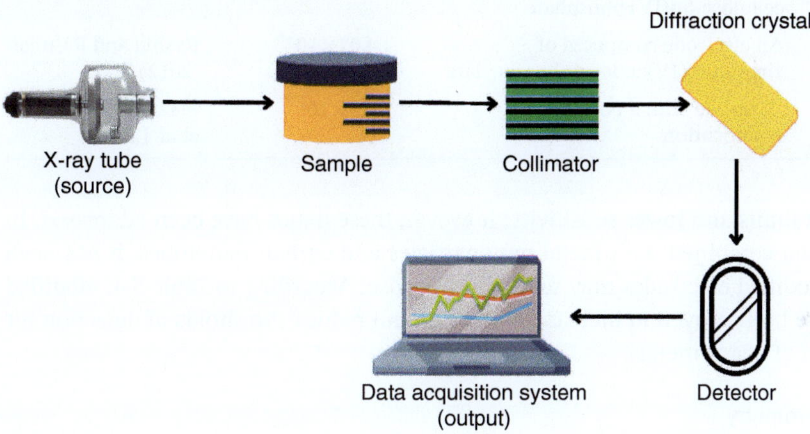

Figure 5.3 The components of an XRFS, which is used to evaluate elemental composition, are shown in a block diagram. These components include the X-ray source, the specimen chamber, the fluorescence detection instrument, the information-gathering system, and the visualizing display. *Source:* Reproduced from Wegrzynek, © 2004/International Atomic Energy Agency (IAEA).

5.3.2.1 Potentiometry

This technique is based on the fact that electromotive force (EMF) measurements may be taken regardless of the need for an electrical current. The quantitative measurement of ions in solutions is a common use of potentiometric methods. But heavy metal ion detection is another area where it excels due to its inexpensive cost, rapid reaction timing, exceptional selectivity, and wide response range. Two disadvantages of this approach include

Table 5.4 Potentiometry for detecting ions of heavy metals.

Metal ion(s)	Electrode transformation	Detectable level ($mol\,l^{-1}$)	Reference
Hg	Zirconium antimonate ionophore-modified screen-printed electrode	5×10^{-8}	Aglan et al. (2018)
Hg	Electrode made of nanoporous silica-modified carbon paste that has been functionalized with thiourea	7×10^{-8}	Javanbakht et al. (2009)
Hg	Electrode made from Kryptofix®5*-modified activated carbon paste in a palm shell	1.0×10^{-7}	Ismaiel et al. (2012)
Hg	PVC membrane electrode with a tribromomercurate-rhodamine B	2.4×10^{-6}	Othman (2006)
Cd	Dithizone-modified electrodes functionalized with multiwall carbon nanotubes	1.0×10^{-7}	Karimi et al. (2012)
Cd	Cation-selective membrane electrode based on a polyaniline Sn(IV) silicate composite membrane	1.0×10^{-7}	Naushad et al. (2015)
Pb	Electrode for a composite membrane made of PVC and carboxymethyl cellulose-Sn(IV) phosphate	1.0×10^{-6}	Inamuddin et al. (2015)
Pb	An electrode composed of zirconium(IV) iodosulphosalicylate	4.07×10^{-6}	Rashid and Rahman (2012)
Pb	Electrode with a polypyrrole modification	7.0×10^{-7}	Mazloum-ardakani et al. (2012)

high detection limits and lower sensitivity; however, these issues have been addressed. In particular, when combined with metal nanoparticles and carbon nanotubes, it has been shown that modified electrodes may serve this purpose. According to Table 5.4, modified electrodes have been proven to increase sensitivity and reduce thresholds of detection for ions composed of heavy metals.

5.3.2.2 Amperometry

Amperometry is a potentiostatic method that involves applying an established potential difference between the electrode that is working and the electrode used as a reference inside the solution containing electroactive species. Heavy metal ion reduction results in very low current flow, which may be measured. These investigations use amperometry techniques to record the current as though it were a time-varying quantity. Because the working electrode maintains a constant voltage, this method may be used to distinguish one metal ion from others that are also electroactive. A faradic reaction in the target analyte is triggered by applying a potential with the right polarity and enough force. Since the working electrode has a smaller surface area, only a small percentage of the analyte responds to such a poor faradic response. Several modifications have been made to the electrode surface in an effort to address this problem.

Table 5.5 Employing several modified electrodes to amperometrically detect heavy metal ions.

Metal ion(s)	Enhancement of electrode	Detectable level (mol l^{-1})	Reference
Hg	Electrode made of platinum modified with poly(vinylferrocenium) (PVF[+])	5×10^{-10}	Çelebi et al. (2009)
Hg	Pt/CeO_2/urease-modified electrode	1.8×10^{-8}	Gumpu et al. (2015)
Hg	Platinum electrode modified with L-tyrosine	1.4×10^{-8}	Majid et al. (2002)
Cd	Tape ion sensor based on microwater/polyvinyl chloride-2-nitrophenylethyl ether (PVC-NPOE) gel interfaces	1.78×10^{-7}	Lee et al. (2009)
As, Cd, Co, Cr, Ni, Pb	Electrode made of cobalt oxide-modified glassy carbon that is functional with glucose oxidase	50[a]	Mugheri et al. (2016)
Pb	Electrode made of a single carbon fiber	1.3×10^{-6}	Li et al. (2007)

[a] Parts per billion (ppb).

Hg^{2+} was detected using an extremely sensitive screen-printed electrode enhanced with a nano-structured coating of carbon black having an acceptable sensitivity range of 5 nM. Amperometric biosensors based on biologically enhanced electrodes have also been used to detect heavy metal ions (Mohammadi et al. 2005). Amperometric metal ion detection has been improved by swapping the electrodes, as seen in Table 5.5

5.3.2.3 Voltammetry

Heavy metal ions are often identified by the use of voltammetry, an electrochemical approach. Voltammetry is the study of EMFs by measuring current at different applied potentials. Voltammetry's outstanding sensitivity, high specificity, and restricted range of detection make it a popular method toward detecting the heavy metal ions. There are several variations of voltammetry, but they all measure current by changing the potential. The different types of voltammetry that may be used to investigate heavy metal ions are described below.

5.3.2.3.1 Cyclic Voltammetry As the name implies, cyclic voltammetry involves repeatedly cycling the direction of the applied voltage. A cyclic voltagram is generated by measuring the forward and backward scanning currents and plotting them in opposition to the input voltage. Some useful information is provided by this cyclic voltagram, such as the peak anodic and cathodic potentials and currents. To selectively detect Hg^{2+}, cyclic voltammetry was used on a glassy carbon electrode (GCE/rGO-SH/Au nano-particles) that included gold (Au) nano-particles thiol-functionalized reduction graphene oxide (TRGO) (Devi et al. 2018). The linear detection relationship for Hg^{2+} was seen in phosphate-buffered saline (PBS) mixture with concentrations from 1 to 10 M, with a detection threshold as low as 0.2 M. When comparing the cyclic voltammetric responsiveness regarding GCE/rGO-SH/Au nano-particles against that for GCE/rGO/Au nano-particles without thiol functionalization, as illustrated in Figure 5.4, a significant shift in the anodic maximum potential emerged. Since the -SH group on the rGO layers does have any significant attraction for the "Au" micron-sized particles, the enhanced anodized currents seen in the GCE/rGO-SH/Au nano-particle electrodes setup could possibly be owing to the larger loading that results from these particles.

5.3.2.3.2 Pulse Voltammetry As a subset of voltammetry, pulse voltammetry is based on the employment of voltage pulses of variable amplitudes and forms. Standard, differential, and reverse pulse voltammetry are only a few examples of the many subtypes of pulse

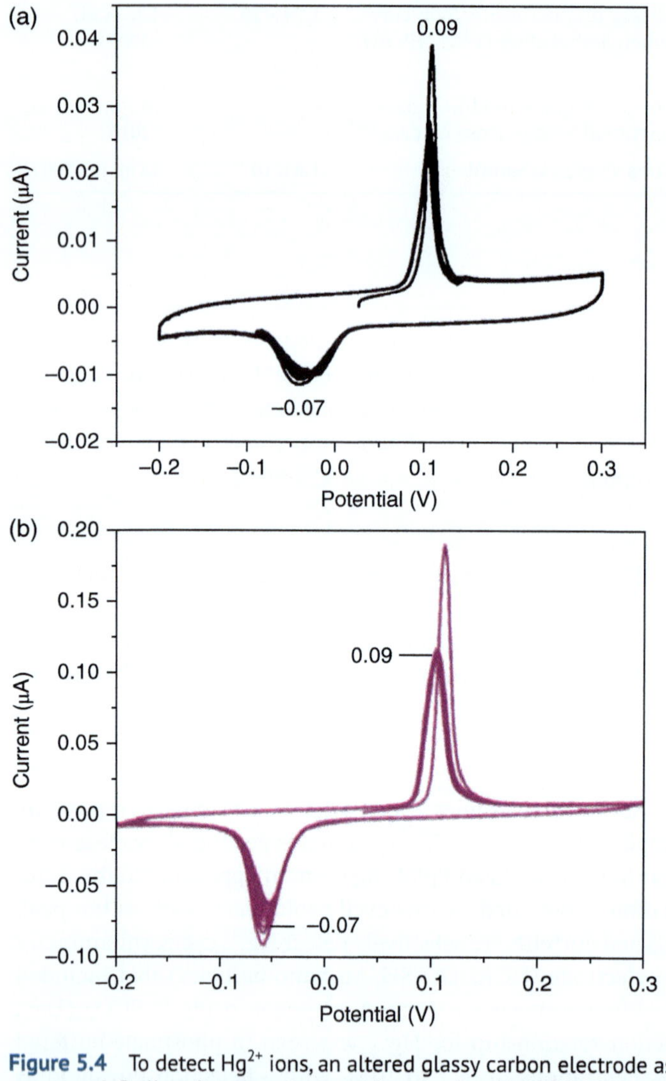

Figure 5.4 To detect Hg^{2+} ions, an altered glassy carbon electrode as well as cyclic voltammetry were used. Cyclic voltagrams were produced using (a) GCE/rGO/Au; (b) nanoparticles of GCE/rGO/-SH/Au were captured while immersed in a 0.1 M KCl solution and scanned at 50 mVs^{-1}. Reproduced from Devi et al. (2018).

voltammetry that may be distinguished. The most common pulse voltammetric technique, differential pulse, is very sensitive to the identification of ions containing heavy metals. Xia et al. (2010) employed differential pulse voltammetry to simultaneously detect lead, copper, and cadmium using an electrode constructed of carbon paste improved by hexagonal mesoporous silica (HMS)-immobilized quercetin (HMS-Qu/CPE). Voltammetric responses were measured for both quercetin-modified carbon paste electrodes (Qu/CPE) and quercetin/ionic liquid-modified carbon paste electrodes (QuIL/CPE) in relation to Cd^{2+}, Cu^{2+}, and Pb^{2+} ions. Significant increases in peak cadmium, copper, and lead currents were seen at HMS-Qu/CPE, which may have resulted from the symbiotic relationship between HMS and Qu (Figure 5.5).

5.3.2.3.3 Stripping Voltammetry The two primary processes of stripping voltammetry are electrodeposition when an analyte mixture is placed on the electrodes from a thoroughly blended solution, and the voltammetric process, wherein the analyte is drawn away and may subsequently be evaluated using a VM method. Stripping voltammetry may be moreover subdivided into ASV and cathodic stripping voltammetry (CSV) based on whether a cathodic potential scan or anodic potential scan is used. Stripping voltammetry requires equipment that is somewhat challenging, costly, and amenable to miniaturization. ASV in an acetate buffer was used to identify zinc ions using a disposable copper-based electrochemical sensor. The sensor could detect and measure the Zn^{2+} content in blood serum with outstanding precision as well as a modest range of detection. Figure 5.6 displays anodic stripping voltammograms for a range of Zn^{2+} ion intensities.

5.3.2.3.4 Stripping Methods Used in Conjunction with Voltammetric Techniques New detection strategies have been developed by combining stripping techniques with different pulse voltammetric methods, among which are linear sweep ASV, differential pulse ASV,

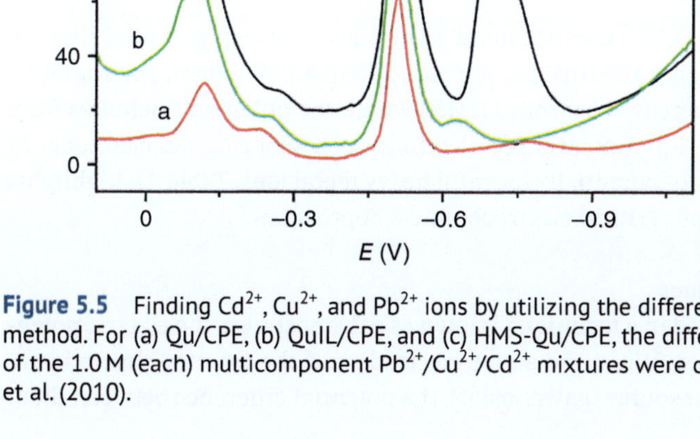

Figure 5.5 Finding Cd^{2+}, Cu^{2+}, and Pb^{2+} ions by utilizing the differential pulse voltammetric method. For (a) Qu/CPE, (b) QuIL/CPE, and (c) HMS-Qu/CPE, the differential pulse voltammograms of the 1.0 M (each) multicomponent $Pb^{2+}/Cu^{2+}/Cd^{2+}$ mixtures were captured. Reproduced from Xia et al. (2010).

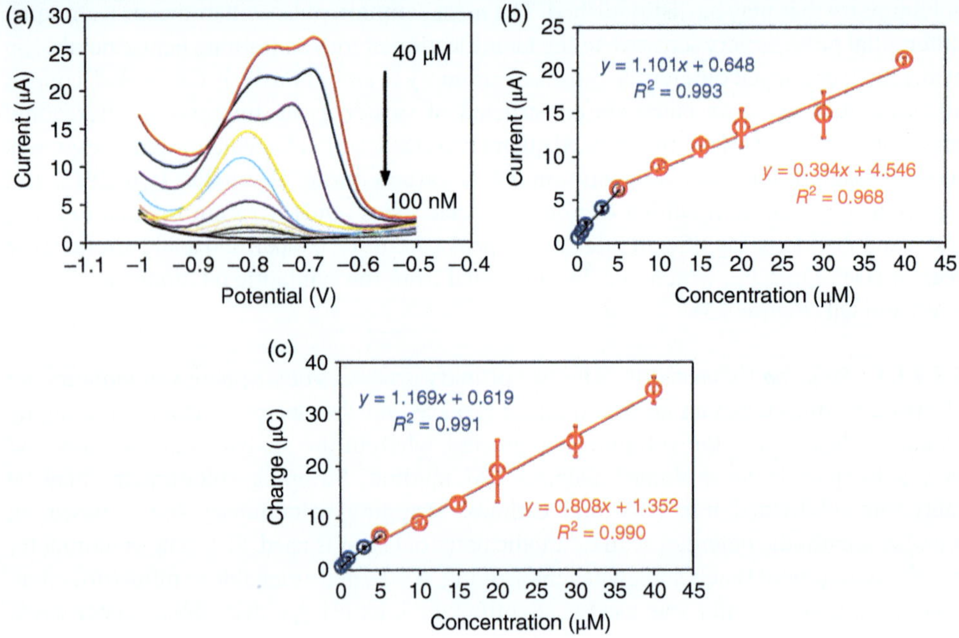

Figure 5.6 To determine the presence of zinc ions, a single-use copper-based electrochemical detector is subjected to an anodically stripping voltammetric analysis. The given images depict (a) Zn^{2+} samples observed by ASV at concentrations ranging from 100 nM to 40 μM, (b) Zn^{2+} curves of calibration based on peaking height, and (c) the regions of peak determined by stripping voltammetry. Adapted from Pei et al. (2014).

square-wave ASV, and so on, which are extremely useful in determining the existence of ionic species. Utilizing square-wave ASV over the cubic electrode and octahedral electrode improved with Fe_3O_4 nanocrystals, ions of heavy metals such as Cd^{2+}, Cu^{2+}, Hg^{2+}, Pb^{2+}, and Zn^{2+} have been discovered. When it comes to electrochemically detecting the tested ions of heavy metals, cubic Fe_3O_4 nanocrystal-modified electrodes are less effective than octahedral Fe_3O_4 nanocrystals, as shown in Figure 5.7. Moreover, octahedral Fe_3O_4 nanocrystals also showed superior electrochemical performance for detecting Pb^{2+} ions, with a higher level of sensing and a weaker detecting threshold than those for detecting other metal ions.

Linear sweep voltammetry (LSV) uses standard electrodes to sweep potential linearly from 10 mV per second to roughly 1000 mV per second to form an I–V curve; polarography uses a polarizable dropping mercury electrode (DME) where the potential fluctuates from its reverse amount as well as a non-polarizable electrode such as a calomel electrode. In order to identify a wide variety of potentially harmful heavy metal ions, Table 5.6 highlights the electrode system and various critical electrochemical approaches.

5.3.2.4 Galvanostatic Techniques

When an electric current is used in a technique known as galvanostatics, the electric potential of that current is measured. When the active electrode and the counter-electrode are connected in series to a current source (galvanostat), the potential difference between them

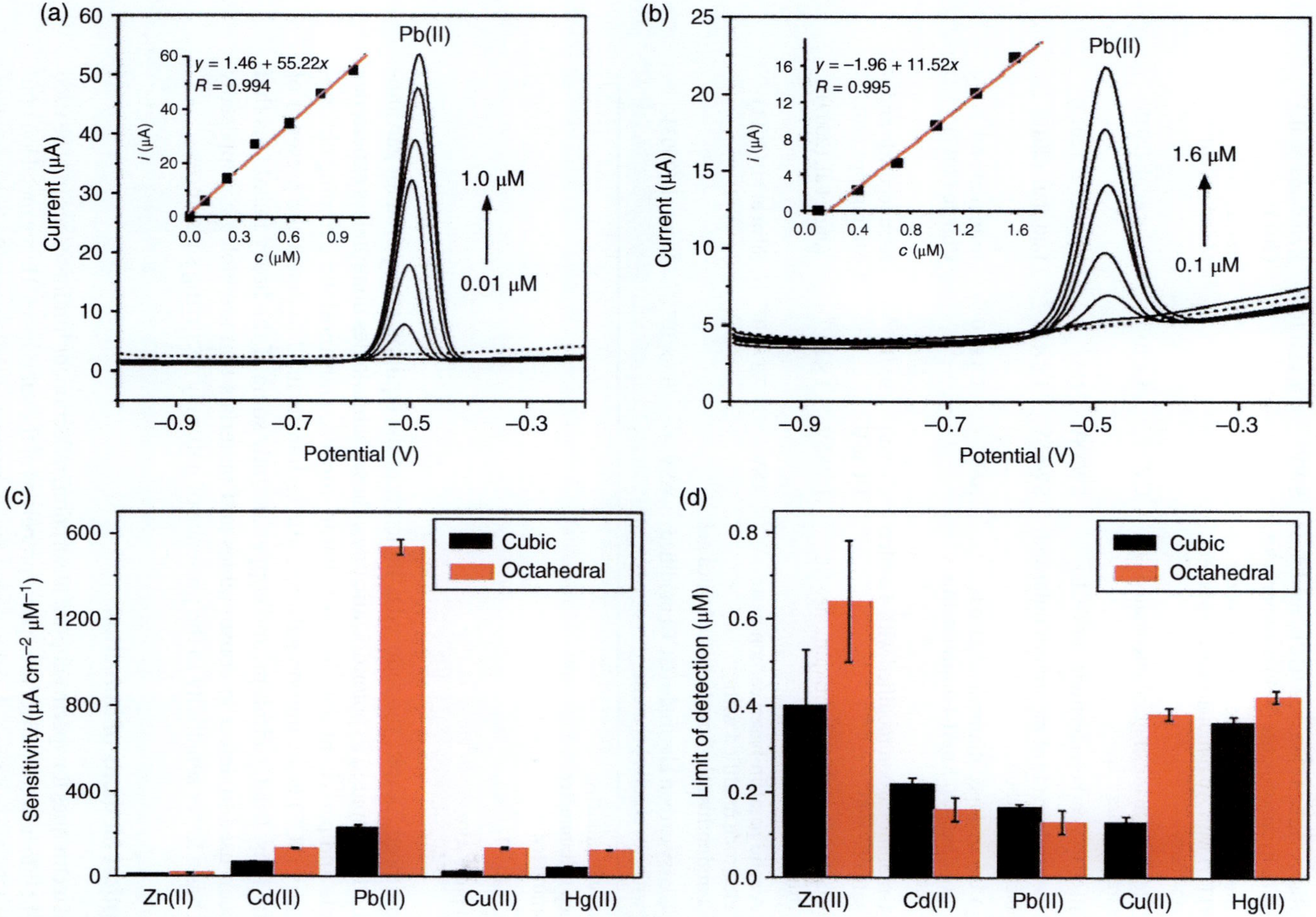

Figure 5.7 SWASV for detecting metal ions of various types. SWASV responses of (a) octahedral and (b) Fe_3O_4 nanocrystals-modified GCE for determining Pb(II), correlation of (c) quality of sensing, and (d) limitations in detection (3σ method) for SWASV detection of Cd^{2+}, Cu^{2+}, Hg^{2+}, Pb^{2+}, and Zn^{2+} at Fe_3O_4 cubic and octahedral nanocrystals (the matching peak current versus concentration linear calibration graphs are shown insets in a and b). Adapted from Yao et al. (2014).

Table 5.6 Various heavy metal ion-detecting voltammetric approaches.

Metal ion	Electrode system	Method	Limit of detection ($mol\,l^{-1}$)	Reference
Hg	Glassy carbon electrode (GCE) modified with hydroxyapatite (HA) nanoparticles	SWV[a]	1.41×10^{-7}	Kanchana et al. (2015)
Hg	Electrodes made of indium tin oxide (ITO) with Au nanoparticles deposited on them	LSV[b]	1×10^{-6}	Ratner and Mandler (2015)
Hg	Electrode made of graphene-coated carbon glass	CV[c]	1×10^{-9}	Talat et al. (2018)
Cd	Bi/glassy carbon electrode (Bi/GCE)	SWASV[d]	NA	Zhao et al. (2017)
Pb	Electrode made of montmorillonite and bismuth carbon	SWASV	1.8×10^{-9}	Luo et al. (2010)
Cd	A boron-doped diamond electrode modified by 0.5 mM 4-aminomethyl benzoic acid	ASV[e]	1.8×10^{-9}	Innuphat and Chooto (2017)
Pb	Graphitic carbon modified with 4-amino salicylic acid	CV and DPASV[f]	9×10^{-8}	Kempegowda and Malingappa (2012)
Pb	Clay nanoparticle anthraquinone–GC thin film	DPASV	1×10^{-9}	Yuan et al. (2009)
Pb	A solid paraffin-based carbon paste electrode modified with 2-aminothiazole-functionalized silica gel	ASV	7.3×10^{-9}	Silva et al. (2011)
Pb	Glassy carbon electrode (GCE) modified with novel calix [4] arene	ASV	6.1×10^{-9}	Lu et al. (2003)

[a] Square-wave voltammetry.
[b] Linear sweep voltammetry.
[c] Cyclic voltammetry.
[d] SWASV.
[e] ASV.
[f] Differential pulse ASV.

may be monitored. Despite the reduced complexity of the equipment required for galvanostatic methods compared to potentiostatic ones, these technologies nonetheless experience considerable and persistent double-layer charge impacts. Galvanostatic stripping chronopotentiometry (SCP) is a common galvanostatic technique used to detect the presence of heavy metal ions. Lead, cadmium, and copper are only some of the heavy metal ions that SCP has been used to detect in consumables and materials that are biologic, supposedly because it has a lower sensitivity to the presence of naturally occurring substances.

5.3.3 Optical Methods of Detection

Traditional techniques for spectroscopy of absorption, reflection, or luminescence may be used to get insight into how they affect the optical properties of the substances. Heavy metal ion optical detection is often performed using optical fibers, integrated optics, specific indicator dyes, capillary-type devices, ionophores, etc. However, there are several drawbacks to optical ion sensing that prevent it from being widely used. Strict pH control is necessary for the precise application of correction factors due to the fact that numerous not specific optical indicators react to various metal ions, and many metal ion indicators combine with hydrogen ion indicators. Indicator dyes' lack of selectivity may be compensated for by using masking agents. To reduce interference from other metal ions, fluoride can potentially be co-immobilized inside the sensor. Alternatively, test strips with reduction substances which transform analytes to distinct varieties or states of oxidization where the selected indicator is supplied may be used.

5.3.3.1 Indicator Dye-Based Sensors

The absorbance or fluorescence that occurs in the binding reagents is altered as a result of the interaction between the heavy-metal ion and the indicator dye in this kind of sensor. In heavy metal ion sensors, the indication functions like a transducer for ionization rather than a visual judgment. Heavy metal ions are used in another kind of indicator as "quenchers," reducing the luminescence of the indicator dye when they interact statically and dynamically. A quencher interacts with a fluorophore during the ground state under static quenching but only in the excited state during dynamic quenching. Dye may be recovered after the dynamic quenching procedure since it is not consumed. High sensitivity and selectivity are possible advantages of fluorescent indicators, but they may also have potential problems. Although indicators are widely employed, they are not always selective, as they may bind to a wide variety of metal ions. Many indicators are not usable for the optical sensing of heavy metal ions due to their lack of stability, inappropriate analytical wavelengths, requirement of additional reagents, or lack of commercial availability in their purest form required for sensing applications. Some indicators can only be kept bound indefinitely or at acidic pH. The main drawback of indication dye-based detectors is that unique dyes and, hence, different diagnostic wavelengths have to be used to detect different heavy metal ions.

5.3.3.2 Ionophore-Based Sensors

Ionophores have been utilized for optical ionization of heavy metals detection because of the limits with indicator reagents, ion-complexing compounds that are organic, and other substances. Due to their ion-carrying, complexing, and ion-binding characteristics, these ionophores may be used for heavy-metal ion selection. Adding chromogenic or fluorogenic moieties to ionophores or coupling them with the appropriate dyes are only two examples of the many sensing approaches that may be used to transform ionophores into reliable metal ion sensors. Another useful technique in order to identify heavy metal ions is the transfer of ions onto membranes using ion carriers. In order to detect heavy metal ions, these optical sensors generate an optical signal using a proton-selective chromo-ionophore and rely on the ions' specific binding with the ionophores. Chromo-ionophores change color or fluorescence when bound to heavy metal ions, which causes a discharge of protons with charges identical to those of the metal ions. It is the coordination bonds between heavy metal ions and many heteroatoms (oxygen, nitrogen, and sulfur) that give rise to ionophores.

5.3.3.3 Review on Optical Sensors

Optically chemical sensors have been used to detect Cd^{2+}, Pb^{2+}, and Hg^{2+} by coating a matrix of polymers with a sensing film of 5,10,15,20-tetra (p-sulphonatophenyl) porphyrin. When these metal ions are complexed with the sensor, new absorbance spectra are created,

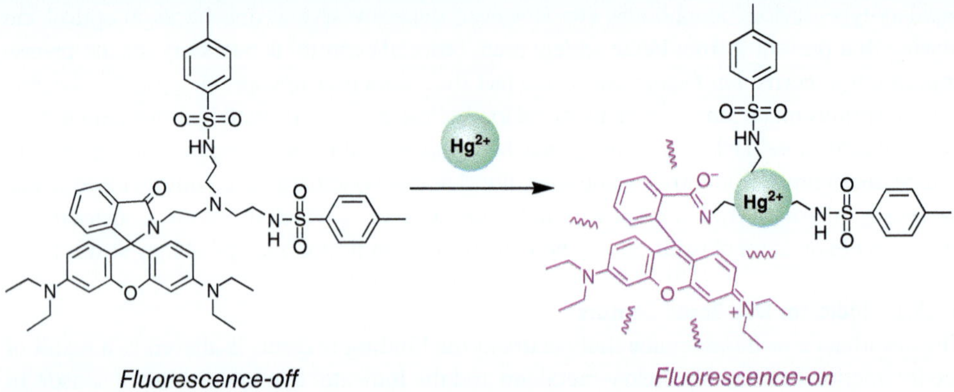

Figure 5.8 The "tren-based tripodal chemosensor" is used in the optical method of Hg^{2+} ion detection to display spectrum advances made by interactions with Hg^{2+} ions. Adapted from Lee et al. (2007).

altering how the sensor reacts to reflected light. With the help of rhodamine and two tosyl groups, a new chemosensor was developed for the selective detection of Hg^{2+} ions (as shown in Figure 5.8). Hg^{2+}'s contact with the sensor not only alters the sensor's color but also significantly increases its fluorescence. The selectivity of the sensor is shown by its capacity to interact with ions of other metals than Hg^{2+} with either no or little spectral shift.

Using a sharp shift in the proportion of intensity corresponding to the monomers (376 nM) to the excimer (465 nM) discharges on metal cation attachment in acetonitrile, an exceptionally sensitive optical sensor for detecting Zn ions has been developed (Ingale and Seela 2012). Two azobenzene compounds containing amino groups are known to react with the Hg^{2+} ion. Because of its coordination, the metal triggers an intramolecular electron transfer, having the independent amino group acting like the provider of electrons. Hg^{2+} coordination causes a reddening of the solution and a shift of 100 nM in the absorbance maxima in bathochromic space. Hg^{2+} ions are responsible for the color change (Fu et al. 2007). An aniline-containing hemicyanine (electron-donating compound) dye was used for the optical identification of Hg^{2+} in pH-neutral solutions of water as well as a benzothiazolium (electron-withdrawing compound) (Tatay et al. 2006). Bis(ferrocenyl) azine, a chromogenic material, was applied to a solid cellulose fiber substrate to enable visual detection of Hg^{2+} ions (Figure 5.9). Upon binding to the metal-specific sensor, the Hg^{2+} ion causes a dramatic change in hue. According to Díez-Gil et al. (2007), this color shift may be used for visual or spectroscopic determination of ion concentration.

When Hg^{2+} is added to such a ligand, the absorption band shifts hypsochromically, changing the hue of the solution from pink to green. Cd^{2+} and Hg^{2+} ions were separately identified using a 1,8-dihydroxyanthraquinone-based dithiacrown ether. As stated by Kadarkaraisamy and Sykes (2007), detection is centered on a rise in the illumination of fluorescence. Kadarkaraisamy and Sykes (2006) employed a fluorescent Pb^{2+} ion sensor made by fusing crown ether to 1,8-dihydroxyanthraquinone. Pb^{2+} binding to form a crown ether modifies the energy of excited states as well as significantly boosts fluorescence. The bathochromic peaks of the emission bands have been discovered to have been displaced due to coordination.

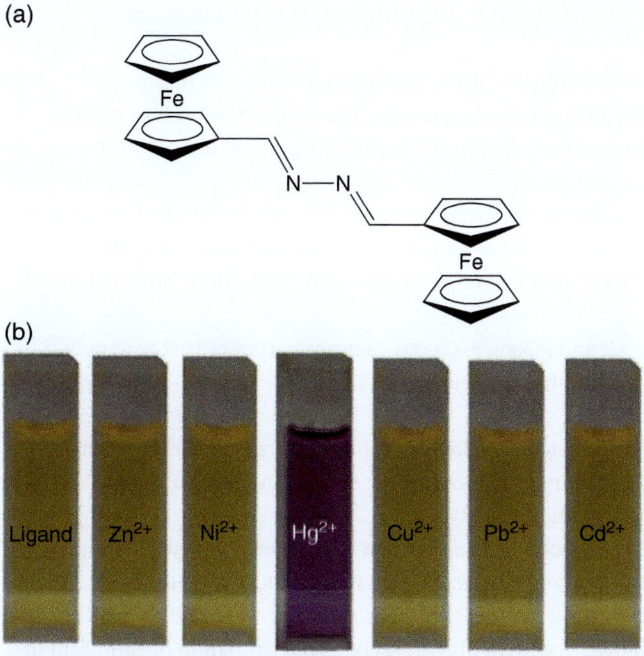

Figure 5.9 To achieve highly specific optical identification of the Hg^{2+} ion, a bis(ferrocenyl) azine was used like the chromogenic molecule. To do this, we (a) used a 1,4-disubstituted azine with two ferrocene groups as a mercury sensor, and (b) watched for color changes when the azine bound to Hg^{2+}. *Source:* Reproduced from Díez-Gil et al., 2007/MDPI.

5.4 Conclusion

Heavy metal ion contamination is a serious issue that has to be addressed. Because of the correlation between heavy metal ion toxicities and a wide range of health problems in living organisms, researchers have developed cutting-edge methods to detect these ions in water and render them safe for human use. There have been a number of effective materials and procedures used in the past for this goal, but they have also come with a number of downsides. Tools and processes are sometimes updated to make them more useful and address any problems they may have. A continual effort to find better-performing, sustainable, easy to use, and cost-effective materials and techniques is making it possible to reduce the levels of the ions of heavy metals to an acceptable point at which their presence is no longer harmful to living organisms.

References

Aglan, R.F., Saleh, H.M., and Mohamed, G.G. (2018). Potentiometric determination of mercury (II) ion in various real samples using novel modified screen-printed electrode. *Applied Water Science* 8: 1–11. https://doi.org/10.1007/s13201-018-0781-z.

Bashir, A., Ahad, S., and Pandith, A.H. (2016). Soft template assisted synthesis of zirconium resorcinol phosphate nanocomposite material for the uptake of heavy-metal ions. *Industrial*

& Engineering Chemistry Research 55 (17): 4820–4829. https://doi.org/10.1021/acs.iecr.6b00208.

Beltrán, B., Leal, L.O., Ferrer, L., and Cerdà, V. (2015). Determination of lead by atomic fluorescence spectrometry using an automated extraction/pre-concentration flow system. *Journal of Analytical Atomic Spectrometry* 30 (5): 1072–1079. https://doi.org/10.1039/C4JA00456F.

Cabrera-Vique, C., Teissedre, P.L., Cabanis, M.T., and Cabanis, J.C. (1997). Determination and levels of chromium in French wine and grapes by graphite furnace atomic absorption spectrometry. *Journal of Agricultural and Food Chemistry* 45 (5): 1808–1811. https://doi.org/10.1021/jf960691b.

Çelebi, M.S., Özyörük, H., Yıldız, A., and Abacı, S. (2009). Determination of Hg2+ on poly (vinylferrocenium)(PVF+)-modified platinum electrode. *Talanta* 78 (2): 405–409. https://doi.org/10.1016/j.talanta.2008.11.028.

Chen, J., Xiao, S., Wu, X. et al. (2005). Determination of lead in water samples by graphite furnace atomic absorption spectrometry after cloud point extraction. *Talanta* 67 (5): 992–996. https://doi.org/10.1016/j.talanta.2005.04.029.

Cui, L., Wu, J., and Ju, H. (2015). Electrochemical sensing of heavy metal ions with inorganic, organic and bio-materials. *Biosensors and Bioelectronics* 63: 276–286. https://doi.org/10.1016/j.bios.2014.07.052.

Cvetković, J., Arpadjan, S., Karadjova, I., and Stafilov, T. (2006). Determination of cadmium in wine by electrothermal atomic absorption spectrometry. *Acta Pharmaceutica* 56 (1): 69–77.

De Gregori, I., Quiroz, W., Pinochet, H. et al. (2007). Speciation analysis of antimony in marine biota by HPLC-(UV)-HG-AFS: extraction procedures and stability of antimony species. *Talanta* 73 (3): 458–465. https://doi.org/10.1016/j.talanta.2007.04.015.

Devi, N.R., Sasidharan, M., and Sundramoorthy, A.K. (2018). Gold nanoparticles-thiol-functionalized reduced graphene oxide coated electrochemical sensor system for selective detection of mercury ion. *Journal of the Electrochemical Society* 165 (8): B3046–B3053. https://doi.org/10.1149/2.0081808jes.

Díez-Gil, C., Caballero, A., Ratera, I. et al. (2007). Naked-eye and selective detection of mercury (II) ions in mixed aqueous media using a cellulose-based support. *Sensors* 7 (12): 3481–3488. https://doi.org/10.3390/s7123481.

Dong, S.F. and Zhu, Z.G. (2002). Determination of the contents of Ca, Mg, Fe, Cu and Zn in suxiao jiuxin pill and the analysis of Ca/Mg and Cu/Zn values. *Guang pu xue yu Guang pu fen xi= Guang pu* 22 (3): 478–479.

Dong, S.F. and Zhu, Z.G. (2003). Determination of the content of inorganic elements in taponin tablet recipe. *Guang pu xue yu Guang pu fen xi= Guang pu* 23 (1): 201–202.

Durmuşkahya, C., Alp, H., Hortooğlu, Z.S. et al. (2016). X-ray fluorescence spectroscopic determination of heavy metals and trace elements in aerial parts of *Origanum sipyleum* L. from Turkey. *Tropical Journal of Pharmaceutical Research* 15 (5): 1013–1015. https://doi.org/10.4314/tjpr.v15i5.16.

El-Bahi, S.M., Sroor, A.T., Arhoma, N.F., and Darwish, S.M. (2013). XRF analysis of heavy metals for surface soil of Qarun Lake and Wadi El Rayan in Faiyum, Egypt. *Open Journal of Metal* 3: https://doi.org/10.4236/ojmetal.2013.32A1003.

Ene, A., Bosneaga, A., and Georgescu, L. (2010). Determination of heavy metals in soils using XRF technique. *Romanian Journal of Physics* 55 (7–8): 815–820.

Falcó, G., Llobet, J.M., Bocio, A., and Domingo, J.L. (2006). Daily intake of arsenic, cadmium, mercury, and lead by consumption of edible marine species. *Journal of Agricultural and Food Chemistry* 54 (16): 6106–6112. https://doi.org/10.1021/jf0610110.

Flora, S.J. (2009). Structural, chemical and biological aspects of antioxidants for strategies against metal and metalloid exposure. *Oxidative Medicine and Cellular Longevity* 2 (4): 191–206. https://doi.org/10.4161/oxim.2.4.9112.

Flora, S.J.S., Mittal, M., and Mehta, A. (2008). Heavy metal induced oxidative stress & its possible reversal by chelation therapy. *Indian Journal of Medical Research* 128 (4): 501–523.

Fu, Y., Li, H., and Hu, W. (2007). Small molecular chromogenic sensors for Hg2+: a strong "push-pull" system exists after binding. *European Journal of Organic Chemistry* 2007: 2459–2463. https://doi.org/10.1002/ejoc.200601086.

Gumpu, M.B., Sethuraman, S., Krishnan, U.M., and Rayappan, J.B.B. (2015). A review on detection of heavy metal ions in water – an electrochemical approach. *Sensors and Actuators B: Chemical* 213: 515–533. https://doi.org/10.1016/j.snb.2015.02.122.

Hafuka, A., Takitani, A., Suzuki, H. et al. (2017). Determination of cadmium in brown rice samples by fluorescence spectroscopy using a fluoroionophore after purification of cadmium by anion exchange resin. *Sensors* 17 (10): 2291. https://doi.org/10.3390/s17102291.

Paul, V., Pandey, R., Ramesh, K. V., & Meena, R. C. (2017). Atomic absorption spectroscopy (AAS) for elemental analysis of plant samples. Manual of ICAR Sponsored Training Programme for Technical Staff of ICAR Institutes on "Physiological Techniques to Analyze the Impact of Climate Change on Crop Plants", 84.

Hutton, L.A., O'Neil, G.D., Read, T.L. et al. (2014). Electrochemical X-ray fluorescence spectroscopy for trace heavy metal analysis: enhancing X-ray fluorescence detection capabilities by four orders of magnitude. *Analytical Chemistry* 86 (9): 4566–4572. https://doi.org/10.1021/ac500608d.

Inamuddin, Naushad, M., Rangreez, T.A., and Alothman, Z.A. (2015). Ion-selective potentiometric determination of Pb (II) ions using PVC-based carboxymethyl cellulose Sn (IV) phosphate composite membrane electrode. *Desalination and Water Treatment* 56 (3): 806–813. https://doi.org/10.1080/19443994.2014.941307.

Ingale, S.A. and Seela, F. (2012). A ratiometric fluorescent on–off Zn2+ chemosensor based on a tripropargylamine pyrene azide click adduct. *The Journal of Organic Chemistry* 77 (20): 9352–9356. https://doi.org/10.1021/jo3014319.

Innuphat, C. and Chooto, P. (2017). Determination of trace levels of Cd (II) in tap water samples by anodic stripping voltammetry with an electrografted boron-doped diamond electrode. *ScienceAsia* 43 (1): 33–41. https://doi.org/10.2306/scienceasia1513-1874.2017.43.033.

Ismaiel, A.A., Aroua, M.K., and Yusoff, R. (2012). Potentiometric determination of trace amounts of mercury (II) in water sample using a new modified palm shell activated carbon paste electrode based on kryptofix 5. *American Journal of Analytical Chemistry* 3: 859–865. https://doi.org/10.4236/ajac.2012.312113.

Javanbakht, M., Divsar, F., Badiei, A. et al. (2009). Potentiometric detection of mercury (II) ions using a carbon paste electrode modified with substituted thiourea-functionalized highly ordered nanoporous silica. *Analytical Sciences* 25 (6): 789–794.

Kadarkaraisamy, M. and Sykes, A.G. (2006). Luminescence detection of transition and heavy metals by inversion of excited states: synthesis, spectroscopy, and X-ray crystallography of

Ca, Mn, Pb, and Zn complexes of 1, 8-anthraquinone-18-crown-5. *Inorganic Chemistry* 45 (2): 779–786. https://doi.org/10.1021/ic051672z.

Kadarkaraisamy, M. and Sykes, A.G. (2007). Selective luminescence detection of cadmium (II) and mercury (II) utilizing sulfur-containing anthraquinone macrocycles (part 2) and formation of an unusual Hg22+-crown ether dimer via reduction of Hg (II) by DMF. *Polyhedron* 26 (6): 1323–1330. https://doi.org/10.1016/j.poly.2006.10.046.

Kanchana, P., Sudhan, N., Anandhakumar, S. et al. (2015). Electrochemical detection of mercury using biosynthesized hydroxyapatite nanoparticles modified glassy carbon electrodes without preconcentration. *RSC Advances* 5 (84): 68587–68594. https://doi.org/10.1039/C5RA11424A.

Kápolna, E., Gergely, V., Dernovics, M. et al. (2007). Fate of selenium species in sesame seeds during simulated bakery process. *Journal of Food Engineering* 79 (2): 494–501. https://doi.org/10.1016/j.jfoodeng.2006.01.075.

Karimi, M., Aboufazeli, F., Hamid, R. et al. (2012). Determination of cadmium (II) ions in environmental samples: a potentiometric sensor. *Current World Environment* 7 (2): 201. https://doi.org/10.12944/CWE.7.2.02.

Kempegowda, R.G. and Malingappa, P. (2012). A binderless, covalently bulk modified electrochemical sensor: application to simultaneous determination of lead and cadmium at trace level. *Analytica Chimica Acta* 728: 9–17. https://doi.org/10.1016/j.aca.2012.03.053.

Kojuncu, Ý., Bundalevska, J.M., Ay, Ü. et al. (2004). Atomic absorption spectrometry determination of Cd, Cu, Fe, Ni, Pb, Zn, and TI traces in seawater following flotation separation. *Separation Science and Technology* 39 (11): 2751–2765. https://doi.org/10.1081/SS-200026751.

Lee, M.H., Wu, J.S., Lee, J.W. et al. (2007). Highly sensitive and selective chemosensor for Hg2+ based on the rhodamine fluorophore. *Organic Letters* 9 (13): 2501–2504. https://doi.org/10.1021/ol0708931.

Lee, H.J., Lagger, G., Pereira, C.M. et al. (2009). Amperometric tape ion sensors for cadmium (II) ion analysis. *Talanta* 78 (1): 66–70. https://doi.org/10.1016/j.talanta.2008.10.059.

Li, X.A., Zhou, D.M., Xu, J.J., and Chen, H.Y. (2007). In-channel indirect amperometric detection of heavy metal ions for electrophoresis on a poly (dimethylsiloxane) microchip. *Talanta* 71 (3): 1130–1135. https://doi.org/10.1016/j.talanta.2006.06.009.

Logar, M., Horvat, M., Akagi, H., and Pihlar, B. (2002). Simultaneous determination of inorganic mercury and methylmercury compounds in natural waters. *Analytical and Bioanalytical Chemistry* 374: 1015–1021. https://doi.org/10.1007/s00216-002-1501-x.

Lu, J., He, X., Zeng, X. et al. (2003). Voltammetric determination of mercury (II) in aqueous media using glassy carbon electrodes modified with novel calix [4] arene. *Talanta* 59 (3): 553–560. https://doi.org/10.1016/S0039-9140(02)00569-6.

Luo, L., Wang, X., Ding, Y. et al. (2010). Voltammetric determination of Pb2+ and Cd2+ with montmorillonite-bismuth-carbon electrodes. *Applied Clay Science* 50 (1): 154–157. https://doi.org/10.1016/j.clay.2010.06.024.

Majid, S., El Rhazi, M., Amine, A., and Brett, C.M. (2002). An amperometric method for the determination of trace mercury (II) by formation of complexes with L-tyrosine. *Analytica Chimica Acta* 464 (1): 123–133. https://doi.org/10.1016/S0003-2670(02)00422-1.

Marguí, E., Kregsamer, P., Hidalgo, M. et al. (2010). Analytical approaches for Hg determination in wastewater samples by means of total reflection X-ray fluorescence spectrometry. *Talanta* 82 (2): 821–827. https://doi.org/10.1016/j.talanta.2010.05.066.

Mazloum-Ardakani, M., Amini, M.K., Dehghan, M. et al. (2012). Nanomolar determination of Pb (II) ions by selective templated electrode. *Journal of the Serbian Chemical Society* 77 (7): 899–910. https://doi.org/10.2298/JSC110830210M.

Minami, T., Atsumi, K., and Ueda, J. (2003). Determination of cobalt and nickel by graphite-furnace atomic absorption spectrometry after coprecipitation with scandium hydroxide. *Analytical Sciences* 19 (2): 313–315. https://doi.org/10.2116/analsci.19.313.

Mohammadi, H., Amine, A., Cosnier, S., and Mousty, C. (2005). Mercury–enzyme inhibition assays with an amperometric sucrose biosensor based on a trienzymatic-clay matrix. *Analytica Chimica Acta* 543 (1–2): 143–149. https://doi.org/10.1016/j.aca.2005.04.014.

Moraes, P.M., Santos, F.A., Cavecci, B. et al. (2013). GFAAS determination of mercury in muscle samples of fish from Amazon, Brazil. *Food Chemistry* 141 (3): 2614–2617. https://doi.org/10.1016/j.foodchem.2013.05.008.

Mugheri, A.Q., Tahira, A., Sherazi, S.T.H. et al. (2016). An amperometric indirect determination of heavy metal ions through inhibition of glucose oxidase immobilized on cobalt oxide nanostructures. *Sensor Letters* 14 (12): 1178–1186. https://doi.org/10.1166/sl.2016.3752.

Muniz-Naveiro, O., Domínguez-González, R., Bermejo-Barrera, A. et al. (2007). Selenium speciation in cow milk obtained after supplementation with different selenium forms to the cow feed using liquid chromatography coupled with hydride generation-atomic fluorescence spectrometry. *Talanta* 71 (4): 1587–1593. https://doi.org/10.1016/j.talanta.2006.07.040.

Naushad, M., Inamuddin, and Rangreez, T.A. (2015). Potentiometric determination of Cd (II) ions using PVC-based polyaniline Sn (IV) silicate composite cation-exchanger ion-selective membrane electrode. *Desalination and Water Treatment* 55 (2): 463–470. https://doi.org/10.1080/19443994.2014.915389.

Odonchimeg, S., Oyun, J., and Javkhlantugs, N. (2016). Determination of plantinum in rocks by graphite furnace atomic absorption spectrometry after separation on sorbent. *International Research Journal of Engineering and Technology* 3: 753–757.

Oliveira, V., Sarmiento, A.M., Gómez-Ariza, J.L. et al. (2006). New preservation method for inorganic arsenic speciation in acid mine drainage samples. *Talanta* 69 (5): 1182–1189. https://doi.org/10.1016/j.talanta.2005.12.034.

Othman, A.M. (2006). Potentiometric determination of mercury (II) using a tribromomercurate–rhodamine B PVC membrane sensor. *International Journal of Environmental Analytical Chemistry* 86 (05): 367–379. https://doi.org/10.1080/03067310500248379.

Palmer, P.T., Jacobs, R., Baker, P.E. et al. (2009). Use of field-portable XRF analyzers for rapid screening of toxic elements in FDA-regulated products. *Journal of Agricultural and Food Chemistry* 57 (7): 2605–2613. https://doi.org/10.1021/jf803285h.

Pei, X., Kang, W., Yue, W. et al. (2014). Disposable copper-based electrochemical sensor for anodic stripping voltammetry. *Analytical Chemistry* 86 (10): 4893–4900. https://doi.org/10.1021/ac500277j.

Radu, T. and Diamond, D. (2009). Comparison of soil pollution concentrations determined using AAS and portable XRF techniques. *Journal of Hazardous Materials* 171 (1–3): 1168–1171. https://doi.org/10.1016/j.jhazmat.2009.06.062.

Radulescu, C., Dulama, I.D., Stihi, C. et al. (2014). Determination of heavy metal levels in water and therapeutic mud by atomic absorption spectrometry. *Romanian Journal of Physics* 59 (9–10): 1057–1066.

Rashid, M. and Rahman, N. (2012). Potentiometric sensor for the determination of lead (II) ion based on zirconium (IV) iodosulphosalicylate. *Science of Advanced Materials* 4 (12): 1232–1237. https://doi.org/10.1166/sam.2012.1391.

Ratner, N. and Mandler, D. (2015). Electrochemical detection of low concentrations of mercury in water using gold nanoparticles. *Analytical Chemistry* 87 (10): 5148–5155. https://doi.org/10.1021/ac504584f.

Rebollo, F. *XRF: X-Ray Fluorescence Spectroscopy*. Alter Technology https://wpo-altertechnology.com/xrf-x-ray-fluorescence-spectroscopy-hi-rel-parts.

Ruiz-Chancho, M.J., Sabé, R., López-Sánchez, J.F. et al. (2005). New approaches to the extraction of arsenic species from soils. *Microchimica Acta* 151: 241–248. https://doi.org/10.1007/s00604-005-0405-9.

Sánchez-Rodas, D., Luis Gómez-Ariza, J., and Oliveira, V. (2006). Development of a rapid extraction procedure for speciation of arsenic in chicken meat. *Analytical and Bioanalytical Chemistry* 385: 1172–1177. https://doi.org/10.1007/s00216-006-0379-4.

Schaeffer, R., Soeroes, C., Ipolyi, I. et al. (2005). Determination of arsenic species in seafood samples from the Aegean Sea by liquid chromatography–(photo-oxidation)–hydride generation–atomic fluorescence spectrometry. *Analytica Chimica Acta* 547 (1): 109–118. https://doi.org/10.1016/j.aca.2005.01.032.

Sevcikova, M., Modra, H., Slaninova, A., and Svobodova, Z. (2011). Metals as a cause of oxidative stress in fish: a review. *Veterinary Medicine* 56 (11): 537–546.

Silva, D.H., Costa, D.A., Takeuchi, R.M., and Santos, A.L. (2011). Fast and simultaneous determination of Pb2+ and Cu2+ in water samples using a solid paraffin-based carbon paste electrode chemically modified with 2-aminothiazole-silica-gel. *Journal of the Brazilian Chemical Society* 22: 1727–1735. https://doi.org/10.1590/S0103-50532011000900015.

Singh, A., Sharma, R.K., Agrawal, M., and Marshall, F.M. (2010). Health risk assessment of heavy metals via dietary intake of foodstuffs from the wastewater irrigated site of a dry tropical area of India. *Food and Chemical Toxicology* 48 (2): 611–619. https://doi.org/10.1016/j.fct.2009.11.041.

Taha, K. (2017). Heavy elements analyses in the soil using X-ray fluorescence and inductively coupled plasma-atomic emission spectroscopy. *International Journal of Advances in Science Engineering and Technology* 5: 118–120.

Talat, M., Tripathi, P., and Srivastava, O.N. (2018). Highly sensitive electrochemical detection of mercury present in the beauty creams using graphene modified glassy carbon electrode. *Innovations in Corrosion and Materials Science* 8 (1): 24–31. https://doi.org/10.2174/2352094908666180109152904.

Tareen, A.K., Sultan, I.N., Parakulsuksatid, P. et al. (2014). Detection of heavy metals (Pb, Sb, Al, As) through atomic absorption spectroscopy from drinking water of District Pishin, Balochistan, Pakistan. *International Journal of Current Microbiology and Applied Sciences* 3 (1): 299–308.

Tatay, S., Gaviña, P., Coronado, E., and Palomares, E. (2006). Optical mercury sensing using a benzothiazolium hemicyanine dye. *Organic Letters* 8 (17): 3857–3860. https://doi.org/10.1021/ol0615580.

Tautkus, S., Steponeniene, L., and Kazlauskas, R. (2004). Determination of iron in natural and mineral waters by flame atomic absorption spectrometry. *Journal of the Serbian Chemical Society* 69 (5): 393–402. https://doi.org/10.2298/JSC0405393T.

Tsade, H.K. (2016). Atomic absorption spectroscopic determination of heavy metal concentrations in Kulufo River, Arbaminch, Gamo Gofa, Ethiopia. *International Journal of Environmental Analytical Chemistry* 3 (177): 2. https://doi.org/10.4172/2380-2391.1000177.

Valko, M.M.H.C.M., Morris, H., and Cronin, M.T.D. (2005). Metals, toxicity and oxidative stress. *Current Medicinal Chemistry* 12 (10): 1161–1208. https://doi.org/10.2174/0929867053764635.

Wegrzynek, D. (2004). *Commercial Applications of X-Ray Spectrometric Techniques*, 1. Copyright. The IAEA Laboratories https://www-pub.iaea.org/MTCD/Publications/PDF/SupplementaryMaterials/TECDOC_1713_CD/template-cd/datasets/papers/Wegrzynek%20Commercial%20applications%20of%20Xray.pdf.

Wierzbicki, T. and Pyrzyñska, K. (2002). Determination of vanadium content in wine by GF AAS. *Chemia Analityczna* 47 (3): 449–456.

Xia, F., Zhang, X., Zhou, C. et al. (2010). Simultaneous determination of copper, lead, and cadmium at hexagonal mesoporous silica immobilized quercetin modified carbon paste electrode. *Journal of Analytical Methods in Chemistry* 2010: 824197. https://doi.org/10.1155/2010/824197.

Yao, X.Z., Guo, Z., Yuan, Q.H. et al. (2014). Exploiting differential electrochemical stripping behaviors of Fe3O4 nanocrystals toward heavy metal ions by crystal cutting. *ACS Applied Materials & Interfaces* 6 (15): 12203–12213. https://doi.org/10.1021/am501617a.

Yuan, X., Ling, K.H., and Keung, C.W. (2009). The analysis of heavy metals in Chinese herbal medicine by flow injection–mercury hydride system and graphite furnace atomic absorption spectrometry. *Phytochemical Analysis* 20 (4): 293–297. https://doi.org/10.1002/pca.1126.

Yuan-Zhen, P.E.N.G., Huang, Y.M., Dong-Xing, Y.U.A.N. et al. (2012). Rapid analysis of heavy metals in coastal seawater using preconcentration with precipitation/co-precipitation on membrane and detection with X-ray fluorescence. *Chinese Journal of Analytical Chemistry* 40 (6): 877–882. https://doi.org/10.1016/S1872-2040(11)60554-9.

Zarazúa, G., Girón-Romero, K., Tejeda, S. et al. (2014). Total reflection X-ray fluorescence analysis of toxic metals in fish tissues. *American Journal of Analytical Chemistry* 05: 805–811. https://doi.org/10.4236/ajac.2014.512089.

Zhao, G., Wang, H., and Liu, G. (2017). Direct quantification of Cd2+ in the presence of Cu2+ by a combination of anodic stripping voltammetry using a Bi-film-modified glassy carbon electrode and an artificial neural network. *Sensors* 17 (7): 1558. https://doi.org/10.3390/s17071558.

Zhong, W.S., Ren, T., and Zhao, L.J. (2016). Determination of Pb (Lead), Cd (Cadmium), Cr (Chromium), Cu (Copper), and Ni (Nickel) in Chinese tea with high-resolution continuum source graphite furnace atomic absorption spectrometry. *Journal of Food and Drug Analysis* 24 (1): 46–55. https://doi.org/10.1016/j.jfda.2015.04.010.

6

Remediation of Heavy Metals in Environmental Resources Using Physical Methods

C. Arun[1], A. Sethupathy[2], R.V. Hemavathy[3], and Chandrasekaran Ramprasad[4]

[1] *Department of Biotechnology, Periyar Maniammai Institute of Science & Technology (Deemed to be University), Thanjavur, Tamil Nadu, India*
[2] *Department of Chemical Engineering, Anjalai Ammal Mahalingam Engineering College Kovilvenni, Thiruvarur, Tamil Nadu, India*
[3] *Department of Biotechnology, Rajalakshmi Engineering College, Thandalam, Chennai, Tamil Nadu, India*
[4] *School of Civil Engineering, Centre for Advanced Research in Environment (CARE), SASTRA Deemed to be University, Thanjavur, Tamil Nadu, India*

6.1 Introduction

Generally, the metals and metalloids group exhibit a density of $4000 \pm 100\,\mathrm{mg\,cm^{-3}}$, are collectively represented in terms of HMs, and are considered worldwide contaminants/pollutants of natural sources (Vardhan et al. 2019; Sivaranjanee et al. 2022). These HMs can be naturally found in soils, sediments, rocks, water sources, and microbes. HMs can be exhibited as actual levels of concentration in the earth's crust region. Some HMs (nickel, silver, copper, zinc, gold, cadmium, lead, mercury, manganese, cobalt, and molybdenum) are generated naturally, which persist in the natural sources for a long period (Stowers et al., 2014). HMs can be used in several modes: cobalt plays a substantial role in propionic acid reduction during liquid fermentation (Hu et al. 2017). Trace HMs such as selenium, zinc, and copper are aids to proceed the biological process (human metabolism). Some HMs (tin, zinc, lead, copper, and tungsten) can be applied for industrial purposes. Artificially derived bio-inorganic catalyst uses HMs as a central atom in chemical transformation. HMs such as lead, chromium, thallium, mercury, and cadmium show destructive impacts at low levels of concentration (Terfassa et al. 2014). The uncontrollable population and industrialization will have a huge impact on natural resources. HMs like nickel, lead, and tin are generated in considerable amounts from printed circuit board production industries. Milling, electrolysis deposition, anodizing cleaning, conversion coating, and electroplating industries are the releasing of HMs such as zinc, chromium, copper, platinum, vanadium, titanium, and silver. Arsenic waste is released from arsenate-chromate copper-based wood treatment. Cadmium, chromium and sulfide substances are released from inorganic pigment production industries. In petroleum-refining industries, conversion

Remediation of Heavy Metals: Sustainable Technologies and Recent Advances, First Edition. Edited by Rangabhashiyam Selvasembian, Binota Thokchom, Pardeep Singh, Ali H. Jawad, and Willis Gwenzi.

catalysts are contaminated with vanadium, chromium, and nickel. The tons of sludge quantity obtained from various industries can be noxious in nature, which leads to effectual wastewater treatment methods needed to eradicate it from the natural resources (Al-Zubaidi and Yang 2020; Fu et al. 2021; Ramprasad et al. 2022; Zhang et al. 2022).

The earth's crust contains a variety of naturally occurring heavy metals. Heavy metal pollution is caused by both natural and anthropomorphic activities such as mining, smelting, industrial production, and the use of metals and metal-containing compounds for domestic and agricultural purposes. Natural activities like volcanic eruption, ocean sprays, and weathering of rocks, dust, and aerosol particles are the key source for heavy metals generation (Rizvi et al. 2020). The eight heavy metals listed as the most prevalent in the environment by the United States Environmental Protection Agency (USEPA) compilation are lead (Pb), chromium (Cr), arsenic (As), zinc (Zn), cadmium (Cd), copper (Cu), mercury (Hg), and nickel (Ni) (Selvi et al. 2019). The high density and high toxicity of heavy metals, even at low concentrations, make them significant environmental contaminants. Various researchers have reported that these sources contribute to human exposure and environmental contamination (Rahman and Bastola 2014; Ramprasad and Philip 2016). The various sources of heavy metals and their potential for environmental contamination are shown in Figure 6.1. Broad classification of heavy metals with examples is tabulated in Table 6.1. Heavy metals are an ill-defined group that is frequently found at contaminated sites. They are distinguished by their long persistence in the natural environment, resulting in serious health consequences in humans, fauna, and flora even at very low concentrations (1 or 2 μg in some cases) (Philip et al. 2019; Ramprasad et al. 2022).

Numerous harmful heavy metals, such as Cr, Cd, Hg, and Pb, that are dumped by industry into the environment are recalcitrant and will not degrade faster and will poison the soil and water to a higher extent. Due to the high propensity of heavy metals, they have a

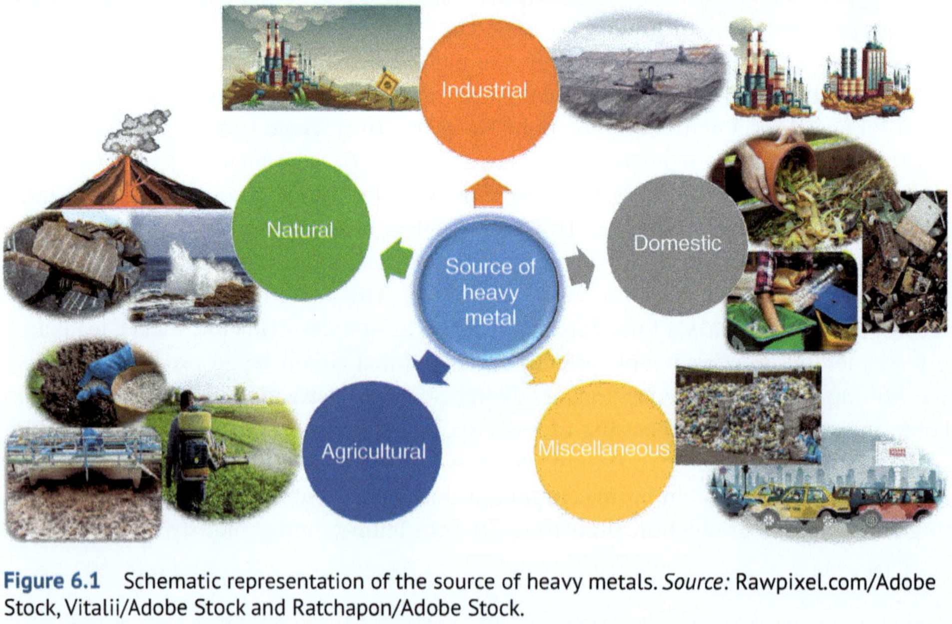

Figure 6.1 Schematic representation of the source of heavy metals. *Source:* Rawpixel.com/Adobe Stock, Vitalii/Adobe Stock and Ratchapon/Adobe Stock.

Table 6.1 Different classes of heavy metals, with their examples and application.

Class of heavy metals	Examples of heavy metals	Application	References
Macronutrients	Cobalt (Co), iron (Fe)	Catalysts, paint dryers, biomedical, steel and rechargeable battery chemicals	Czarnek et al. (2015), Ali et al. (2016), and Sharma et al. (2019)
Micronutrients	Copper (Cu), nickel (Ni), chromium (Cr), manganese (Mn)	Automotive, textile, leather, biomedical, electrical, and electronic	Duarte-Nass et al. (2020) and Haque et al. (2021)
Highly toxic	Mercury (Hg), cadmium (Cd), arsenic (As), Tin (Sn), zinc (Zn)	Medical, batteries, electronic, metallurgy	Bradberry (2016), Ali et al. (2016), and Rahman and Singh (2020)
Radioactive	Uranium, thorium, radium	Nuclear reactor, electronic, medical appliances; military	Ding et al. (2019) and Singh et al. (2023)
Rare earth	Lanthanum, yttrium, cerium	Metallurgy, mineralogy, electronics, medical	Balaram (2019) and Ramprasad et al. (2022)
Precious	Gold, platinum, silver, palladium	Jewelry, electronic, and electrical	Kwak et al. (2018) and Korium et al. (2021)

Table 6.2 Indian and European (EU) standards for heavy metals in soil, food, and drinking water.

Heavy metal	Indian standard			European Union standard in soils ($\mu g\,g^{-1}$)
	Soil ($\mu g\,kg^{-1}$)	Food ($mg\,kg^{-1}$)	Water ($mg\,l^{-1}$)	
Cadmium (Cd)	3.0–6.0	1.5	0.01	3.0
Chromium (Cr)	NA	20	0.05	150
Copper (Cu)	135–270	30	0.05	140
Nickel (Ni)	75–150	1.5	0.03	75
Lead (Pb)	250–500	2.5	0.1	300
Zinc (Zn)	300–600	50	5.0	300

tendency to accumulate in various environmental matrices, leading to concentrations that are deceptively greater than the recommended average safety threshold as shown in Table 6.2 (Rzymski et al. 2014; Li et al. 2019). The Comprehensive Environmental Response Compensation and Liability Act, United States, specifies the following as the maximum permissible level of heavy metals in aqueous medium: Cr – $0.01\,mg\,l^{-1}$, Ar – $0.01\,mg\,l^{-1}$, Cd – $0.05\,mg\,l^{-1}$, Hg – $0.002\,mg\,l^{-1}$, Pb – $0.015\,mg\,l^{-1}$, and Ag – $0.05\,mg\,l^{-1}$, respectively (Jaishankar et al. 2014). A significant number of human life-threatening disorders, including atherosclerosis, cancer, Alzheimer's disease, and Parkinson's disease, can be caused by heavy metal concentrations that are higher than that are advised (Muszynska and Hanus-Fajerska 2015). Due to this, numerous researchers have researched many physical, chemical, and biological remediation techniques to lower these contaminants' levels in the environment below the allowable regulatory limit.

Many treatment systems have been adopted for the removal of heavy metals, such as physical, chemical, and biological methods. There were several advantages and disadvantages of each that are tabulated in Table 6.3. In situ remediation is the treatment of the pollutant in the original place, without moving the contaminants, while ex situ remediation involves excavation and removal of the pollutants for treatment. However, the selection of the most appropriate heavy metal remediation method depends on the site characteristics, types of pollutants to be removed, contaminant concentration, and the final use of the contaminated medium (Ramprasad et al. 2022). It was observed from the table that the physical methods for the removal of heavy metals had slighted advantages over the other methods. Hence, the further sections discuss in detail the various physical methods for the removal of heavy metals.

Table 6.3 Advantages and limitations of various treatment technologies for remediation of heavy metals.

Method	Remediation technique	Advantages	Disadvantages	Application and status of usage	References
Physical	Filtration, adsorption, membrane filtration, biosorption, flocculation, encapsulation, and landfilling	Easy to install and faster recovery rate, low installation cost, higher security	Limited to small area and pollutant targeted one, high cost on waste disposal and storage, more maintenance cost	In situ and ex situ applications possible, and widely practiced	Song et al. (2017), Liu et al. (2018), Ramprasad and Philip (2018), Mao et al. (2020), Selvi and Aruliah (2018)
Chemical	Chemical filtration, electrocoagulation, stabilization, solidification, soil flushing, vitrification, and advanced oxidation process	High efficiency, fast to treatment process, low to high contamination level can be treated	Higher cost for installation and maintenance, trace of pollutants remains in treated water, more sludge generation, and time consuming	In situ and moderate ex situ applications, widely practiced	
Biological	Biodegradation by microbes (aerobic and anaerobic) and phytodegradation	Low cost, simple to implement and higher public acceptance, more sustainable	More time consuming, shock loads cannot be sustained, generates sludge, lower efficiency for high concentration	In situ only preferred, very rare for ex situ	

6.2 Toxicity of HMs

The toxicity of the pollutants or heavy metals is characterized by their half-life time or persistence of pollutants in ambient atmospheric conditions, residence time, acute or chronic effects, lethality effects of metals, bioaccumulation, biomagnification, and carcinogenic properties (Manzetti et al. 2014). Aquatic species are pretentious by the influence of HMs in the water sources. Generally, microbial species mineralize the HMs and are utilized by phytoplankton, move into the stages of biomagnification, transient toward high trophic level of ecosystem and disturb aquatic species in morphological and physiological terms. HMs severely affects human metabolism, nervous system, and cell growth. Autoimmune diseases such as kidney failure and rheumatoid arthritis are caused by lead and mercury. Some HMs (vanadium, mercury, antimony, cobalt, and nickel) can be carcinogenic. Arsenic and vanadium can be mutagenic, which cause lung cancer, skin cancer, and hair fall. Thallium causes behavioral alteration and neurological disorders, especially in children, and also causes central nervous system disorder and heartbeat disturbance in adults. Osteoporosis and bone marrow disorder are caused by cadmium at low-level concentrations. Nephrotoxicity is caused by mercury, heptavalent manganese, cadmium, and lead, which cause immune system disorders (Ebrahimi et al., 2020; Barros et al., 2021). To reduce human and natural resource exposure to noxious substances, waste management steps need to be improved. Environmental protection agency (EPA) of the United States implemented limits on contaminant levels and toxicity release from heavy metals, which are reported in Table 6.4. According to Matulik et al. (2017), sewage containing high doses of mercury which discarded by Chisso Corporation of Japan. The mercury level accrues in sea creatures and via biomagnification results in mercury intoxication in population. Unusual symptoms reported in the residents of Minamata such as vision problem, convulsions and numbness which leads to loss of human lives.

6.3 Physical Methods for Remediation of HMs from Wastewater

Figure 6.2 indicates the physical methods for remediation of HMs. The physical methods, such as coagulation, flocculation, ion exchange, adsorption, and membrane filtration, are explained as follows:

Table 6.4 Maximum contaminant level standards for noxious HMs.

Types of HMs	Maximum contaminant level ($g\,l^{-1}$)	Toxicity	Reference
Cu	0.25×10^{-3}	Insomnia, liver damage	Ahmad et al. (2020), Jehan et al. (2020)
Zn	0.8×10^{-3}	Thirsty, depression and neurological disorder	
As	0.05×10^{-3}	Visceral cancer	
Cd	0.01×10^{-3}	Kidney failure, human carcinogen	
Cr	0.05×10^{-3}	Carcinogenic, vomiting, headache	

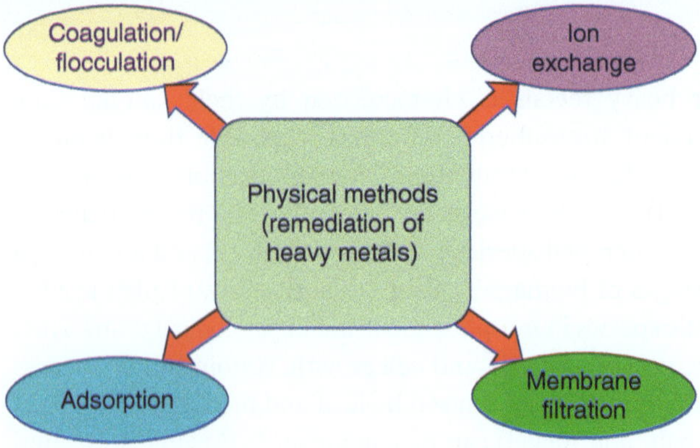

Figure 6.2 Physical methods for remediation of heavy metals (HMs).

6.4 Coagulation and Flocculation

Coagulation can reduce the net surface charge of colloidal matters for stabilization by means of electrostatic repulsion. Generally, the coagulation process requires coagulant for effective particle stabilization state (Ramirez Arenas et al. 2020; Marichamy et al. 2021). The coagulants such as polyaluminum chloride, aluminum sulfate, magnesium chloride, polyethylenimine, and aluminum hydroxide oxide. In flocculation method, the rise of particle size by means of effective mixing, interaction, and collision with organic and inorganic polymers. Then, the smaller particles become larger ones (agglomeration state) and these can be separated using straining or flotation and filtration (Zhang et al. 2023). Begum et al. (2021) reported that the biopolymer (chitosan) acts as a good coagulation and flocculation agent. The combined coagulation with flocculation method was effective for decontamination of wastewater from various industries and drinking water treatment. As per US EPA, the implication of coagulation (alum) and lime softening for eradicating 98% of chromium, cadmium, and lead content in wastewater. Zainol et al. (2019) carried out the investigation in river water as a source of heavy metal removal with alum and ferric sulfate. For ferric sulfate coagulation, chromium removal (98%) was attained in the pH range of 6.4–9.4. For alum, 99% of lead removal was attained in the pH of 6–8. Carolin et al. (2017) reported that ferric chloride is an effective coagulant for the removal of metals such as lead, copper, nickel, and zinc from synthetic wastewater.

6.5 Ion Exchange

Ion exchange method can be used for removal of HMs at low concentrations in aqueous solution. It is a cost-effective method using inexpensive substances to remove solubilized ions from liquid state to solid state. Though, this method is more sensitive pH of aqueous solution. Soliman and Moustafa (2020) reported that naturally derived inorganic zeolite/synthetic organic resin with anion or cation was used to eradicate aqueous solution HMs. It occurs in the form of reversible reaction by exchange of ions from aqueous solution. Organic resin exhibits high molecular weight polyelectrolytes for exchange of mobile ions that set the more number of exchanges per resin. Synthetic resins are a solid matter which are insoluble in water content to absorb negative or positive ions in an electrolyte solution (Koohsaryan et al. 2020; Soliman and Moustafa 2020). Then, it releases the remaining ions of the similar charge in the aqueous solution in equal amounts. Resins are broadly divided into weak or strong acid/base anion/cation exchangers. Ion exchange method is applied in different industrial effluents for HMs removal. Lee et al. (2020) reported that ion exchange resins are sensitive to high concentrations of HMs in aqueous solution. Liquid ion exchange method removes HMs such as chromium, zinc, and cadmium. Metal finishing operation uses acid, base, and plating chemicals that are ionized in an aqueous solution which can be removed using ion exchange method. This step indicates the last step in pollution control in which HMs or cyanide complexes are removed before disposal into the natural source (Prelot et al. 2014; Wang et al. 2014; Saravanan et al. 2023).

6.6 Adsorption

The release of HMs increased day by day through industrial activities. It indicates the necessary methods to be developed to eradicate the metal ions concentration in wastewater. Adsorption is one of the simplest methods to eliminate HMs or adsorbate from polluted wastewater. During adsorption process, HMs can be adhered to on the adsorbent surface rather than move into the internal matrix (Yin and Deng 2015; Agarwal et al. 2021). This method is more feasible to utilize in natural biological, physical, and chemical systems. Generally, adsorption is a mass transfer process in which contaminants are attached to the solid support with chemical/physical interaction by means of hydrogen bonding or van der Waals force of attraction. It is a widely approachable method when compared to ion exchange, oxidative, membrane, and ozonation (Yadav et al. 2019). The merits of adsorption method such as minimize waste biosolids, eco-friendly nature, high efficacy, cost-effective and the used adsorbents can be regenerated (Gunatilake 2015). Aslam et al. (2022) reported that adsorption method mechanism depends on mass transfer rate and solid-liquid equilibrium. Adsorbents are generally derived from waste byproducts, agricultural waste, modified biopolymers, natural material and surface-modified magnet adsorbents, and it can be utilized for removal of HMs from wastewater (Ramprasad et al. 2020). Literature analysis indicates removal of metal ions with adsorbents like red mud, coal, nanoparticles, waste biosolid, zeolite, fertilizer waste (Yao et al. 2015; Karna et al. 2017). Agriculture waste products such as water hyacinth, orange peel, sunflower, coir pith, and marine algae have been investigated for HMs removal (Renu et al. 2017). Islam et al. (2019) discussed the Cr^{6+} removal with sawdust as adsorbent and it attained 50–80% of efficiency. Božić et al. (2013) reported that removal of nickel, copper, cadmium, and zinc by beech sawdust and also linden trees and poplar sawdust were investigated for the removal of copper, cadmium, zinc, manganese, and nickel.

6.7 Membrane Filtration

For wastewater treatment, thin film composite and porous membranes are effectively used. The membrane process can be regulated by driving forces such as low pressure (distillation, ultrafiltration, and microfiltration), osmotic pressure (electrodialysis, forward osmosis, and liquid membrane), and high pressure (reverse osmosis and nanofiltration). The key factors influencing the membrane process are pore distribution, pore size, surface charge, functional groups, and degree of hydrophilicity. These factors play a substantial role in wastewater treatment and HMs removal. The selectivity and flux rate are used to determine the efficacy of membrane process. Porous membranes such as polyamide, cellulose acetate, and polysulfone are applied in solution flow where the pressure level is low. Size exclusion, adsorptive ability, steric hindrance, and Donnan exclusion are governed in membrane technology. Abdullah et al. (2019) reported that membrane filtration is highly recognized for inorganic effluent treatment and also reduces the organic substances and suspended solids. It gives more attention to HMs removal because it exhibits cheap and high efficacy (Liu et al., 2021). Nanofiltration and reverse osmosis can be effectively removed from the metal ions (Ni^{2+}, Cu^{2+}, Cr^{6+}) (Taghizadeh et al. 2013). For separation process, ultrafiltration method can be executed with help of chemicals to attain effectual removal of HMs. This process is also known as polymer-enhanced ultrafiltration and micellar-enhanced ultrafiltration (Huang et al. 2016). Micellar-enhanced ultrafiltration method exhibits the merits as high flux, low energy cost, and high HMs removal efficacy and demerits as more operating cost. Polymer-enhanced ultrafiltration method uses water-soluble polymers, which cannot be recovered by ultrafiltration method (Huang et al. 2016).

6.8 Conclusion

The rapid rise of HMs load on the ecosystem will cause critical menace to human population and natural ecosystem. Hence, the eradication of HMs is a major concern. HMs can be removed by using physical, chemical, and biological methods. The choosing of the method depends on feasible steps and economic and technical applicability. Physical methods were creating high impact on removal of heavy metals from various contaminated sources. Membrane filtration is a flexible method which consumes minimum chemical agent and generates low level of solid content with high efficacy of HMs removal but it has a limit such as minimum flow rate. Flocculation and coagulation methods attained high-level sludge settling and dewatering but these require heavy load of chemicals. For ion exchange method, the merits such as process stability, chemical safety, and pH of the sample will not be affected but it has a demerit of fouling problem. Adsorption method was effective removal of heavy metals from wastewater using natural and synthetic adsorbents. The obtained waste from the adsorption process can be reused as adsorbent for further removal of HMs. This method can be selected depending on load of HMs, economical input, ecological impact, and easy-way operation.

References

Abdullah, N., Yusof, N., Lau, W.J. et al. (2019). Recent trends of heavy metal removal from water/wastewater by membrane technologies. *Journal of Industrial and Engineering Chemistry* 76: 17–38.

Agarwal, A., Singh, A., Banerjee, B.D. et al. (2021). Exotic hydrogel matrix as an efficient platform for sustainable production of biomass and lipid from chlorella sorokiniana. *ACS Applied Bio Materials* 4 (8): 6304–6315.

Ahmad, S., Pandey, A., Pathak, V.V. et al. (2020). Phycoremediation: algae as eco-friendly tools for the removal of heavy metals from wastewaters. In: *Bioremediation of Industrial Waste for Environmental Safety: Volume II: Biological Agents and Methods for Industrial Waste Management*, 53–76. Singapore: Springer.

Ali, M.M., Ali, M.L., Islam, M.S., and Rahman, M.Z. (2016). Preliminary assessment of heavy metals in water and sediment of Karnaphuli River, Bangladesh. *Environmental Nanotechnology, Monitoring & Management* 5: 27–35.

Al-Zubaidi, I. and Yang, C. (2020). Waste management of spent petroleum refinery catalyst: mechanism review. *European Journal of Engineering and Technology Research* 5 (8): 938–947.

Aslam, M.M.A., Den, W., and Kuo, H.W. (2022). Elucidating the mass transfer mechanism of CrVI adsorption by encapsulated chitosan-carbon nanotubes-iron beads in packed-bed columns. *Journal of Water Process Engineering* 46: 102586.

Balaram, V. (2019). Rare earth elements: a review of applications, occurrence, exploration, analysis, recycling, and environmental impact. *Geoscience Frontiers* 10 (4): 1285–1303.

Barros, A., Tejada-Tovar, C., and Gonzalez-Delgado, A.D. (2021). Comparative assessment of Al_2O3 modified biomasses from agricultural residues for nickel and cadmium removal. *Journal of Water and Land Development* 49: 29–34.

Begum, S., Yuhana, N.Y., Saleh, N.M. et al. (2021). Review of chitosan composite as a heavy metal adsorbent: material preparation and properties. *Carbohydrate Polymers* 259: 117613.

Božić, D., Gorgievski, M., Stanković, V. et al. (2013). Adsorption of heavy metal ions by beech sawdust–kinetics, mechanism and equilibrium of the process. *Ecological Engineering* 58: 202–206.

Bradberry, S.M. (2016). Metals (cobalt, copper, lead, mercury). *Medicine* 44 (3): 182–184.

Carolin, C.F., Kumar, P.S., Saravanan, A. et al. (2017). Efficient techniques for the removal of toxic heavy metals from aquatic environment: a review. *Journal of Environmental Chemical Engineering* 5 (3): 2782–2799.

Czarnek, K., Terpiłowska, S., and Siwicki, A.K. (2015). Selected aspects of the action of cobalt ions in the human body. *Central European Journal of Immunology* 40 (2): 236–242.

Ding, Y., Xia, G., Ji, H., and Xiong, X. (2019). Accurate quantitative determination of heavy metals in oily soil by laser induced breakdown spectroscopy (LIBS) combined with interval partial least squares (IPLS). *Analytical Methods* 11 (29): 3657–3664.

Duarte-Nass, C., Rebolledo, K., Valenzuela, T. et al. (2020). Application of microbe-induced carbonate precipitation for copper removal from copper-enriched waters: challenges to future industrial application. *Journal of Environmental Management* 256: 109938.

Ebrahimi, M., Khalili, N., Razi, S. et al. (2020). Effects of lead and cadmium on the immune system and cancer progression. *Journal of Environmental Health Science and Engineering* 18: 335–343.

Fu, Q., Tan, X., Ye, S. et al. (2021). Mechanism analysis of heavy metal lead captured by natural-aged microplastics. *Chemosphere* 270: 128624.

Gunatilake, S.K. (2015). Methods of removing heavy metals from industrial wastewater. *Methods* 1 (1): 14.

Haque, S., Tripathy, S., and Patra, C.R. (2021). Manganese-based advanced nanoparticles for biomedical applications: future opportunity and challenges. *Nanoscale* 13 (39): 16405–16426.

Hu, B.F., Jia, X.L., Hu, J. et al. (2017). Assessment of heavy metal pollution and health risks in the soil-plant-human system in the Yangtze River Delta, China. *International Journal of Environmental Research and Public Health* 14: 1042.

Huang, Y., Wu, D., Wang, X. et al. (2016). Removal of heavy metals from water using polyvinylamine by polymer-enhanced ultrafiltration and flocculation. *Separation and Purification Technology* 158: 124–136.

Islam, M.A., Angove, M.J., and Morton, D.W. (2019). Recent innovative research on chromium (VI) adsorption mechanism. *Environmental Nanotechnology, Monitoring & Management* 12: 100267.

Jaishankar, M., Tseten, T., Anbalagan, N. et al. (2014). Toxicity, mechanism and health effects of some heavy metals. *Interdisciplinary Toxicology* 7 (2): 60.

Jehan, S., Khattak, S.A., Muhammad, S. et al. (2020). Ecological and health risk assessment of heavy metals in the Hattar industrial estate, Pakistan. *Toxin Reviews* 39 (1): 68–77.

Karna, R.R., Luxton, T., Bronstein, K.E. et al. (2017). State of the science review: potential for beneficial use of waste by-products for in situ remediation of metal-contaminated soil and sediment. *Critical Reviews in Environmental Science and Technology* 47 (2): 65–129.

Koohsaryan, E., Anbia, M., and Maghsoodlu, M. (2020). Application of zeolites as non-phosphate detergent builders: a review. *Journal of Environmental Chemical Engineering* 8 (5): 104287.

Korium, M.S., Roozbahani, H., Alizadeh, M. et al. (2021). Direct metal laser sintering of precious metals for jewelry applications: process parameter selection and microstructure analysis. *IEEE Access* 9: 126530–126540.

Kwak, N.W., Jeong, S.J., Seo, H.G. et al. (2018). In situ synthesis of supported metal nanocatalysts through heterogeneous doping. *Nature Communications* 9 (1): 4829.

Lee, J.C., Hong, H.J., Chung, K.W., and Kim, S. (2020). Separation of platinum, palladium and rhodium from aqueous solutions using ion exchange resin: a review. *Separation and Purification Technology* 246: 116896.

Li, C., Zhou, K., Qin, W. et al. (2019). A review on heavy metals contamination in soil: effects, sources, and remediation techniques. *Soil and Sediment Contamination: An International Journal* 28 (4): 380–394.

Liu, S., Wang, X., Guo, G., and Yan, Z. (2021). Status and environmental management of soil mercury pollution in China: aa review. *The Journal of Environmental Management* 277: 111442.

Liu, H., Xu, F., Xie, Y. et al. (2018). Effect of modified coconut shell biochar on availability of heavy metals and biochemical characteristics of soil in multiple heavy metals contaminated soil. *Science of the Total Environment* 645: 702–709.

Manzetti, S., van der Spoel, E.R., and van der Spoel, D. (2014). Chemical properties, environmental fate, and degradation of seven classes of pollutants. *Chemical Research in Toxicology* 27 (5): 713–737.

Mao, R., Lang, M., Yu, X. et al. (2020). Aging mechanism of microplastics with UV irradiation and its effects on the adsorption of heavy metals. *Journal of Hazardous Materials* 393: 122515.

Marichamy, M.K., Kumaraguru, A., and Jonna, N. (2021). Particle size distribution modeling and kinetic study for coagulation treatment of tannery industry wastewater at response surface optimized condition. *Journal of Cleaner Production* 297: 126657.

Matulik, A.G., Kerstetter, D.W., Hammerschlag, N. et al. (2017). Bioaccumulation and biomagnification of mercury and methylmercury in four sympatric coastal sharks in a protected subtropical lagoon. *Marine Pollution Bulletin* 116 (1): 357–364.

Muszynska, E. and Hanus-Fajerska, E. (2015). Why are heavy metal hyperaccumulating plants so amazing? *BioTechnologia. Journal of Biotechnology Computational Biology and Bionanotechnology* 96 (4): 265–271.

Philip, L., Ramprasad, C., and Krithika, D. (2019). Sustainable wastewater management through decentralized systems: case studies. In: *Water Scarcity and Ways to Reduce the Impact: Management Strategies and Technologies for Zero Liquid Discharge and Future Smart Cities*, 15–45. Cham: Springer.

Prelot, B., Ayed, I., Marchandeau, F., and Zajac, J. (2014). On the real performance of cation exchange resins in wastewater treatment under conditions of cation competition: the case of heavy metal pollution. *Environmental Science and Pollution Research* 21: 9334–9343.

Rahman, P.K. and Bastola, S. (2014). Biological reduction of iron to the elemental state from ochre deposits of Skelton Beck in Northeast England. *Frontiers in Environmental Science* 2: 22.

Rahman, Z. and Singh, V.P. (2020). Bioremediation of toxic heavy metals (THMs) contaminated sites: concepts, applications and challenges. *Environmental Science and Pollution Research* 27: 27563–27581.

Ramirez Arenas, L., Ramseier Gentile, S., Zimmermann, S., and Stoll, S. (2020). Coagulation of TiO2, CeO2 nanoparticles, and polystyrene nanoplastics in bottled mineral and surface waters. Effect of water properties, coagulant type, and dosage. *Water Environment Research* 92 (8): 1184–1194.

Ramprasad, C. and Philip, L. (2016). Surfactants and personal care products removal in pilot scale horizontal and vertical flow constructed wetlands while treating greywater. *Chemical Engineering Journal* 284: 458–468.

Ramprasad, C. and Philip, L. (2018). Contributions of various processes to the removal of surfactants and personal care products in constructed wetland. *Chemical Engineering Journal* 334: 322–333.

Ramprasad, C., Alekhya, D., Bhishmitha, C., and Deepika, C.S. (2020, November). Precipitation of struvite by sustainable waste materials and use as slow release fertilizer–a circular economy approach. In: *IOP Conference Series: Materials Science and Engineering*, vol. Vol. 955, No. 1, 012088. IOP Publishing.

Ramprasad, C., Gwenzi, W., Chaukura, N. et al. (2022). Strategies and options for the sustainable recovery of rare earth elements from electrical and electronic waste. *Chemical Engineering Journal* 442: 135992.

Renu, B., Agarwal, M., and Singh, K. (2017). Methodologies for removal of heavy metal ions from wastewater: an overview. *Interdisciplinary Environmental Review* 18 (2): 124–142.

Rizvi, A., Zaidi, A., Ameen, F. et al. (2020). Heavy metal induced stress on wheat: phytotoxicity and microbiological management. *RSC Advances* 10 (63): 38379–38403.

Rzymski, P., Poniedzialek, B., Niedzielski, P. et al. (2014). Cadmium and lead toxicity and bioaccumulation in *Microcystis aeruginosa*. *Frontiers of Environmental Science & Engineering* 8: 427–432.

Saravanan, A., Yaashikaa, P.R., Kumar, P.S. et al. (2023). Environmental sustainability of toxic arsenic ions removal from wastewater using electrodeionization. *Separation and Purification Technology* 317: 123897.

Selvi, A. and Aruliah, R. (2018). A statistical approach of zinc remediation using acidophilic bacterium via an integrated approach of bioleaching enhanced electrokinetic remediation (BEER) technology. *Chemosphere* 207: 753–763.

Selvi, A., Rajasekar, A., Theerthagiri, J. et al. (2019). Integrated remediation processes toward heavy metal removal/recovery from various environments-a review. *Frontiers in Environmental Science* 7: 66.

Sharma, N., Bisht, S.S., Gupta, S. et al. (2019). Nutraceutical evaluation of horse gram (*Macrotyloma uniflorum*) cultivated in high altitudes of Uttarakhand Himalaya, India. *Indian Journal of Pure & Applied Biosciences* 7 (4): 190–202.

Singh, G., Bhadange, S., Bhawna, F. et al. (2023). Phytoremediation of radioactive elements, possibilities and challenges: special focus on agricultural aspects. *International Journal of Phytoremediation* 25 (1): 1–8.

Sivaranjanee, R., Kumar, P.S., and Mahalaxmi, S. (2022). A review on agro-based materials on the separation of environmental pollutants from water system. *Chemical Engineering Research and Design* 181: 423–457.

Soliman, N.K. and Moustafa, A.F. (2020). Industrial solid waste for heavy metals adsorption features and challenges; a review. *Journal of Materials Research and Technology* 9 (5): 10235–10253.

Song, B., Zeng, G., Gong, J. et al. (2017). Evaluation methods for assessing effectiveness of in situ remediation of soil and sediment contaminated with organic pollutants and heavy metals. *Environment International* 105: 43–55.

Stowers, C.C., Cox, B.M., and Rodriguez, B.A. (2014). Development of an industrializable fermentation process for propionic acid production. *Journal of Industrial Microbiology & Biotechnology* 41 (5): 837–852.

Taghizadeh, M., Asgharinezhad, A.A., Pooladi, M. et al. (2013). A novel magnetic metal organic framework nanocomposite for extraction and preconcentration of heavy metal ions, and its optimization via experimental design methodology. *Microchimica Acta* 180 (11–12): 1073–1084.

Terfassa, B., Schachner, J.A., Traar, P. et al. (2014). Oxorhenium (V) complexes with naphtholate-oxazoline ligands in the catalytic epoxidation of olefins. *Polyhedron* 75: 141–145.

Vardhan, H., Nafady, A., Al-Enizi, A.M., and Ma, S. (2019). Pore surface engineering of covalent organic frameworks: structural diversity and applications. *Nanoscale* 11 (45): 21679–21708.

Wang, B., Guo, X., and Bai, P. (2014). Removal technology of boron dissolved in aqueous solutions–a review. *Colloids and Surfaces A: Physicochemical and Engineering Aspects* 444: 338–344.

Yadav, M., Das, M., Savani, C. et al. (2019). Maleic anhydride cross-linked β-Cyclodextrinconjugated magnetic nanoadsorbent: an ecofriendly approach for simultaneous adsorption of hydrophilic and hydrophobic dyes. *ACS Omega* 4 (7): 11993–12003.

Yao, Z.T., Ji, X.S., Sarker, P.K. et al. (2015). A comprehensive review on the applications of coal fly ash. *Earth-Science Reviews* 141: 105–121.

Yin, J. and Deng, B. (2015). Polymer-matrix nanocomposite membranes for water treatment. *Journal of Membrane Science* 479: 256–275.

Zainol, N.A., Mohamad, N.A.N., Ghani, A. et al. (2019). Use of alum and ferric sulphate for treating landfill leachate via coagulation process: a comparative study. *Journal of Engineering Research and Education* 11: 35–44.

Zhang, Q., Cheng, Y.F., Huang, B.C., and Jin, R.C. (2022). A review of heavy metals inhibitory effects in the process of anaerobic ammonium oxidation. *Journal of Hazardous Materials* 429: 128362.

Zhang, D., Ma, F., and Tao, Y. (2023). Study on effect of nanobubble on ultra-fine flake graphite (UFG) flotation. *Particulate Science and Technology* 1–9.

7

Chemical Approaches to Remediate Heavy Metals

Ayushi Singhal[1,2]*, Arpana Parihar*[1]*, Vedika Khare*[3]*, Gagan Kant Tripathi*[3]*, and Raju Khan*[1,2]

[1] *CSIR – Advanced Materials and Processes Research Institute (AMPRI), Bhopal, Madhya Pradesh, India*
[2] *Academy of Scientific and Innovative Research (AcSIR), Ghaziabad, Uttar Pradesh, India*
[3] *School of Nanotechnology, UTD, RGPV Campus, Bhopal, Madhya Pradesh, India*

7.1 Introduction

Anthropogenic activities cause the accumulation of heavy metals and toxic metalloids in water bodies, soil, and air for a long-time, which contaminate the environment and, thereby, ecosystem (Atieh et al. 2017; Masindi and Muedi 2018). Most of these heavy-metal-contaminated locations are in wealthy countries, which include the United States, European Union Member States, Australia, and China. For instance, heavy metals/metalloids have contaminated 6000 km^2 of farmland in the United States; 250,000 sites in the European Union have been poisoned; and 810,000 km^2 of farmland in China have been contaminated so far (Khalid et al. 2016; Tu et al. 2020). These metal contaminants are nonbiodegradable, get accumulated in soil, and damage the natural system and living resources, directly or indirectly, increasing health hazards (He et al. 2015; Masindi and Muedi 2018). Heavy metal toxic pollutants such as mercury, cobalt, lead, chromium, nickel, arsenic, uranium, cadmium, tin, copper, and others can be found in wastewater discharged by commercial industries such as textile, nonferrous metal refineries, leather industry, tanning, dye, paint, electroplating, printing, surface coating, and nuclear power plants (Zeraatkar et al. 2016). Heavy metal exposure causes multiple health side effects including damage to organs such as the kidney, liver, lungs, and eyes, as well as the biological organ system that includes the endocrine, nervous system, immune system, circulatory system, reproductive system, digestive system, and respiratory system, leading to diseases such as inflammatory disorders, cancer, anemia, hypersensitivity, organ failure, autoimmune disorders, bone disorders, and tumorigenesis (He et al. 2015; Zeraatkar et al. 2016). Some of these components, such as insecticides, herbicides, and fertilizers, are essential for biological metabolism, but because of their high dose, they

Remediation of Heavy Metals: Sustainable Technologies and Recent Advances, First Edition. Edited by Rangabhashiyam Selvasembian, Binota Thokchom, Pardeep Singh, Ali H. Jawad, and Willis Gwenzi.

penetrate the food chain and accumulate in living beings, thus are highly harmful to flora and fauna (Zeraatkar et al. 2016; Arora 2019). The extent of these contaminants in soils can be varied and depends upon the type of soil, environmental conditions, and the distance from the pollutant source. There have been numerous methods discovered to date which can be categorized as physical, biological, and chemical processes. Several strategies for removing heavy metals from the water have also been developed, which help to alleviate the problem of water scarcity by allowing it to be reused. To remove harmful metals from sludge, physical procedures such as crushing, drying, ball milling, filtering, sedimentation, sieving, and metal ion absorption process from wastewater sludge are used (Sharma and Rana 2016). Although this is a low-cost approach, its efficiency (50%) and ability to operate on a big scale remain a major concern (Khoei et al. 2019). Biological methods, in addition to the physical process, have opened the way for the removal of toxic heavy metals and the technique called bio-absorption (Kumar et al. 2019). Bio-absorption is a physiochemical process in which living tissues are capable of chemically interacting with pollutants and eliminating them from the environment. The plant residue that contains lignocellulosic debris, minute aquatic plants, and microorganisms such as algae, yeasts, bacteria, and fungi are all involved in this process. Microalgae exhibit a greater ability to absorb harmful metals from wastewater. Its effectiveness is primarily determined by metal ion concentrations, temperature, pH, chemical oxygen demand (COD), and the chemical constituents of an aqueous medium (Kumar 2014). Biological agents such as Anabaena, nostoc, cyanobacteria, Callitriche Lusitania, and *Azolla caroliniana* are used to extract Ar, Cu, Mt, and Dz from sewage and sludge (Real et al. 2007; Jasrotia et al. 2015; Chen et al. 2020). Although biological technologies for metal extraction are the cheapest, the reliance on microorganisms to survive in different environments is poor and the only disadvantage of this method. Chemical precipitation, adsorption, electroplating, stabilization, reverse osmosis, and ion exchange are a few of the chemical remediation technologies that have been identified so far to remove metal pollutants from industrial and home sludge in several studies (Kumar et al. 2020; Palansooriya et al. 2020). Chemical procedures are extremely efficient (98–99%), but the development of untreated wastes and the use of costly and harmful substances in this process makes it a lower-priority method (Renu et al. 2017; Yang et al. 2020). To achieve optimal efficiency, all of these processes require some changes to become a better fit to remove heavy metals. Aside from these chemical methods, adsorption is a process in which a solid substance, called an adsorbent, is used to attract other particles, ions atoms, and molecules to its surface, primarily through properties of adhesion, ion exchange, and electrostatic attraction, which are thus used for heavy metal contaminants' remediation. Microbial, waste biomass, lignocellulosic, industrial waste, and more advanced chemical adsorbents such as hybrid adsorbents, nano adsorbents, and metal–organic frameworks are all examples of some common adsorbents (Ma et al. 2015; Ndimele 2017; Yuanan et al. 2020). The present chapter provides information about the sources of heavy metal contaminants that contribute to environmental pollution, with a special focus on chemical methods for heavy metal remediation. Further, current challenges and future research directions in the field of heavy metal remediation have also been discussed.

7.2 Sources of Heavy Metal

Heavy metals are found abundantly in the environment and are vital for life, but they may also be hazardous if they accumulate in the food cycle. Mercury, lead, and cadmium belong to the most persistent heavy metals. Heavy metals can be obtained from both natural sources and man-made (anthropogenic) sources. Some of the potent sources of heavy metal are shown in Figure 7.1.

7.2.1 Natural Sources

Many different natural sources have been identified to contain heavy metals. Natural heavy metal emissions occur under a variety of environmental episodes. These natural sources include volcanic explosions, large fires in the forest, weathering of rocks, biological toxic sources, and contaminated soil. Metals discharged from their indigenous spheres into other environmental divisions will undergo weathering by natural processes. Heavy metals are not found in the free form; instead, they are found in complex forms such as oxides and phosphates.

7.2.1.1 Rocks

7.2.1.1.1 Magmatic Rocks Heavy metals have been absorbed into the crystal lattice of the primary minerals that develop during the expansion of the magma throughout the cooling phase. As heavy metals substitute other atoms during crystallization, this is known as isomorphic substitution. Inradius, ion charge, and electronegativity of the substituting element and the main element determine the amount of isomorphic substitution (Bradl 2005).

7.2.1.1.2 Sedimentary Rocks Sedimentary rocks originate from previously existing stones or fragments of extinct creatures. They are generated by the accumulation of tiny rock fragments on the surface of the earth. Rock disintegrates into particles, which are termed sediments, as a result of physical weathering. Chemical weathering causes the rock to disintegrate into ions. These sedimentary particles are transported to certain basins, where they are deposited. A process known as diagenesis or lithification occurs when fresh sediment accumulates. The loose particles are subsequently linked together by chemical changes in the pore fluid, and solid rock gradually emerges from the loose particles (Warf 2010). Compaction squeezes pore fluids out of the underlying sediments, resulting in the development of pore cement, which binds the loose particles together. Sedimentary rocks may hold liquids like water, gas, and oil because of their porous nature. Additionally, they enable liquids to travel through. Because of these two characteristics, sedimentary rocks are particularly important for the provision of water and energy. They could have ore deposits of multiple heavy metals if ore-beating hydrothermal fluids get to them.

7.2.1.1.3 Metamorphic Rocks The third type of rock is metamorphic. Increased temperature and pressure conditions can cause chemical changes in any magmatic, sedimentary, or metamorphic rock (for example, during plate movement into deeper

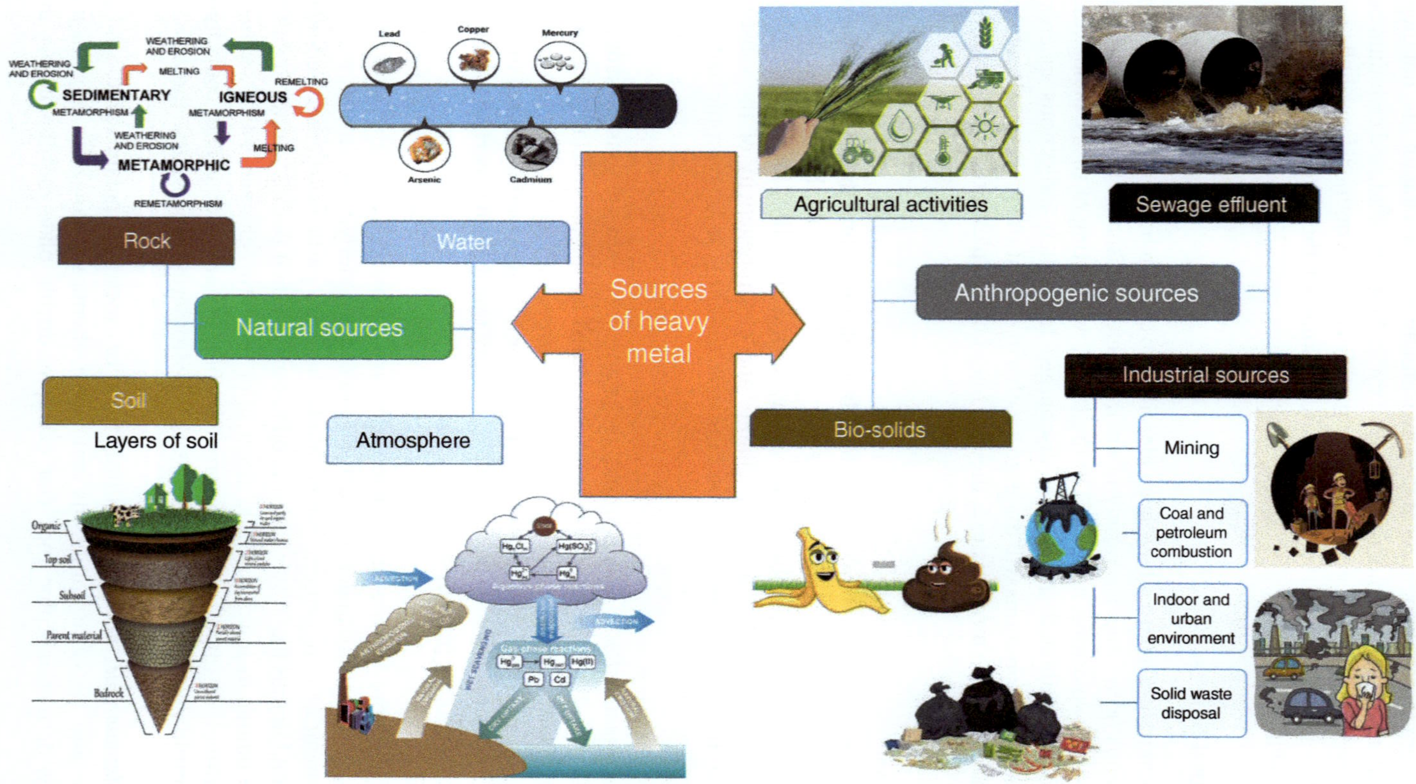

Figure 7.1 Sources of heavy metal. *Source:* Reddogs/Adobe Stock, Africa Studio/Adobe Stock.

portions of the earth's crust). Metamorphosis is the name given to the process of forming metamorphic rocks. Metamorphic rocks are created by the rearranging and recrystallization of minerals that constitute rocks. Both contact or static metamorphosis, which happens when molten lava collides with cemented rock, and dynamic or spatial metamorphosis, which takes place in crystalline schists, are possible. Insight into the geology of metamorphic processes throughout varying physicochemical circumstances, such as temperature, fluid activity, and stress, during the solid stage in between diagenesis and massive melting, is provided by metamorphic rocks (Song and Cao 2021).

7.2.1.2 Soil

Soil is a multifaceted mixed medium made up of dense phases consisting of minerals and organic materials, as well as fluid phases that relate with one another and ions inflowing the soil system (Totsche et al. 2018). Climate, soil organisms, terrain, parent rock type, and time, all impact soil formation or pedogenesis. A soil profile is made up of various layers or horizons that form throughout the evolution of the soil. The first layer, known as O-horizons, is made up primarily of decomposing organic materials including leaves, twigs, and other humic compounds (Bradl 2005).

The underlying horizon is made up of mineral and organic debris and is susceptible to rainwater filtration-induced leaching. Eluviation is the process by which clay-size particles and dissolved chemical components (such as Fe, Ca, Mg, and heavy metals) are transported into the underlying B horizon.

This material collects in the B horizon (alluviation). Clay minerals and iron oxyhydroxides, which are capable of absorbing toxic ions, are abundant in the B horizon. This horizon also provides the majority of plant nutrients. Finally, the somewhat weathered parent rock in the C horizon underneath the B horizon; after that, the undamaged parent rock is present. The soil constitution differs greatly depending on the kind of parental rock, temperature, period, and soil organisms. Its chemical makeup is identical to that of the parent rock. Heavy metal background concentrations in the natural or geogenic environment differ greatly from one location to the next. Soils include three basic components, which are the fundamental ingredients in determining a soil's unique qualities. Organic matter (OM), clay minerals, and iron, aluminum, and manganese oxides are among the constituents (Sahoo et al. 2020).

7.2.1.3 Water

The type of soil and rocks that the water travels over or through has a big impact on the surface chemistry of waterways including water brooks, waterways, springs, ponds, and lakes. The temperature, pH, redox potential, adsorption and desorption processes from suspended particles or bottom sediments, cation exchange, dilution, and evaporation are some of the physical, chemical, and biological factors that affect the identity of water. For instance, according to the research (2020), industrialized regions had lower heavy metal concentrations and a higher percentage of heavy metals overall than emerging ones (Zhou et al. 2020). Progressively, secondary metal waste disposal has replaced primary mining and manufacturing as the main cause of heavy metal pollution in lakes and global water supplies. The primary causes of heavy metal contamination varied throughout the five continents as well. For instance, the most frequent causes of pollution in Asia were mining and manufacturing along with rock weathering, whereas the most frequent in North America were mining and manufacturing along with fertilizers and pesticides. Groundwater near industrial sites that use a variety of chemicals in the manufacturing of batteries, paints, medicines, leather processing, agrochemicals, and other things frequently contains trace amounts of heavy metals. These businesses release partly or untreated wastewater into the sea in coastal areas as well as into surface water bodies including rivers, lakes, and ponds. Around landfills, contamination from a range of heavy metals happens often. Another significant cause of contamination is the excessive use of agrochemicals, which are retained in the unsaturated zone and infiltrate groundwater via irrigation return flow. The accumulation of bio-toxic heavy metals in crops and their subsequent movement down the food chain endangers human health. As a result, several studies in this area have been conducted (Vetrimurugan et al. 2017).

7.2.2 Anthropogenic Sources

Although the word "contaminants" (and contamination rather than pollution) is chosen for soil and land, heavy metals from anthropogenic sources might be categorized as pollutants or contaminants. A study in the United States on the soil demonstrates the relevance of contamination (Burt et al. 2003).

7.2.2.1 Agricultural Activities

Due to the need to utilize a lot of land for food production to feed an ever-growing world population, soil amendments, pesticides, and fertilizers must be used often. Fertilizers are added to the soil to either increase the number of nutrients available to crops or to increase their accessibility by altering factors such as pH. Pesticides are used to safeguard crops (Bradl 2005). Sewage sludge, animal dung, and dredged sediments from harbors and rivers are among the most prevalent sources of soil additions. Heavy metals can be mobilized from these sediments during dredging because the condition of reduction changes to an oxidizing one, enabling heavy metals to be remobilized (Selvi et al. 2019).

Heavy metals placed in soils can impact surface waters via infiltration, as soil, surface, and groundwater are all connected. Irrigation may also release heavy metals.

7.2.2.1.1 Use of Phosphatic Fertilizers The common heavy metal sources in soil are inorganic and biological fertilizers. Inorganic fertilizers, particularly phosphate fertilizers, contain different concentrations of Cd, Cr, Ni, Pb, and In, depending on where they originate. Cadmium is a major problem in plants because it persists and builds up at extreme levels in leaves. This can be ingested by animals or humans (Sharma and Agrawal 2005).

Fertilizers have enabled farmers to boost crop productivity and cultivate nutrient-deficient soils. However, contamination of farmed areas by trace metals naturally contained in the phosphate rocks consumed to create the fertilizers might be a detrimental consequence of phosphate fertilizers (Lambert et al. 2007; He et al. 2021; Rahimi et al. 2021; Recena et al. 2021; Yazici et al. 2021).

7.2.2.1.2 Use of Pesticides Pesticides are used to control insects and illnesses in high-production agriculture and can be applied as seed treatments, sprays, dust, or soil applications. Metal-based pesticides are no longer used; however, their previous use increased the buildup of heavy metals. Hg from methyl mercurial, and As, and Pb from lead arsenate are the most common in the soil and the groundwater (Bradl 2005). Malathion is a broad-spectrum, nonsystemic pesticide used to control insects on fruits, vegetables, and crops, as well as domestic insects and animal parasites (Low et al. 2013). Heavy metals, like pesticides, are contaminants found in the aquatic environment that are both common and persistent (Abdel-Razek et al. 2019).

Heavy metals are often transported into aquatic bodies through wastewater, which includes industrial effluents. Lead, cadmium, and nickel are among the most harmful heavy metals typically found in water bodies, according to reports (Sun et al. 2017).

7.2.2.2 Sewage Effluents

Over 70% of the world's freshwater in all forms is locked up in the ice sheets of Antarctica and Greenland, while the bulk of the remaining freshwater is either too deep to reach or is locked up in soil moisture. Freshwater makes up about 3% of the world's total water (Ashraf et al. 2017). Sewage water from home, industrial, and commercial sources is increasingly being used for urban and periurban agriculture (Singh et al. 2012). Copper, zinc, cadmium, and lead concentrations in sewage effluent were found to be greater than the permitted threshold for wastewater used for irrigation in one study. The discharge of sewage effluent into the river water resulted in visible heavy metal contamination of the river Rupingazi. The higher concentrations of copper, zinc, cadmium, and lead in sewage effluent-irrigated soil revealed that continued use of sewage effluent for irrigation could lead to a gradual accumulation of heavy metals in the soil, which could eventually lead to increased heavy metal uptake into various parts of growing vegetables (Tyagi et al. 2018).

7.2.2.3 Bio-Solids

Solid waste from the sewage, municipality, and industries are all considered "bio-solids."

These bio-solids are utilized for soil augmentation because of their higher nutrient and OM content. They're full of nutrients, heavy metals, and pathogens like *Escherichia coli*. The alarming heavy metals in sewage sludge include Ni, Zn, Ag, Cu, Pb, Se, Mo, Cr, As, and Cd (Navarro et al. 2008; Khanh Nguyen et al. 2021).

7.2.2.4 Industrial Activities

The most typical industrial operations that emit heavy metals into the environment are mining, fuel combustion, wastewater streams, and trash disposal. Preventing the entry of these poisonous and dangerous compounds into the environment has received little attention in the past. In the meanwhile, increased laws regulating the collection and treatment of pollutants have resulted in a significant reduction in heavy metal contamination of ecosystems in many nations. Unfortunately, this is not a worldwide phenomenon (Bradl 2005).

7.2.2.5 Mining

Due to the release and dispersion of mine waste into neighboring agricultural soils, food crops, and stream systems, elevated levels of heavy metals have been seen in and around closed metalliferous mines (Anju and Banerjee 2012).

There are several concerns associated with heavy metal soil contamination from mining and smelting in many contexts. Some of the detrimental environmental repercussions of contaminated mining sites include human health concerns, phytotoxicity, water and soil contamination, and ecotoxicity (Sah et al. 2019).

7.2.2.6 Coal and Petroleum Combustion

The combustion of fossil fuels is mostly to blame for the emission of heavy metals, especially into the atmosphere. The primary sources of Ni and Cu emissions have been recognized as petroleum and coal burning (Tian et al. 2012). The origins of the heavy metals were mainly human activities like waste incineration, steel manufacturing, coal fly ash, polluted soil, and vehicle emissions (Al Hejami et al. 2020). Road dust tainted by the coating of specific automotive parts, such as the brakes, can be blamed for the presence of Zn and Pb (Tian et al. 2012).

7.2.2.7 Indoor and Urban Environments

The world's megacities and huge metropolitan centers are continually expanding. These areas are characterized by extremely high population densities, excessive energy consumption, substantial industrial and transportation activities, and usually by very low standards for housing, water and power supply, as well as for waste collection and disposal. These elements result in severe contamination of the environment (Bradl 2005). Heavy metals are found in indoor settling dust from both internal and external sources, and their abundances in interior settings are significantly higher than their crustal abundance. Heavy metals linked with materials, such as soil, road dust, and industrial and vehicle particulates that are airborne or dragged in on clothing and footwear, might be external sources (Sarma Bora et al. 2021). Automobile emissions are the most significant external resources of heavy metals (Wang et al. 2021).

7.2.2.7.1 Solid Waste Disposal Thousands of millions of tonnes of solid garbage are created every year across the world. Heavy metals are mostly found in waste from industrial operations, particularly energy generation, mining, agricultural activities (animal dung), and household garbage. These wastes are frequently disposed of without being properly treated at waste disposal facilities that do not fulfill the standards for safe disposal (Bradl 2005).

Rapid industry, urbanization, and population boom are all factors that contribute to increased solid waste output in India. Improper municipal solid waste management has resulted in a variety of environmental consequences. In India, innovative waste disposal processes such as composting, incineration, palletization, and bio-methanation have been utilized for just 67% of solid waste, with the remaining solid waste being disposed of on rubbish dumping sites. Dumping waste on land produces toxic leachate that may pollute groundwater and surface water, posing a major environmental threat (Kaur et al. 2019).

7.3 Chemical Remediation Technique for Heavy Metal Contamination in the Environment

One of the major concerns of the environment is heavy metal pollution. Due to wide anthropogenic activities, as mentioned in the previous section, heavy metal pollution is affecting largely. As of now, there are many efficient methods available for the remediation of heavy metals. Among them, chemical precipitation is efficient and by far the most popular industrial process since it is easy to use and reasonably priced to run (Wu 2019). Chemicals and heavy metal ions interact during precipitation processes to create insoluble precipitates. Filtration or sedimentation can be used to separate the developing precipitates from the water. After that, the purified water is decanted and properly released or recycled. Sulfide precipitation and hydroxide precipitation are two examples of common chemical precipitation processes (Pohl 2020). It is a useful method for getting rid of heavy metals from wastewater. Although other techniques like sulfide and carbonate precipitation are occasionally utilized, hydroxides are the most common way to precipitate metals (Wang et al. 2005). In some circumstances, it is necessary to oxidize or decrease the chemical species to a valence that may subsequently be precipitated directly. Fluorine can be removed using calcium chloride, while phosphate can be removed by precipitation as iron or aluminum salts (Renu et al. 2017). Several industrial wastewater contaminants have been successfully removed using chemical precipitation. It is suitable for automatic control and functions under ambient circumstances (Fu and Wang 2011). In this section, the chemical techniques used for the remediation of heavy metals from various sources are enlisted and described.

7.3.1 Chemical Precipitation

The chemical precipitation method includes adding chemical reagents which are followed by the removal of heavy metals in the form of precipitated solid. Commonly, it is used for the removal of cations, but it can also be used for the removal of the anions such as fluorides, cyanide, phosphate, and some organic molecules such as phenols (Wang et al. 2005). Generally, precipitation of the solid material is obtained by the addition of the coagulants. Coagulants such as alum, lime, salts, and some polymers are added to the wastewater, which forms the removal of the precipitant and heavy metals as a solid mass. Some of the techniques used for the removal of heavy metals are ion exchange, chemical precipitation, flocculation, membrane method, osmosis, electrodialysis, etc. (Arora 2019). Chemical precipitation can be used to extract and remove metal ions such as aluminum, antimony, arsenic, cadmium, copper, cobalt, iron, etc. It can also be used to remove fluorides, phosphates, and sulfides.

Chemical precipitation includes hydroxide precipitation, sulfide precipitation, carbonate precipitation, cyanide precipitation, phosphorous precipitation, lime soda/ash precipitation, heavy metal chelating precipitation, and precipitation combined with other techniques (Wang et al. 2005; Fu and Wang 2011).

7.3.1.1 Hydroxide Precipitation

Hydroxide precipitation uses an alkaline agent, which raises the pH of the solution and causes the metal ions to reduce solubility and form precipitate. The heavy metals are precipitated as hydroxide, which is then removed by sedimentation or filtration.

Generally, alkaline compounds such as caustic soda or lime are used as precipitants.

In a recent study, magnesium hydroxy carbonate having low-grade magnesite was used to treat heavy metal-polluted wastewater. The VO^{2+}, Cr^{3+}, and Fe^{3+} concentrations decreased to the limit of discharge upon the treatment with the magnesium hydroxycarbonate. The precipitate formed is composed of Fe_2O_3, V_2O_5, and Cr_2O_3, which can be further used. In summary, the use of magnesium hydroxycarbonate as a precipitating agent is an efficient and economical way to remove heavy metals (Zhang and Duan 2020).

Sulfide precipitation: Sodium sulfide and hydrogen sulfide are used as precipitants in this method to remove the heavy metal ions as insoluble metal sulfides. The sulfide precipitation process is somehow the same as hydroxide precipitation. Upon the addition of sulfide precipitant to the heavy metal polluted, it forms sludge with the heavy metal, which is then removed by filtration or sedimentation. The hydrogen and sodium sulfides both cause toxicity; to minimize the risk, gaseous ferrous sulfides are used. In another method, metal sulfide precipitation was done for the remediation of heavy metals. The sulfide precipitation methods offer a high rate of metal removal in a shorter period (Pohl 2020). In a study, for the exclusion of lead from the recycling battery, a chemical precipitation technique was employed. Using 1,3-benzenediamidoethanethiol dianion (BDET2-) acts as a ligand and binds soft metal to form a metal–ligand complex. The method successfully removed 99.4% of the lead within 15 minutes (Wu 2019).

Heavy metal-chelating precipitation: To removal the chelated heavy metals from the industrial effluents, a novel strategy using the zero-valent iron and hydrogen peroxide. The zero-valent iron and hydrogen peroxide degraded the chelated compound and further precipitated the heavy metals. The process was used for the removal of nickel ions and it reduces the COD value. The efficiency for the removal of heavy metal (Ni) was found to be 98.4% (Fu et al. 2012). For the removal of heavy metal complexes, Cu-EDTA chemical precipitation was employed. The complex is first oxidized via nonthermal plasma oxidation, followed by precipitation. The copper release occurs due to the plasma oxidation. The experimental setup for the nonthermal plasma oxidation is shown in Figure 7.2a. The copper removal efficiency with varying treatment times at different inputs is shown in Figure 7.2b. The energy efficiency for copper removal at various inputs is shown in Figure 7.2c. As shown in the figure, the energy efficiency first increases with the increasing energy input, but after a certain extent, it decreases with increasing energy input. The copper removal efficiency was found to be 78.2% (Wang et al. 2019).

Despite all the advantages of the chemical precipitation method, some limitations are there that limit the use of this method. The chemical precipitation method is only used when the contamination is high; with the low amount of contamination, it cannot be used. Moreover, it is costly and produces a large amount of sludge that needs to be maintained.

7.3.2 Coagulation

Coagulation or flocculation is used largely for water treatment, either drinking water or wastewater (Sabir et al. 2021). For coagulation to happen, coagulants are added to the wastewater. Coagulants are chemicals which, upon the addition of the colloidal material, form small aggregate particles called flocs. These flocs then attract the suspended material in the water and grow to form a larger size mass (Akbal and Camci 2010). Coagulants can be both natural and synthetic (Verma et al. 2012). As per the chemical composition, chemical coagulants can be classified as organic materials, inorganic materials, and hybrid materials (Tang et al. 2016). The most common coagulants are aluminum, ferrous sulfate, and

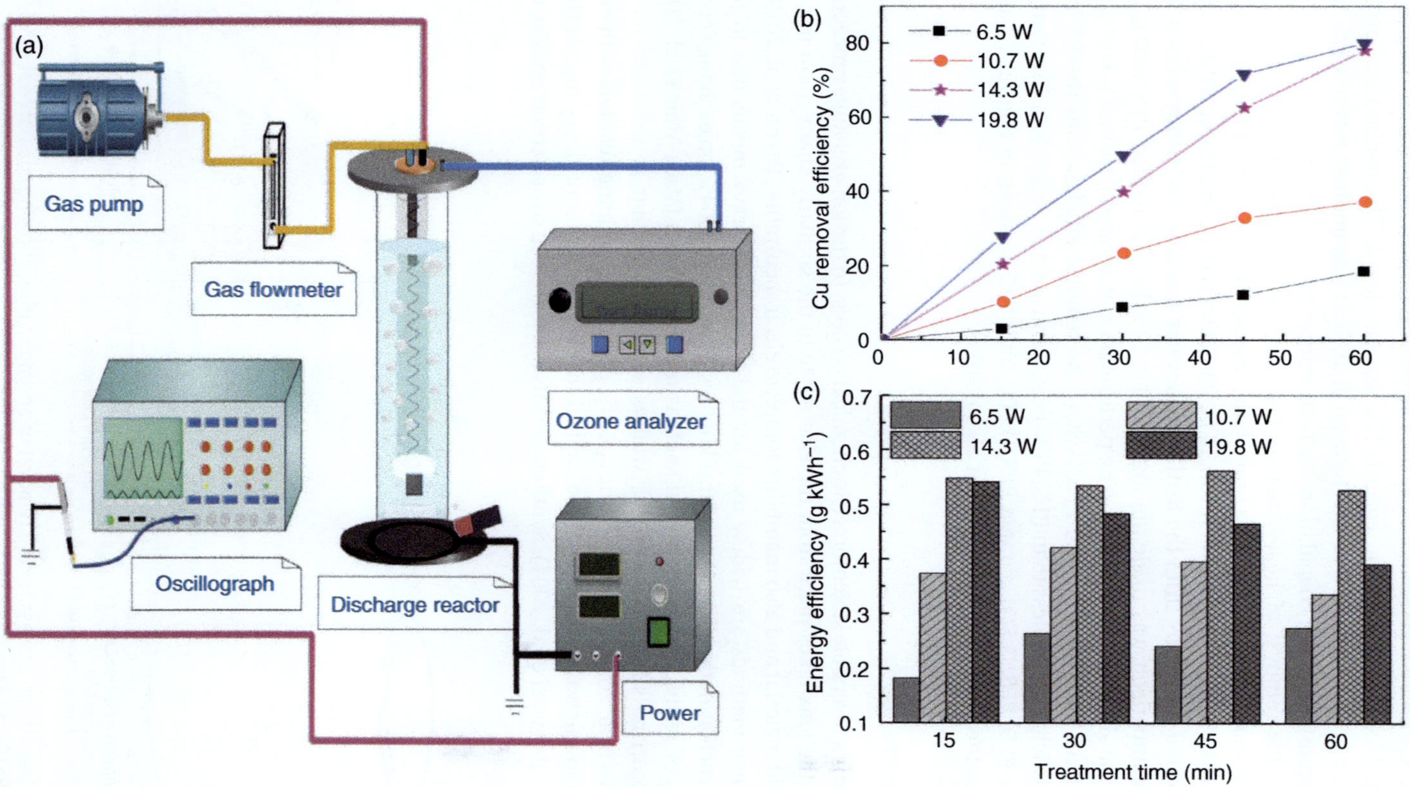

Figure 7.2 (a) Experimental arrangements for the removal of EDTA-chelated copper using nonthermal plasma oxidation, copper release, and removal at different energy inputs: (b) removal efficiency and (c) energy efficiency. *Source:* Wang et al. (2019)/with permission of Elsevier.

ferric chloride. Coagulation can be employed for a wide range of heavy metals (Charerntanyarak 1999; El Samrani et al. 2008; Johnson et al. 2008). For the removal of heavy metals (Cu and Ni) generated from the real plating effluent, the hybrid technique of Fenton ozone-based oxidation and coagulation was employed. The method successfully removed 99.7% of Ni and 99.72% of Cu (Nguyen et al. 2021). In another method, the combined techniques oxidation, coagulation, and adsorption techniques were utilized at optimum pH for the removal of heavy metals. The method proved to be economical and efficient. The method successfully removed the metals in the order Cd (79.0%) ≪ Co (94.8%) < Ni (94.4%) ≪ Cu (98.0%) < Cr (98.3%) < Pb (99.5%) (Bora and Dutta 2019).

The removal of antimony from the textile wastewater was achieved using aerated PFS/$FeSO_4$. The comparison was made among the PFS, PFS/$FeSO_4$, aerated PFS/$FeSO_4$, and PAC, one of the aluminum coagulants. The performance of aerated PFS/$FeSO_4$ was found to be better than the PAC, PFS, and PFS/$FeSO_4$. The removal efficiency for the PAC was below 20%, which was very less than PFS, PFS/$FeSO_4$, and aerated PFS/$FeSO_4$. The removal efficiency for PFS, PFS/$FeSO_4$, aerated PFS/$FeSO_4$, and PAC is shown in Figure 7.3, along with the basic mechanism involved (Liu et al. 2019b).

7.3.3 Ion Exchange

Ion exchange is a reversible stoichiometric chemical process. In the ion exchange process, the exchange of the ions takes place from the electrolyte to the similarly charged ion into the insoluble solid material and also maintains complete electroneutrality (Bashir et al. 2019).

Ion exchange methods are widely used for the removal of heavy metals due to their advantages, such as high efficiency and fast kinetics (Kang et al. 2004). Ion exchange occurs with the use of resins. It can be natural such as natural zeolites (Vaca Mier et al. 2001), naturally occurring silicate minerals as well as synthetic. Biochar is also used for ion exchange (Wu et al. 2021a). They have the specific property to exchange cations with metal ions present in the wastewater. It can be applied to a large number of heavy metals (Feng et al. 2000; Dąbrowski et al. 2004). Synthetic resins are preferred over natural ones as they

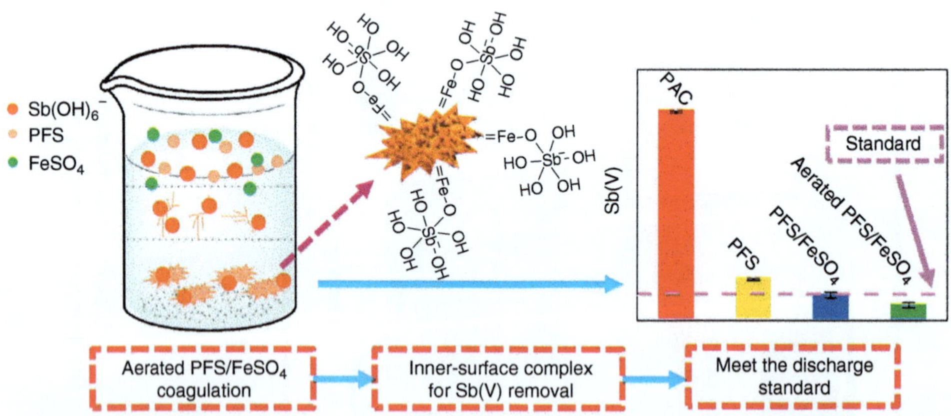

Figure 7.3 Coagulation process for the removal of antimony, inset showing the removal efficiency for PFS, PFS/$FeSO_4$, aerated PFS/$FeSO_4$, and PAC. *Source:* Liu et al. (2019b)/with permission of Elsevier.

are easy to use and remove heavy metals. Ion exchange resins can be regenerated and used again (Verbych et al. 2004). Most cation exchangers are supposed to be sulfonic groups and carboxylic groups.

For lead removal from contaminated water, cation exchange resins were used. The removal efficiency was found to be >90%. After the removal, the regeneration of the exhausted resin was done with the use of $Ca(NO_3)_2$ and was tested using an electrochemical workstation as shown in Figure 7.4. The efficiency was found to be decreased with the increased flow rate. The maximum removal was found at pH ranges from 2.4 to 4.8 and increases with the increasing bed height (Lalmi et al. 2018).

The removal of heavy metals was achieved in a study using RUB-15 nanoplates as adsorbents. RUB-15 was synthesized using tetramethylammonium hydroxide (TMAOH) and tetraethyl orthosilicate (TEOS) through the hydrothermal process. Although the RUB-15 was originally synthesized for the uranyl ions' removal, it presents good adsorption capacity for the heavy metal ions (Pb and Cd). The adsorption capacity was 338 $mg\,g^{-1}$ and 190 $mg\,g^{-1}$ for lead and cadmium ions, respectively. The low-cost synthesis, wide applicability, and easy regeneration made RUB-15 a promising adsorbent for the remediation of heavy metal and radioactive elements (Chen et al. 2017). For the elimination of copper ions, the activated carbon was prepared from resins exchanged with zinc ions. The preparation was done by heat treatment followed by carbon dioxide activation. This activated carbon showed high adsorption capacity than the commercial ones and can be used to prepare electrode material for the removal of heavy metals (Wu et al. 2021b). In a study by Zeng et al., the mercury ions were efficiently removed using nano humboldtine decorated almandine (NHDA). The adsorption capacity was found to be 575.17 $mg\,g^{-1}$ (Zeng et al. 2021), which is the highest among the similar studies conducted (Asiabi et al. 2018; Liu et al. 2019a). In another study, the removal of lead ions ion-exchange approach was employed. The process was done using an iron oxide decorated organic framework hybrid. The adsorption capacity was found to be 610 $mg\,g^{-1}$, with a 93% removal efficiency. The adsorption shows high selectivity even in the presence of coexisting positively charged ions (Goyal et al. 2021). Despite the advantages of the ion exchange method, there are some disadvantages, such as high cost, being used for smaller batches only, and can be used for a limited number of ions (Pan and An 2019). Moreover, the regeneration of the resins can also cause pollution as they need a large number of acids and alkalis (Verbych et al. 2004).

7.3.4 Electrochemical Method

Electrochemical processes may include electrodialysis, electroreduction, electron ionization, and electrocoagulation. In the electrocoagulation process, the flocculating agent is prepared in situ with the help of electrochemical dissolution of the sacrificial anode. The electrochemically synthesized agents show 100-fold more removal efficiency than the one that is pre-precipitated. Electrochemical techniques are effective, economical processes with no use of chemicals and minimal sludge. Electrochemical techniques are rapid, easy, and ecofriendly (Jin et al. 2016).

For the removal of heavy metal from the surface finishing wastewater, the electrocoagulation or electrochemical Fenton process was used with iron anodes. The setup designed for the electrochemical process is shown in Figure 7.5a. Using the electrochemical process, the whole heavy metal was removed and the COD value was also removed. These results were

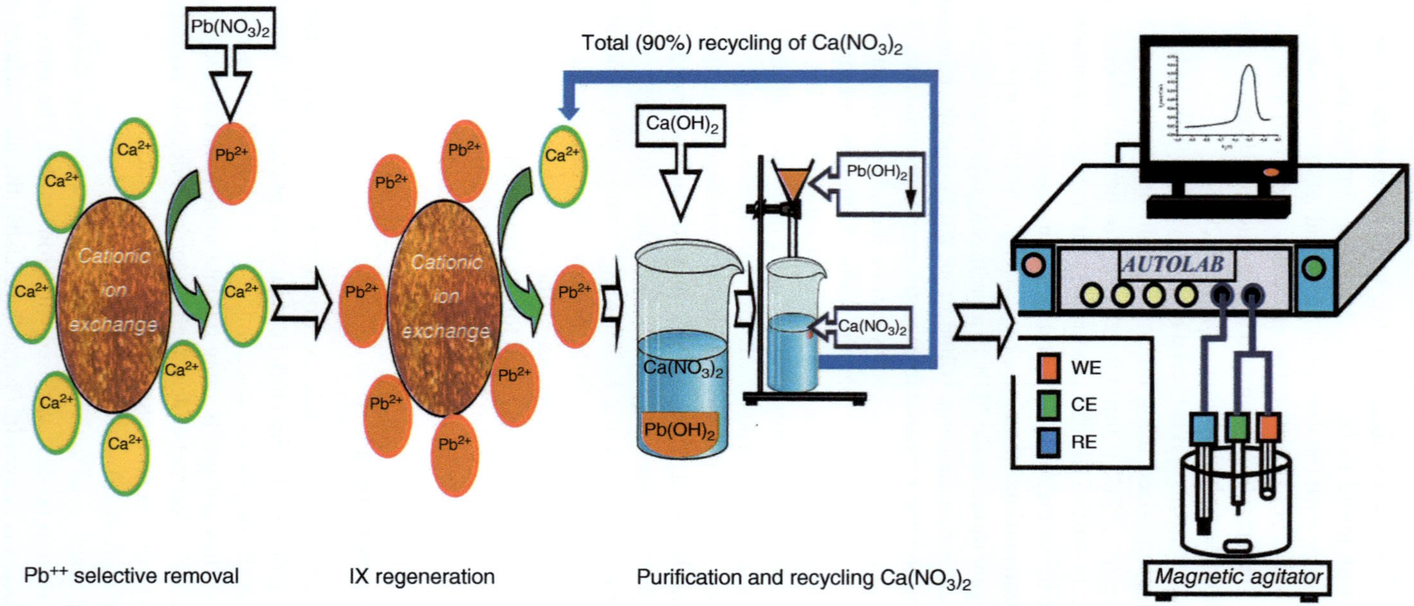

Figure 7.4 Ion exchange mechanism for the lead removal using ion exchange resins and its regeneration. *Source:* Adapted from Lalmi et al. (2018).

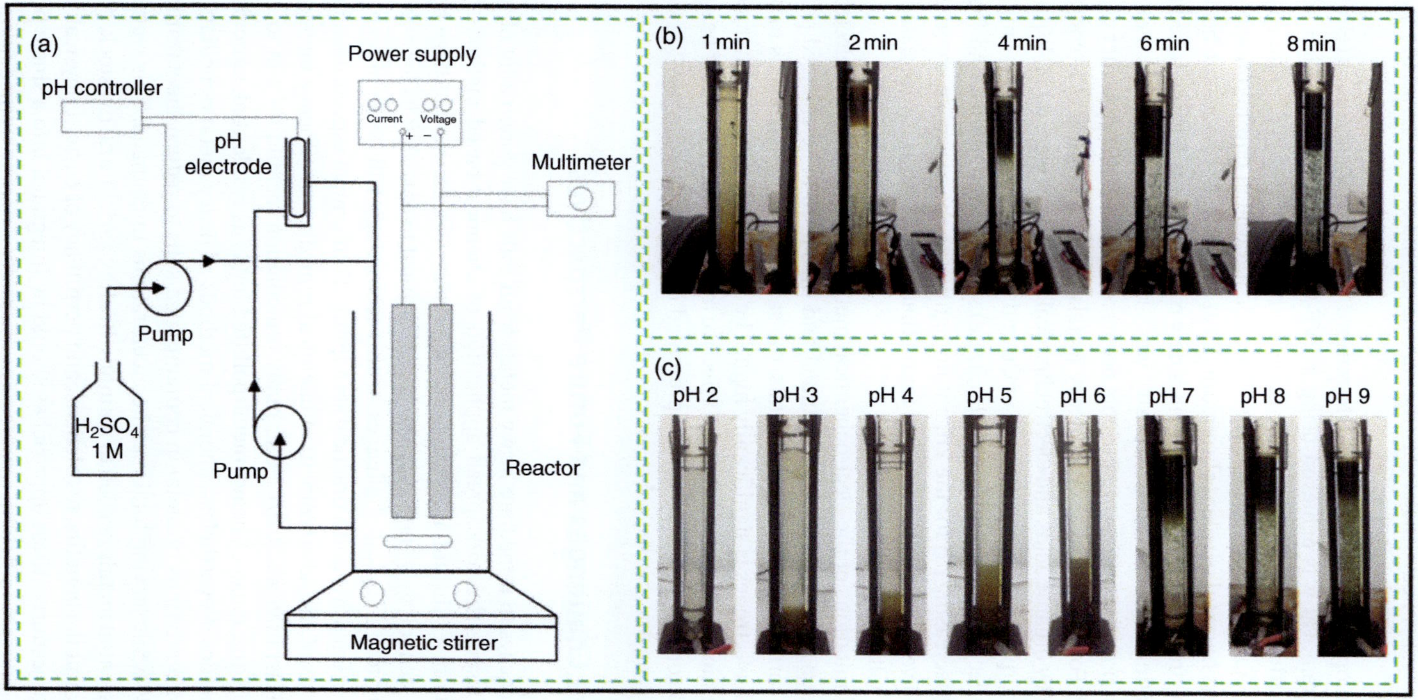

Figure 7.5 (a) Experimental setup design. Effect of varying (b) time and (c) pH on the removal process. *Source:* Ya et al. (2018)/with permission of Elsevier.

achieved at a low current density of $24.2\,\text{mA}\,\text{cm}^{-2}$ within the time of four minutes. To increase the COD removal, the electrochemical Fenton process was further applied. The effect on the removal of heavy metal with the varying pH and time is shown in Figure 7.5b,c (Ya et al. 2018).

A redox metallopolymer was employed to remove heavy metal oxyanions (HMOs) from the water. The removal was achieved using a PVF-CNT electrode with high adsorption capacity. The method is advantageous not only for the toxic removal (Cr and As) but also for reducing the toxicity of the by-products. The toxic hexavalent chromium was adsorbed and delivered less harmful trivalent chromium (Su et al. 2018). To remove the heavy metals (Cu, Pb, and Cd) from soil-washing effluent, a new remediation technique based on asymmetrical alternating current electrochemistry was utilized. In the proposed method, after the remediation process, no loss of nutrients occurred and no toxic byproduct was formed (Xu et al. 2019). For the remediation of heavy metals (Cu and Zn) from the contaminated soil, adsorption by organic acids and electrochemical techniques were done. With the electrochemical adsorption, the concentration of Cu and Zn reached near about $1\,\text{mg}\,\text{l}^{-1}$ in the leachate. The effect of pH on organic acid (citrate) solution was studied and found to be decreased with decreasing pH for the Zn, but no significant changes were observed for Cu. The effect of cell voltage was observed and found that removal capacity decreased with the decreasing cell voltage (Yang et al. 2020).

Some of the limitations of the electrochemical methods include that the electrodes used for the production of coagulants are consumed and thus need to be changed periodically and due to the corrosive nature of the metals, the sacrificial electrode needs to be coated. The cost of such electrodes is significantly high. The integration of some other techniques with electrochemical techniques will be an effective means for the removal of heavy metals.

The details of some of the literature are presented in Table 7.1.

7.4 Current Challenges and Future Perspectives

Chemical approaches to remediate heavy metals from soils have gotten a lot of interest in recent years. Despite the widespread application of chemical-based methods for heavy metal remediation, a substantial future concern persists. The efforts that have been made to broaden the technology's application sector are critical to ensuring its long-term viability. The existing technological challenges of this technology will be significant research goals from both a scientific and a practical standpoint. Four crucial aspects need immediate attention in terms of the advancement of chemical remediation of heavy metals from the environment (Hazrati et al. 2020). First, many remediation techniques aim to encourage the regeneration and reuse of heavy metal-polluted lands as residential, agricultural, and industrial properties. Nevertheless, chemical methods of heavy metal remediation reduce bioavailable concentrations of metal in contaminated soils to reduce the hazardous effect of heavy metals while leaving total metal concentrations unchanged. This would cause a deficiency of essential metals needed to maintain healthy agricultural livestock. As a result, taking into account the benefits, restrictions, and pertinency of current physical, biological and chemical procedures, these approaches should be integrated into a single platform to create a novel, highly efficient, and environmentally friendly remediation strategy.

Table 7.1 Different remediation techniques, their removal efficiency, and time taken for heavy metal removal.

S No.	Technique(s)	Heavy metal	Precipitation reaction duration	Removal efficiency	Reference
1.	*Chemical precipitation*				
	Hydroxide precipitation	Vo Cr Fe	20 min	—	Zhang and Duan (2020)
	Sulfide precipitation	Pb Ca Cu	15 min	99.4%	Wu (2019)
	Chelating precipitation	Ni Cu	60 min	98.4% 78.2%	Fu et al. (2012) Wang et al. (2019)
2.	*Coagulation*	Cu Ni	50 min	99.72% 99.7%	Nguyen et al. (2021)
		Cd Co Ni Cu Cr Pb	240 min (4 h)	79.0% 94.8% 94.4% 98.0% 98.3% 99.5%	Bora and Dutta (2019)
		PFS PFS/$FeSO_4$ PFS/$FeSO_4$ (aerated) PAC	30 min	27.0–32.1% 42.3–47.1% 93.8% 20%	Liu et al. (2019b)
3.	*Ion exchange*	Pb	—	≥90%	Lalmi et al. (2018)
			0–24 h	93%	Goyal et al. (2021)
		Pb Cd	48 h	99%	Chen et al. (2017)
		Cu	160 min	—	Wu et al. (2021b)
4.	*Electrochemical method*	Fe (II) Fe (III)	4 min	>67%	Ya et al. (2018)
		Cr As	120 min (2 h)	>80%	Su et al. (2018)
		Cu Pb Cd	30 min to 6 h	—	Xu et al. (2019)
		Cu Zn	one to seven days	50%	Yang et al. (2020).

Meanwhile, before implementing a newer method for the cleanup of metal-polluted sites, it is necessary to analyze its price and technical viability. Secondly, in the soil environment, heavy metal contamination is frequently characterized by high concentrations, coupled effluence, persistence, and complexity. Furthermore, the concurrent stability of numerous heavy metals in damaged soil remnants is a major task due to their distinct characteristics. Furthermore, recycling and repurposing natural, industrial and agricultural wastes as cheaper and environmentally beneficial stabilizers of soil may ultimately reduce hazardous waste-dumping pressure and the cost of remediation. Third, in the vast majority of cases, stabilized soils are classed as hazardous wastes because they contain high levels of toxic heavy metal contaminants. Chemically stable heavy metals in polluted soils are particularly vulnerable to release when local environmental circumstances change, which include microbial activity, acid depositions, and temperature changes. Furthermore, the widespread application of chemical methods in the remediation of heavy metals from soil pays to an upsurge in stabilized soil formation. Due to high clearance costs and probable ecofriendly implications, technical treatment and regulation of these chemically stabilized soils became a major blockage. As a solution to the aforementioned issues, stabilized soils should be proposed as a viable choice for long-term, profitable, and naturally friendly civil engineering resources. Horticultural soil, dam, road construction, and sanitary landfills are all common disposal techniques for stabilized soils. To minimize secondary pollution and preserve ecological security, the metal stabilization routine in soils must be long-standing checked and more appraised based on these specific disposal situations. As a result, standardized toxic discharge methodologies should be developed to assess the determined leaching of heavy metals in stabilized muds under various disposal scenarios, taking into account environmental variables. Furthermore, while the majority of studies emphasize small-scale evaluations of heavy metal stabilization routines using numerous artificially augmented aging procedures, the outcome of these assessments is seldom backed up by lasting monitoring data from large-scale stabilization treatments. Another worry is that there is currently no comprehensive assessment system or documented technical guideline for chemical-based heavy metal remediation, making it difficult to provide adequate direction for creating control programs and carrying out remediation actions. In context to this, a combined assessment system should be constructed for measuring heavy metal remediation efficacy onsite, using physicochemical and biological indicators, and longstanding monitoring studies of the field should be done to build a uniform methodological guideline for chemical method-based metal removal.

7.5 Conclusions

This chapter discussed the most frequently used chemical methods that have been used to date to remove harmful heavy metals or compounds such as chromium, lead, zinc, nickel, arsenic, mercury, and others. Adsorption is one of the strategies that has proved successful in overcoming the drawbacks of physical, chemical, and biological procedures. In the present situation, progressive water treatment machinery is crucial for ensuring good-quality water, reducing biological and chemical pollutants, and boosting industrial manufacturing processes. The majority of scientific reports have so far been based on lab tests. The absence of knowledge about main systems is the most significant shortcoming of using cost-effect analysis. The ease of availability, quiet process, selectivity, high reactivity, and reusability are all critical factors to consider while applying adsorption. Furthermore, this technique has been remarkably cooperative in repurposing leftover, undesirable, or discarded objects such as lignocellulosic materials, microorganisms, flash, agricultural waste, sludge, and industrial waste by physical and chemical pretreatment and producing them into an adsorptive planned to remove heavy metals from wastewater. Recently, various nanomaterials have been used for sewage treatment; for instance, nano adsorbents such as oxides Zn, Fe, Mn, Mg, and carbon nanotubes, electrocatalysts, photocatalysts, and nanofilms are currently being exploited. Chemical stabilization technology has been widely employed for decades in the rehabilitation of polluted soils due to its high effectiveness, ease of use, broad applicability, and low cost. Yet, based on the existing situation, there is still a significant technological hurdle that must be addressed to ensure its long-term viability. To considerably promote the further development of this technology, more research is required to address some essential and critical challenges. Because there is a typical absence of long-term field monitoring data, one significant task in stabilization techniques is its longstanding effectiveness. The long-term evaluation of overall metal stabilization performance is a crucial component of heavy metal removal efforts. However, several studies are primarily based on an understanding of heavy metal leachability, bioavailability, bioaccessibility, and mobility in soils, making it difficult to establish effective disposal plans and improve risk management of stabilized soils. Many polluted areas throughout the world have been and are now being cleaned up using one of the strategies listed above. However, as previously stated, there is a scarcity of data on large-scale operations in the literature. A desirable outcome is the filling of the data gap on real-world heavy metal remediation, which is aided by closer collaboration between research and industry. This will aid future soil clean-up efforts, which is an urgent need in many developed countries. At present, an integrated remediation system that includes physical, chemical, and biological methods will be needed to assess heavy metal suppleness, soil quality enhancement, and ecosystem preservation purposes after chemical stabilization treatment.

Acknowledgments

The authors express their sincere thanks to Director, CSIR-AMPRI, for his support and encouragement in this work. Raju Khan would like to acknowledge SERB for providing funds in the form of the IPA/2020/000130 project.

References

Abdel-Razek, M.A. et al. (2019). Bioremediation of a pesticide and selected heavy metals in wastewater from various sources using a consortium of microalgae and cyanobacteria. *Slovenian Veterinary Research* 56: 61–73. https://doi.org/10.26873/SVR-744-2019.

Akbal, F. and Camci, S. (2010). Comparison of electrocoagulation and chemical coagulation for heavy metal removal. *Chemical Engineering and Technology* 33 (10): 1655–1664. https://doi.org/10.1002/ceat.201000091.

Al Hejami, A. et al. (2020). Heavy metals in indoor settled dusts in Toronto, Canada. *Science of the Total Environment* 703: 134895. https://doi.org/10.1016/j.scitotenv.2019.134895.

Anju, M. and Banerjee, D.K. (2012). Multivariate statistical analysis of heavy metals in soils of a Pb-Zn mining area, India. *Environmental Monitoring and Assessment* 184 (7): 4191–4206. https://doi.org/10.1007/s10661-011-2255-8.

Arora, R. (2019). Adsorption of heavy metals-a review. *Materials Today: Proceedings* 18 (1): 4745–4750. https://doi.org/10.1016/j.matpr.2019.07.462.

Ashraf, M. et al. (2017). *Challenges and Opportunities for Using Wastewater in Agriculture: A Review*, vol. vol. 2(2), 1–20. http://Jaab.Uaar.Edu.Pk.

Asiabi, H. et al. (2018). Functionalized layered double hydroxide with nitrogen and sulfur co-decorated carbondots for highly selective and efficient removal of soft Hg2+ and Ag+ ions. *Journal of Hazardous Materials* 357: 217–225. https://doi.org/10.1016/J.JHAZMAT.2018.05.055.

Atieh, M. *et al.* (2017) 'Metals in the Environment: Toxic Metals Removal', http://hindawi.com, 2017. 10.1155/2017/4309198.

Bashir, A. et al. (2019). Removal of heavy metal ions from aqueous system by ion-exchange and biosorption methods. *Environmental Chemistry Letters* 17 (2): 729–754. https://doi.org/10.1007/s10311-018-00828-y.

Bora, A.J. and Dutta, R.K. (2019). Removal of metals (Pb, Cd, Cu, Cr, Ni, and Co)from drinking water by oxidation-coagulation-absorption at optimized pH. *Journal of Water Process Engineering* 31 (March): 100839. https://doi.org/10.1016/j.jwpe.2019.100839.

Bradl, H.B. (2005). Chapter 1 sources and origins of heavy metals. *Interface Science and Technology* 6 (C): 1–27. https://doi.org/10.1016/S1573-4285(05)80020-1.

Burt, R. et al. (2003). Major and trace elements of selected Pedons in the USA. *Journal of Environmental Quality* 32 (6): 2109–2121. https://doi.org/10.2134/jeq2003.2109.

Charerntanyarak, L. (1999). Heavy metals removal by chemical coagulation and precipitation. *Water Science and Technology* 39 (10–11): 135–138. https://doi.org/10.1016/S0273-1223(99)00304-2.

Chen, Z. et al. (2017). Layered silicate RUB-15 for efficient removal of UO22+ and heavy metal ions by ion-exchange. *Environmental Science: Nano* 4 (9): 1851–1858. https://doi.org/10.1039/c7en00366h.

Chen, X.X. et al. (2020). Health risk assessment associated with heavy metal accumulation in wheat after long-term phosphorus fertilizer application. *Environmental Pollution* 262: 114348. https://doi.org/10.1016/j.envpol.2020.114348.

Dąbrowski, A. et al. (2004). Selective removal of the heavy metal ions from waters and industrial wastewaters by ion-exchange method. *Chemosphere* 56 (2): 91–106. https://doi.org/10.1016/j.chemosphere.2004.03.006.

El Samrani, A.G., Lartiges, B.S., and Villiéras, F. (2008). Chemical coagulation of combined sewer overflow: heavy metal removal and treatment optimization. *Water Research* 42 (4–5): 951–960. https://doi.org/10.1016/j.watres.2007.09.009.

Feng, D., Aldrich, C., and Tan, H. (2000). Treatment of acid mine water by use of heavy metal precipitation and ion exchange. *Minerals Engineering* 13 (6): 623–642. https://doi.org/10.1016/S0892-6875(00)00045-5.

Fu, F. and Wang, Q. (2011). Removal of heavy metal ions from wastewaters: a review. *Journal of Environmental Management* 92 (3): 407–418. https://doi.org/10.1016/j.jenvman.2010.11.011.

Fu, F. et al. (2012). Application of a novel strategy-advanced Fenton-chemical precipitation to the treatment of strong stability chelated heavy metal containing wastewater. *Chemical Engineering Journal* 189–190: 283–287. https://doi.org/10.1016/j.cej.2012.02.073.

Goyal, P., Tiwary, C.S., and Misra, S.K. (2021). Ion exchange based approach for rapid and selective Pb(II) removal using iron oxide decorated metal organic framework hybrid. *Journal of Environmental Management* 277: 111469. https://doi.org/10.1016/j.jenvman.2020.111469.

Hazrati, S. et al. (2020). Mitigation in availability and toxicity of multi-metal contaminated soil by combining soil washing and organic amendments stabilization. *Ecotoxicology and Environmental Safety* 201: 110807. https://doi.org/10.1016/j.ecoenv.2020.110807.

He, Z. et al. (2015). Heavy metal contamination of soils: sources, indicators and assessment. *Journal of Environmental Indicators* 9: 17–18.

He, H. et al. (2021). Correction to: Strong phosphorus (P)-zinc (Zn) interactions in a calcareous soil-alfalfa system suggest that rational P fertilization should be considered for Zn biofortification on Zn-deficient soils and phytoremediation of Zn-contaminated soils. *Plant and Soil* 461 (1–2): 135. https://doi.org/10.1007/s11104-021-04966-1.

Jasrotia, S. et al. (2015). *Performance of Aquatic Plant Species for Phytoremediation of Arsenic-Contaminated Water*. Springer [Preprint]. 10.1007/s13201-015-0300-4.

Jin, W. et al. (2016). Electrochemical processes for the environmental remediation of toxic Cr(VI): a review. *Electrochimica Acta* 191 (Vi): 1044–1055. https://doi.org/10.1016/j.electacta.2016.01.130.

Johnson, P.D. et al. (2008). Enhanced removal of heavy metals in primary treatment using coagulation and flocculation. *Water Environment Research* 80 (5): 472–479. https://doi.org/10.2175/106143007x221490.

Kang, S.Y. et al. (2004). Competitive adsorption characteristics of Co2+, Ni2+, and Cr3+ by IRN-77 cation exchange resin in synthesized wastewater. *Chemosphere* 56 (2): 141–147. https://doi.org/10.1016/J.CHEMOSPHERE.2004.02.004.

Kaur, R., Sharma, S., and Kaur, N. (2019). Heavy metals toxicity and the environment. *Journal of Pharmacognosy and Phytochemistry* 8 (1S): 247–249.

Khalid, S. et al. (2016). *A Comparison of Technologies for Remediation of Heavy Metal Contaminated Soils*, 182. Elsevier 10.1016/j.gexplo.2016.11.021ï.

Khanh Nguyen, V. et al. (2021). Review on pretreatment techniques to improve anaerobic digestion of sewage sludge. *Fuel* 285: 119105. https://doi.org/10.1016/j.fuel.2020.119105.

Khoei, A.J. et al. (2019). Application of physical and biological methods to remove heavy metal, arsenic and pesticides, malathion and diazinon from water. *Turkish Journal of Fisheries and Aquatic Sciences* 19 (1): 21–28. https://doi.org/10.4194/1303-2712-v19_1_03.

Kumar, M. (2014). Utilization of agro residue corncob for production of acetone-butanol using *Clostridium acetobutylicum* and process optimization through RSM. *Journal of Microbial and Biochemical Technology* s8 (01): 1–5. https://doi.org/10.4172/1948-5948.s8-005.

Kumar, M. et al. (2019). Valorization of coal fired-fly ash for potential heavy metal removal from the single and multi-contaminated system. *Heliyon* 5 (10): e02562. https://doi.org/10.1016/j.heliyon.2019.e02562.

Kumar, M., Singh, A.K., and Sikandar, M. (2020). Biosorption of Hg (II) from aqueous solution using algal biomass: kinetics and isotherm studies. *Heliyon* 6 (1): e03321. https://doi.org/10.1016/j.heliyon.2020.e03321.

Lalmi, A. et al. (2018). Removal of lead from polluted waters using ion exchange resin with Ca(NO3)2 for elution. *Hydrometallurgy* 178: 287–293. https://doi.org/10.1016/j.hydromet.2018.05.009.

Lambert, R., Grant, C., and Sauvé, S. (2007). Cadmium and zinc in soil solution extracts following the application of phosphate fertilizers. *Science of the Total Environment* 378 (3): 293–305. https://doi.org/10.1016/j.scitotenv.2007.02.008.

Liu, F. et al. (2019a). A novel monolith ZnS-ZIF-8 adsorption material for ultraeffective Hg (II) capture from wastewater. *Journal of Hazardous Materials* 367: 381–389. https://doi.org/10.1016/J.JHAZMAT.2018.12.098.

Liu, Y. et al. (2019b). Coagulation removal of Sb(V) from textile wastewater matrix with enhanced strategy: comparison study and mechanism analysis. *Chemosphere* 237: 124494. https://doi.org/10.1016/j.chemosphere.2019.124494.

Low, V.L. et al. (2013). Current susceptibility status of malaysian culex quinquefasciatus (Diptera: Culicidae) against ddt, propoxur, malathion, and permethrin. *Journal of Medical Entomology* 50 (1): 103–111. https://doi.org/10.1603/ME12068.

Ma, W. et al. (2015). Synthesis of macroporous silica biomass nanocomposite based on XG/MgSiO3 for the removal of toxic ions. *Bioresource Technology* 186: 356–359. https://doi.org/10.1016/j.biortech.2015.03.133.

Masindi, V. and Muedi, K.L. (2018). Environmental contamination by heavy metals. In: *Heavy Metals*. IntechOpen 10.5772/intechopen.76082.

Navarro, M.C. et al. (2008). Abandoned mine sites as a source of contamination by heavy metals: a case study in a semi-arid zone. *Journal of Geochemical Exploration* 96 (2–3): 183–193. https://doi.org/10.1016/j.gexplo.2007.04.011.

Ndimele, P.E. (2017). *The Political Ecology of Oil and Gas Activities in the Nigerian Aquatic Ecosystem*. Academic Press 10.1016/c2015-0-05649-0.

Nguyen, M.K. et al. (2021). Fenton/ozone-based oxidation and coagulation processes for removing metals (Cu, Ni)-EDTA from plating wastewater. *Journal of Water Process Engineering* 39: 101836. https://doi.org/10.1016/j.jwpe.2020.101836.

Palansooriya, K.N. et al. (2020). Soil amendments for immobilization of potentially toxic elements in contaminated soils: a critical review. *Environment International* 134: 105046. https://doi.org/10.1016/j.envint.2019.105046.

Pan, Z.F. and An, L. (2019). Removal of heavy metal from wastewater using ion exchange membranes. In: *Applications of Ion Exchange Materials in the Environment*, 25–46. Springer 10.1007/978-3-030-10430-6_2.

Pohl, A. (2020). Removal of heavy metal ions from water and wastewaters by sulfur-containing precipitation agents. *Water, Air, and Soil Pollution* 231 (10): 1–17. https://doi.org/10.1007/S11270-020-04863-W/TABLES/2.

Rahimi, M. et al. (2021). Assessing the distribution of cadmium under different land-use types and its effect on human health in different gender and age groups. *Environmental Science and Pollution Research* 28 (35): 49258–49267. https://doi.org/10.1007/s11356-021-12881-2.

Real, F. et al. (2007). Removal of diazinon by various advanced oxidation processes. *Wiley Online Library* 82 (6): 566–574. https://doi.org/10.1002/jctb.1702.

Recena, R., García-López, A.M., and Delgado, A. (2021). Zinc uptake by plants as affected by fertilization with Zn sulfate, phosphorus availability, and soil properties. *Agronomy* 11 (2): 390. https://doi.org/10.3390/agronomy11020390.

Renu, N.A., Agarwal, M., and Singh, K. (2017). Methodologies for removal of heavy metal ions from wastewater: an overview. *Interdisciplinary Environmental Review* 18 (2): 124. https://doi.org/10.1504/ier.2017.087915.

Sabir, A. et al. (2021). *Agricultural Waste Absorbents for Heavy Metal Removal*, 195–228. Springer 10.1007/978-3-030-47400-3_8.

Sah, D. et al. (2019). Pollution characteristics, human health risk through multiple exposure pathways, and source apportionment of heavy metals in PM10 at Indo-Gangetic site'. *Urban Climate* 27: 149–162. https://doi.org/10.1016/j.uclim.2018.11.010.

Sahoo, P.K. et al. (2020). Regional-scale mapping for determining geochemical background values in soils of the Itacaiúnas River Basin, Brazil: the use of compositional data analysis (CoDA). *Geoderma* 376: 114504. https://doi.org/10.1016/j.geoderma.2020.114504.

Sarma Bora, M. et al. (2021). Pollution and health risk assessment of toxic metals in solid waste dumping site soil and its impact on groundwater: a case study. *International Journal of Environmental Analytical Chemistry* 103: 1–21. https://doi.org/10.1080/03067319.2020.1862813.

Selvi, A. et al. (2019). Integrated remediation processes toward heavy metal removal/recovery from various environments-a review. *Frontiers in Environmental Science* 7: 66. https://doi.org/10.3389/fenvs.2019.00066.

Sharma, R.K. and Agrawal, M. (2005). Biological effects of heavy metals: An overview. *Journal of Environmental Biology* 26 (2 Suppl): 301–313.

Sharma, S. and Rana, S.K. (2016). Physical, chemical and phytoremediation technique for removal of heavy metals. *Journal of Heavy Metal Toxicity and Diseases* 1 (2): 10. https://doi.org/10.21767/2473-6457.100010.

Singh, P.K., Deshbhratar, P.B., and Ramteke, D.S. (2012). Effects of sewage wastewater irrigation on soil properties, crop yield and environment. *Agricultural Water Management* 103: 100–104. https://doi.org/10.1016/j.agwat.2011.10.022.

Song, S. and Cao, Y. (2021). Textures and structures of metamorphic rocks. In: *Encyclopedia of Geology*, 375–388. Elsevier 10.1016/b978-0-08-102908-4.00052-7.

Su, X. et al. (2018). Electrochemically-mediated selective capture of heavy metal chromium and arsenic oxyanions from water. *Nature Communications* 9 (1): 4701. https://doi.org/10.1038/s41467-018-07159-0.

Sun, C. et al. (2017). Preparation of a cellulosic adsorbent by functionalization with pyridone diacid for removal of Pb(II) and Co(II) from aqueous solutions. *Cellulose* 24 (12): 5615–5624. https://doi.org/10.1007/s10570-017-1519-z.

Tang, X. et al. (2016). Chemical coagulation process for the removal of heavy metals from water: a review. *Desalination and Water Treatment* 57 (4): 1733–1748. https://doi.org/10.1080/19443994.2014.977959.

Tian, H. et al. (2012). Temporal and spatial variation characteristics of atmospheric emissions of Cd, Cr, and Pb from coal in China. *Atmospheric Environment* 50: 157–163. https://doi.org/10.1016/j.atmosenv.2011.12.045.

Totsche, K.U. et al. (2018). Microaggregates in soils. *Journal of Plant Nutrition and Soil Science* 181 (1): 104–136. https://doi.org/10.1002/jpln.201600451.

Tu, C. et al. (2020). Biochar and bacteria inoculated biochar enhanced Cd and Cu immobilization and enzymatic activity in a polluted soil. *Environment International* 137: 105576. https://doi.org/10.1016/J.ENVINT.2020.105576.

Tyagi, V.K. et al. (2018). Anaerobic co-digestion of organic fraction of municipal solid waste (OFMSW): progress and challenges. *Renewable and Sustainable Energy Reviews* 93: 380–399. https://doi.org/10.1016/j.rser.2018.05.051.

Vaca Mier, M. et al. (2001). Heavy metal removal with mexican clinoptilolite: multi-component ionic exchange. *Water Research* 35 (2): 373–378. https://doi.org/10.1016/S0043-1354(00)00270-0.

Verbych, S. et al. (2004). Ion exchange extraction of heavy metal ions from wastewater. *Separation Science and Technology* 39 (9): 2031–2040. https://doi.org/10.1081/SS-120039317.

Verma, A.K., Dash, R.R., and Bhunia, P. (2012). A review on chemical coagulation/flocculation technologies for removal of colour from textile wastewaters. *Journal of Environmental Management* 93 (1): 154–168. https://doi.org/10.1016/j.jenvman.2011.09.012.

Vetrimurugan, E. et al. (2017). Human exposure risk to heavy metals through groundwater used for drinking in an intensively irrigated river delta. *Applied Water Science* 7 (6): 3267–3280. https://doi.org/10.1007/s13201-016-0472-6.

Wang, L.K. et al. (2005). Chemical precipitation. In: *Physicochemical Treatment Processes*, vol. vol. 3, 141–197. Humana Press 10.1385/1-59259-820-x:141.

Wang, Q. et al. (2019). Non-thermal plasma oxidation of Cu(II)-EDTA and simultaneous Cu(II) elimination by chemical precipitation. *Journal of Environmental Management* 248: 109237. https://doi.org/10.1016/J.JENVMAN.2019.07.008.

Wang, Y.L. et al. (2021). Estimation of soil and dust ingestion rates from the stochastic human exposure and dose simulation soil and dust model for children in Taiwan. *Environmental Science and Technology* 55 (17): 11805–11813. https://doi.org/10.1021/acs.est.1c00706.

Warf, B. (2010). Sedimentary rock. In: *Encyclopedia of Geography*. Thousand Oaks, Calif.: SAGE https://doi.org/10.4135/9781412939591.n1015.

Wu, R. (2019). Removal of heavy metal ions from industrial wastewater based on chemical precipitation method. *Ekoloji* 28 (107): 2443–2452.

Wu, J. et al. (2021a). A novel modified method for the efficient removal of Pb and Cd from wastewater by biochar: enhanced the ion exchange and precipitation capacity. *Science of the Total Environment* 754: 142150. https://doi.org/10.1016/j.scitotenv.2020.142150.

Wu, S. et al. (2021b). ZnCl2 enabled synthesis of activated carbons from ion-exchange resin for efficient removal of Cu2+ ions from water via capacitive deionization. *Chemosphere* 264: 128557. https://doi.org/10.1016/j.chemosphere.2020.128557.

Xu, J. et al. (2019). Remediation of heavy metal contaminated soil by asymmetrical alternating current electrochemistry. *Nature Communications* 10 (1): 1–8. https://doi.org/10.1038/s41467-019-10472-x.

Ya, V. et al. (2018). Electrochemical treatment for simultaneous removal of heavy metals and organics from surface finishing wastewater using sacrificial iron anode. *Journal of the Taiwan Institute of Chemical Engineers* 83: 107–114. https://doi.org/10.1016/j.jtice.2017.12.004.

Yang, X. et al. (2020). Remediation of heavy metal contaminated soils by organic acid extraction and electrochemical adsorption. *Environmental Pollution* 264: 114745. https://doi.org/10.1016/j.envpol.2020.114745.

Yazici, M.A. et al. (2021). Reduced root mycorrhizal colonization as affected by phosphorus fertilization is responsible for high cadmium accumulation in wheat. *Plant and Soil* 468 (1–2): 19–35. https://doi.org/10.1007/s11104-021-05041-5.

Yuanan, H. et al. (2020). Quantitative source apportionment of heavy metal(loid)s in the agricultural soils of an industrializing region and associated model uncertainty. *Journal of Hazardous Materials* 391: 122244. https://doi.org/10.1016/j.jhazmat.2020.122244.

Zeng, Q. et al. (2021). Efficient removal of Hg2+ from aqueous solution by a novel composite of nano humboldtine decorated almandine (NHDA): ion exchange, reducing-oxidation and adsorption. *Journal of Hazardous Materials* 404 (PA): 124035. https://doi.org/10.1016/j.jhazmat.2020.124035.

Zeraatkar, A. et al. (2016). *Potential Use of Algae for Heavy Metal Bioremediation, A Critical Review*. Elsevier [Preprint]. 10.1016/j.jenvman.2016.06.059.

Zhang, Y. and Duan, X. (2020). Chemical precipitation of heavy metals from wastewater by using the synthetical magnesium hydroxy carbonate. *Water Science and Technology* 81 (6): 1130–1136. https://doi.org/10.2166/wst.2020.208.

Zhou, Q. et al. (2020). Total concentrations and sources of heavy metal pollution in global river and lake water bodies from 1972 to 2017. *Global Ecology and Conservation* 22: e00925. https://doi.org/10.1016/j.gecco.2020.e00925.

8

Carbon-Based Absorption Materials for Heavy Metal Removal

Ching T. Moi[1] *and Ruhima Khan*[2]

[1] *Department of Chemistry, Indian Institute of Technology Guwahati, Guwahati, Assam, India*
[2] *Department of Chemistry, Gauhati University, Guwahati, Assam, India*

8.1 Introduction

With the increase in urbanization and industrialization over the past century, large amounts of untreated heavy metal ions are released into soil and water (Chai et al. 2021). Heavy metal pollution is a threat to the environment across the globe, causing serious detrimental effects on human well-being and biological systems. Metals with an atomic density greater than $4 \pm 1\,g\,cm^{-3}$ are termed heavy metals (Mariana et al. 2021). The most abundant heavy metals include copper (Cu), zinc (Zn), lead (Pb), nickel (Ni), cadmium (Cd), arsenic (As), chromium (Cr), and mercury (Hg) (Duan et al. 2020). They originate from human activities, such as mining, painting, battery, chemical manufacturing, the use of pesticides and fertilizers, etc. The lethal effect of heavy metals can cause symptoms like insomnia, depression, weakness, deprived concentration, mood swings, high blood pressure, memory loss, etc. While some heavy metals like iron (Fe), Cu, manganese (Mn), and Zn are required in the body in trace amounts, their presence in huge amounts could be hazardous (Ihsanullah et al. 2016).

Numerous treatment technologies are employed for heavy metal removal, such as adsorption, coagulation, chemical precipitation, oxidation, ion exchange, electrochemical processes, and membrane filtration (Fu and Wang 2011; Yang et al. 2019). Considering the setback of most of these treatment methods due to energy consumption, the complexity of operating conditions, and producing secondary contaminants, the adsorption method could be the potentially preferred alternative (Yu et al. 2021). The adsorption technique offers selectivity, design flexibility, and the recovery of the adsorbent (Agarwal and Singh 2017).

Presently, nanomaterials with multiple applications are state of the art in materials science research and have gained a wide range of applications in clean energy production, energy storage, environmental remediation, water treatment, basic science, and daily life (Xu et al. 2018). From the reported literature, researchers have worked on numerous metal

Remediation of Heavy Metals: Sustainable Technologies and Recent Advances, First Edition. Edited by Rangabhashiyam Selvasembian, Binota Thokchom, Pardeep Singh, Ali H. Jawad, and Willis Gwenzi.

oxides (Zeng et al. 2019), layered double hydroxides (LDHs) (Rahman et al. 2018), clay (Gu et al. 2019), and mesoporous carbon (Hu et al. 2022) as suitable adsorbents of heavy metal ions. Biological waste materials, which are environmentally benign, easily available, and present in bulk quantities mainly from agricultural byproducts, may have a potential application as a low-cost adsorbent based on ecological and commercial standpoints. Agricultural waste materials such as rice husk (Elhafez et al. 2017), banana peels (Joseph et al. 2019), and biomass like seaweed, fallen leaves, etc. (Jha et al. 2021), which are processed into activated carbon (AC) can act as low-cost and non-conventional adsorbents for the elimination of heavy metal ions (Jha et al. 2021).

For decades, adsorption using carbon-based nanomaterials (CNMs) has existed at the forefront of nanotechnology (Xu et al. 2018). Nanomaterials have been used as adsorbents due to their low cost, large surface area, and eco-friendly for the elimination of specific heavy metal ions from aqueous solutions. CNMs, such as graphene oxide (GO), AC, carbon black (CB), CNTs, and CNFs, are utilized for the adsorption of various environmental pollutants. Carbonaceous materials have the benefits of an increased surface area with high chemical durability (Qasem et al. 2021; Hu et al. 2022).

8.2 Sources of Heavy Metal in Water

Soil erosion, geological weathering, and airborne dirt naturally cause heavy metals in water bodies. The anthropogenic activities include atmospheric precipitation, industrial waste, mining activities, agronomy, and automobile exhaust. The rapid advances in technology and industrialization come at the price of environmental pollution with heavy metals (Mariana et al. 2021). Figure 8.1 shows the sources of heavy metal contamination in wastewater.

8.2.1 Human Health and Heavy Metal Toxicity

The term "heavy metals" refers to elements based on their atomic mass and their toxicity to a living organism. Meanwhile, lighter elements such as beryllium and lithium can also be lethal. Some heavy metal ions, like Fe, Zn, Cu, etc., are indispensable for human health in

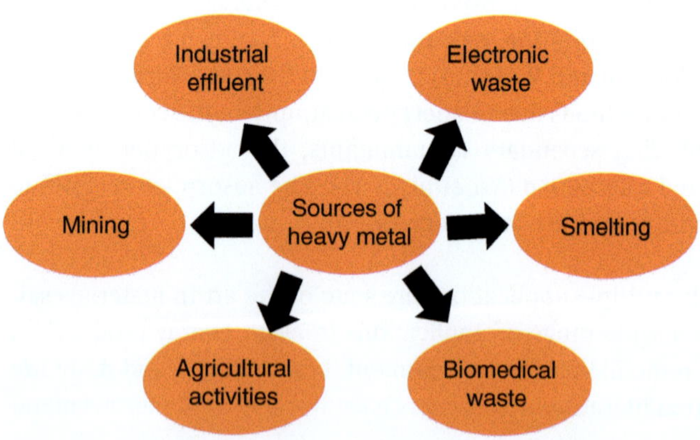

Figure 8.1 Sources of heavy metal pollution in wastewater.

trace amounts (Mariana et al. 2021). Most lethal and ordinary heavy metal ions include Pb, Hg, Cd, As, Mn, and radioactive metals. The toxic impact of heavy metal ions is the fact that they can cause dysfunction and complication in human metabolic activities and other biological processes. The following section discusses heavy metal toxicity on human health (Nadeem et al. 2006).

8.2.2 Toxicity of Mercury

Hg can exist in different forms, like elemental, organic, and inorganic. Their degrees of toxicity and their effects on the human system differ accordingly. Hg is among the top elements considered by the World Health Organization (WHO) as a major threat to public health. Hg is a d-block element and exists in liquid form under normal conditions (Baby et al. 2019). Major contributors to Hg contamination are agricultural waste, paper and pulp industries, pharmaceutical industries, etc. A human being is exposed to methyl-mercury by eating fish, and Hg poisoning causes a disease known as pink disease or acrodynia. Growing fetuses are most susceptible to Hg exposure as it can harmfully affect the developing brain and nervous system. A high dose of Hg exposure can be fatal as it can sabotage the immune, nervous, and digestive systems and cause kidney and lung failure (Baby et al. 2019). The compounds methyl-mercury and mercuric chloride are termed carcinogenic by the environmental protection agency. The inhalation, ingestion, or dermal exposure of different forms of Hg compounds can cause skin and eye diseases, hair loss, insomnia, etc. The permissible limit of Hg concentration is 0.01 $mg\,l^{-1}$ for drinking water as set by the WHO. Kabiri et al. (2016) stated the removal of Hg^{2+} from contaminated water using an As-synthesized 3D porous graphene membrane functionalized with a thiol group. They obtained a removal efficiency close to 100% for low (4 $mg\,l^{-1}$) and high (120 $mg\,l^{-1}$) Hg(II) concentrations (Kabiri et al. 2016). Another group reported the use of functionalized CNT like MSWCNTs-CoS composite (1666.0 $mg\,g^{-1}$), KA-CNTs (250.5 $mg\,g^{-1}$), and MWCNTs@mSiO_2-MPTS (349.7 $mg\,g^{-1}$) indicating a high adsorption affinity for Hg^{II} removal from aqueous solution (Gupta and Gupta 2018).

8.2.3 Toxicity of Lead

Pb is a non-biodegradable heavy metal element with an atomic mass of 207.2 u, and which appears silvery bluish in color and changes to gray upon oxidation. Pb is termed a carcinogenic element to human health and is hazardous to flora and fauna of the environment due to its extensive application (Baby et al. 2019). The main sources of Pb contamination come from the effluent of battery manufacturing, manure and pesticides, dyes, fuel additives, toys, and jewelry. Exposure to Pb can take place through environmental and professional sources. Pb mainly enters the human body by ingestion of food or drinks contaminated with Pb or by inhalation due to the burning of Pb-containing material. High levels of Pb exposure can attack the central nervous system and can cause cancer, mental health problems, kidney failure, hypertension, and sterility (Hiraoka et al. 2006). Infants and pregnant women are more vulnerable to Pb poisoning. In the bloodstream, Pb concentrations as low as 5 $\mu g\,dl^{-1}$ can affect the intellect in children and cause miscarriage and underweight infants in pregnant women (Nadeem et al. 2006; Ghorbani et al. 2020). Based on experimental evidence proved by adsorption, the affinity of the Pd(II) ion with GO is controlled by pH. The studies showed that at low pH values, Pd(II) ions adsorb on the surface of GO by ionic exchange of Pb^{2+} with H^+/Na^+. Meanwhile, at higher pH, the GO surface gets deprotonated by the presence of oxygen-containing functional groups. Thus, the electrostatic force is created between the negatively charged GO surface and Pb(II) ion (Wang et al. 2015; Ghorbani et al. 2020).

8.2.4 Toxicity of Arsenic

As is a transition element with an atomic mass of 74.92 u and is usually present along with other metals like Fe, sodium, calcium (Ca), and sulfur and also in metallic form. The underground water of many countries is contaminated with As. Man gets As poisoning mainly through drinking water contaminated by As from poor disposal of chemicals, textiles, glass, pesticides, dyes, and pigment. In humans, long-term exposure to As can cause skin cancer, pigmentation, and pulmonary and cardiovascular diseases (Mohan and Pittman 2007; Baby et al. 2019). As can exist in various oxidation states of −3, 0, +3, and +5 and numerous forms, like As and arsenious acids, dimethylarsinic acid, etc. Among them, arsenate ($AsO_4{}^{3-}$) in As^{5+} and arsenite ($AsO_3{}^{3-}$) in As^{3+} are present in water and toxic to humankind and living organisms. As^{3+} exists predominantly in groundwater and As^{5+} is mainly stable in the aerobic surrounding. The allowed As concentration in drinking water, as adopted by the WHO guidelines, is 10 ppb (0.01 $mg\,l^{-1}$) (Xu et al. 2018). As(V) and As(III) removal from the water is studied using a char carbon (CC) derived from fly ash (FA) and AC. At pH 2.2, As(V) adsorption by CC is 34.5 $mg\,g^{-1}$ as compared to 30.5 $mg\,g^{-1}$ by AC at pH 7. Simultaneously, As(III) removal by CC is 89.2 $mg\,g^{-1}$ (pH 2.2) and 29.9 $mg\,g^{-1}$ by AC (pH 7.0) (Pintor et al. 2020).

8.2.5 Toxicity of Chromium

Cr is considered the sixth abundant transition element with a mass of 51.99 u, with lustrous, brittle and steel gray. Cr can exist in the multivalent oxidation state of Cr^{2+}, Cr^{4+}, Cr^{5+}, and Cr^{6+}. The trivalent and hexavalent states of Cr are the most stable forms. Cr in +3 oxidation state is a basic necessity for mineral supplements in humankind and animals (Baby et al. 2019). The potential source of Cr is mainly from textile industries, chemical and metallurgical industries, electroplating, leather tanning, and chromate preparation, and the effluents from industries are released into the air, water, and soil polluting the environment. Cr in the +6 oxidation state released from industrial waste is highly stable, lethal, and carcinogenic to mankind and it can exist in the shape of chromate ($CrO_4{}^{2-}$) and dichromate ($Cr_2O_7{}^{2-}$) ions. Cr^{6+} poisoning can cause severe health problems like kidney and liver failure, diarrhea, vomiting, etc. The set limit of Cr concentration in water is 50 $\mu g\,l^{-1}$ by the WHO (Wang et al. 2013). ZnO tetrapods and AC nanomaterial composite are documented for the removal of Cr^{6+} ions. Their high surface area, porosity, high stability and nontoxicity with rich functional groups favor the adsorption and improved elimination capacity of Cr^{6+} ions. The adsorption efficiency of Cr^{6+} is widely affected by pH. At optimum pH of 2, the ZnO-T/AC composite shows the highest removal efficiency of 97% in comparison to bare AC and ZnO-T, which show 43% and 55%, respectively (Sharma et al. 2019).

8.2.6 Toxicity of Cadmium

Cd is a group 12 transition element with an atomic number of 48, and it has similar properties to Hg and Zn and is bluish-white. The Agency for Toxic Substances and Disease Registry (ATSDR) classified Cd as the seventh most toxic metal. Cd contamination mainly comes from dyes and pigments, nuclear plants, electroplating, plastic and batteries manufacturing, and textile industries (Baby et al. 2019). Mostly, a low level of Cd is present in the environment. However, effluents from manufacturing industries play a great role in increasing Cd concentration. The acceptable limit of Cd concentration in drinking water is $0.003\,\mathrm{mg\,l^{-1}}$, as recommended by the WHO (Wang et al. 2013). Cd ions have a long biological lifetime of 10–35 years in humans and are found to be present mostly in kidneys. Cd toxicity can cause severe health destruction like respiratory and kidney malfunctioning, carcinogenesis, anemia, and cough. The removal efficiency of Cd(II) ions is studied by Ihsanullah et al. (2016) using CNM (such as CNTs, AC, CNFs and FA). Based on the studies, they found that surface modification plays a crucial role in the adsorption capacity. CNTs have the highest adsorption capacities of $2.02\,\mathrm{mg\,g^{-1}}$, as compared to AC ($1.98\,\mathrm{mg\,g^{-1}}$), CNFs ($1.22\,\mathrm{mg\,g^{-1}}$), and FA ($1.58\,\mathrm{mg\,g^{-1}}$) (Al-Khadi Ihsanullaha et al. 2015).

8.3 Effects of Water Environmental Chemistry on Heavy Metal Removal

The parameters for measuring the quality of water concerning heavy metal remediation are temperature, pH of the aqueous medium, and ionic strength. Mostly, in the developing world, the source of heavy metal pollution mainly comes from industrial effluents. Meanwhile, reports have shown the existence of heavy metal pollutants in domestic wastewater. The nature and quality of water affect the adsorption ability of metal ions by the adsorbent surface. These sections represent the chemical properties of heavy metal ions and the result of water quality on heavy metal remediation (Elhafez et al. 2017; Xu et al. 2018).

8.3.1 Temperature

Temperature is an essential measure affecting the behavior of heavy metal ions and has substantial influences on their subsequent removal. The reaction rate for the removal of heavy metals is improved by raising the temperature for endothermic reactions, and it decreases for exothermic reactions with rising temperature (Xu et al. 2018). From the reported literature, the efficiency of Ni(II) ions' removal using tea waste has been increased by 22% with increases in temperature from 25 °C to 60 °C. The enhancement in removal efficiency can be credited to increases in the mobility of heavy metal ions with rising temperature and breaking of bonds, which further enhance the adsorption sites (Malkoc and Nuhoglu 2005). Typically, an increase in temperature causes a reduction in mass transfer resistance and accelerates diffusion. For example, temperature increases caused an enhancement in heavy metal ions' (such as Pb^{2+}, As^{3+}, and As^{5+}) removal by $MnFe_2O_4$ and GO-$MnFe_2O_4$ NPs (Kumar et al. 2014). At higher temperatures, the adsorption procedure is more effective and spontaneous, with an increase in calculated negative ΔG^0 values (Jin et al. 2015). Besides, active mass transport and the thickness of the boundary layer get reduced with the rise in temperature, which further enhances the number of the surface-active site on the material and also permits metal ions to be easily removed from the liquid medium (Gupta and Rastogi 2008). However, the effect of temperature is irrelevant for a reaction involving small activation energy where the reaction rate is independent of diffusion. On the other hand, in exothermic reactions, adsorption increases with a decrease in temperature (Hadavifar et al. 2014).

8.3.2 pH Value

Based on solution chemistry, the pH has a substantial effect on heavy metal ions present in an aqueous medium. So, it is an important parameter for heavy metal elimination. Normally, heavy metals are present in their cationic states at acidic to neutral pH and are easily transferable and soluble in an aqueous medium. With an increase in pH, the formation of complexes takes place between anion and hydroxide present in water. The pH also affects the adsorbent's ionic concentration present in the functional group, state of ionization, and surface charge of the adsorbent materials. Studies have shown that the mobility and durability of Cu ions increase with decreasing pH (Kumpiene et al. 2008; Joseph et al. 2019). On the other hand, as the pH rises, the oxidation state of metal ions changes due to complex formation with the hydroxide present in an alkaline solution. Cr show changes in oxidation state with increasing pH from Cr(III) to more stable and lethal Cr(IV) (Pantsar-Kallio et al. 2001). The variation in solution pH affects the accumulation and ionization state of the functional groups attached to the surface of CNMs like graphene, SWNT, and MWNT, thus reducing the binding capacity of the pollutants. Based on experimental evidence, pH $\geq$6 MWNT exists in a negative charge, which prevents the formation of aggregation and decreases the micro- and nanopore volumes (Kabiri et al. 2016). Furthermore, at low pH, heavy metal ions' removal by adsorption processes is slow. The high concentration of H^+ ions restricts the surface contact between the soluble heavy metal ions and adsorbent by competing for surface adsorption sites, which decreases the removal processes (Chen et al. 2010). Although heavy metal ions' adsorption is minimal at lower pH, there is also a report on an increase in adsorption with a decrease in pH. However, the surface of the adsorbent becomes negatively charged, which can react easily with positively charged metal ions (Vimala and Das 2009).

In general, pH is a deciding factor, given the nature and potential removal of heavy metal ions. Based on the reported literature, acidic pH (<4) is known to inhibit heavy metal adsorption, whereas pH values between 5 and 7 are found to be the most efficient. Considering the above point, in developing countries, to get maximum adsorption capacity, the pH of a contaminated water source should be maintained at neutral pH (Joseph et al. 2019).

8.3.3 Ionic Strength and Coexisting Ions

The ionic strength defined the concentration of the electrolyte solution. The ability to remove heavy metal ions by nanomaterials having high surface area and good adsorption capacity is strongly influenced by ionic strength, which affects the thickness of the electric double layer and interface potential. The upsurge in ionic strength greatly affects the electrostatic interaction between metal ions and other surfaces and hence their removal (Ferraz and Lourençlo 2000). The types of adsorption taking place in the nanomaterial show a high effect on the concentration of the electrolyte solutions. The water source containing chlorine ions can form negatively charged or neutral metal-chloride complexes, which are soluble in the electrolyte solution and difficult to remove (Ferraz and Lourençlo 2000). The ability to remove Cu(II) and Ni(II) ions decreases with upsurges in ionic strength. The adsorption affinity is reduced with an increase in the formation of heavy metal-chloride complexes (Villaescusa et al. 2004). Zhang et al. reported the heavy metal (such as Cu(II), Pb(II), and Zn(II)) removal from the compost pit, and found a decrease in the removal efficiency with the upsurge in ionic strength. Moreover, studies have shown an increase in heavy metal ions' elimination with ionic strength. With the participation of inner-sphere surface complexation, the elimination efficiency of As(III) and Ni(II) increases by about 25% with an increase in the strength of ionic solution from 0.01 to 1.0 M Cl^- (Joseph et al. 2019).

8.4 Carbon-Based Nanomaterials

With the advance in nanotechnology, carbonaceous adsorbents with high surface chemistry, excellent electrical conductivity, and high adsorption capacity are studied for heavy metal adsorption. However, the main setback of CNMs is their sluggish kinetics and post-extraction, which restrict the practical applications of contemporary CNMs. Figure 8.2 shows the CNM for heavy metal adsorption from wastewater.

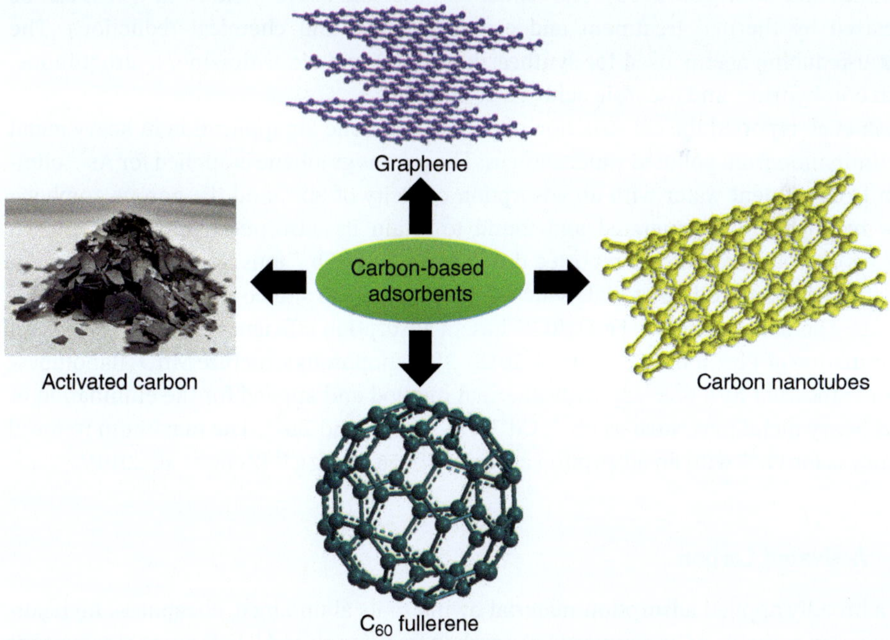

Figure 8.2 Carbon-based nanomaterial (CNM) for heavy metal adsorption from wastewater.

8.4.1 Graphene and Derivatives

Graphene is a single-sheet CNM consisting of a two-dimensional (2D) hexagonal network. Graphene was first introduced by Andre Geim and Sir Konstantin Novoselov in 2004, who received the Nobel Prize in 2010 for their outstanding discovery (Ali et al. 2019). Graphene constitutes the basic starting material which can be enclosed into fullerenes (0D), rolled to form CNTs (1D), and assembled into graphite (3D). Graphene (2D sheet) is widely applied in several fields, such as semiconductors, energy storage, batteries, solar cells, LCDs, water remediation, and biological sciences. Graphene has desirable properties like high surface area, tuneable surface chemistry, thermal mobility, mechanical strength, and electrical conductivity (Qasem et al. 2021). GO is obtained by the oxidation of graphene through chemical or thermal reduction processes, which can be synthesized by both bottom-up (such as the unfolding of CNT or reduced GO (rGO)) and top-down (e.g. electrochemical and ultrasonic exfoliation and acidic treatment) (Xu et al. 2018). The hypothetical precise surface area of graphene is about $2630\,m^2\,g^{-1}$, and it offers a planar sheet for molecule adsorption on both sides. The physicochemical properties of graphene sheets, such as hydrophobicity, high surface area, and surface chemistry, restrict their dispersibility in water, which results in the formation of agglomeration (Li et al. 2021). The chemical or thermal reduction of GO yields rGO, which shows properties of both pristine graphene (i.e. high mechanical strength and high surface area) and GO (i.e. readily dispersible in an aqueous medium). rGO, as an oxygenated derivative of graphene, has rich oxygen-containing functional groups of epoxy, hydroxyl and carboxyl groups. Hence, graphene planar sheets are highly hydrophilic and dispersible. The superior properties of GO come from its rich surface chemistry, which offers better remediation for various heavy metal ions (Smith and Rodrigues 2015). The reduction processes of GO yield rGO, which can be synthesized by thermal treatment and electrochemical and chemical reductions. The chemical-reducing agents used for synthesizing (rGO) include hydrazine, hydroquinone, sodium borohydride, and ascorbic acid (Yu et al. 2021).

Tabish et al. reported the construction of porous graphene for application in heavy metal ions' elimination from polluted water sources. The porous graphene is applied for As^{3+} elimination from effluent water with an adsorption capacity of 80%, and the porous graphene can be recycled and regenerated and found to retain its adsorption properties (Tabish et al. 2018). Another researcher reported the elimination of Pb^{2+} ions from an aqueous solution using nanocomposite material, which is fabricated by in situ co-precipitation of Fe_3O_4 with rGO. The nanocomposite Fe_3O_4/rGO has an adsorption efficiency of $373.14\,mg\,g^{-1}$ for the elimination of Pb^{2+} ions (Tabish et al. 2018). 3D nanoporous structure MnO_2 nanotubes/rGO is synthesized by a one-step hydrothermal method and applied for the elimination of various heavy metal ions, such as Pb^{2+}, Cd^{2+}, Ag^{+}, Cu^{2+}, and Zn^{2+}. The maximum removal efficiency is for Pb^{2+} with an adsorption capacity of $356.37\,mg\,g^{-1}$ (Zeng et al. 2019).

8.4.2 Activated Carbon

AC is a broadly applied adsorption material owing to its abundance, cheapness, high surface area, and porosity. AC is composed of a carbon framework and it belongs to mesoporous material. Based on the synthetic procedure, AC could be both micro- and mesopores (Chai et al. 2021). According to the reported literature, AC is produced by physical activation (PA) and chemical activation (CA) methods. PA involves carbonization at low temperatures under anaerobic conditions, and CA involves the use of chemical agents (e.g. KOH, H_3PO_4, and $ZnCl_2$), thus requiring minimum synthesis duration. CA is considered a better synthesis procedure due to its enhanced porosity, although it is not environmentally benign (Mariana et al. 2021).

8.4.3 Carbon Nanotubes

The improved electronic, thermal, and mechanical properties of CNTs make them favorable materials for various applications. The CNT is made up of two types: single-wall CNT (SWCNT), which consists of one graphene sheet rolled effortlessly to give a cylindrical shape with a diameter of 1 nm and length of 1 cm; and the multi-wall CNTs (MWCNTs), which consist of arrays of cylinders formed by more than one graphene sheet wrapped around themselves with a diameter of 100 nano-meter (nm) and length of more than 20 μm (Coleman et al. 2006). The CNT material surface is connected with numerous functional groups which favor the interaction between organic and inorganic molecules. These interactions are proven by the presence of hydrogen bonding, π–π stacking, and van der Waals forces. CNT has been demonstrated to possess improved adsorption ability compared to outdated adsorbent material (Hoang et al. 2022).

8.4.4 SWCNTs in the Purification of Heavy Metal-Contaminated Water

SWCNTs consist of numerous applications with excellent physicochemical properties and are composed of 1-D carbon nanomaterials with a hollow tube of one atom thickness. The unique properties of SWCNTs include high surface area, porosity, good surface chemistry, and nanosize, which make them favorable for application in numerous fields such as chemical and biological sensors, electronic, semiconductor, energy storage, and environmental remediation (Hoang et al. 2022). The three main techniques mostly used for SWCNT synthesis are laser ablation, electric arc, and chemical vapor deposition (CVD), in which the CVD method is considered the most efficient commercial approach for SWCNT synthesis (Hiraoka et al. 2006). Fabrication of SWCNT with magnetite cobalt sulfide has been designed by Alijani et al. for the elimination of Hg^{2+}. The nanocomposite shows an adsorption capability of >99.56% within seven minutes' duration as compared to pristine SWCNTs, which show 45.39% Hg^{2+} removal efficiency (Alijani and Shariatinia 2018). Molecular dynamic simulation is conducted by Anitha et al. for studying the adsorption properties of SWCNTs and their functionalized counterparts (SWCNTs-OH, SWCNTs-NH_2, and SWCNTs-COOH) for heavy metal ions' (such as Cd^{2+}, Cu^{2+}, Pb^{2+}, and Hg^{2+}) removal. The simulation studies show that carbonyl-functionalized SWCNTs-COOH has the highest adsorption capacities of about 150–230% as compared to pristine SWCNTs. Meanwhile, SWCNTs-OH and SWCNTs-NH show a low adsorption ability of 10–47% higher than pristine SWCNTs (Anitha et al. 2015).

8.4.5 MWCNTs in the Purification of Heavy Metal-Contaminated Water

MWCNTs attract numerous applications because of their exceptional properties of tremendous thermal and electrical conductivity, easy flexibility, high tensile strength, and good surface area. The exclusive physicochemical properties of MWCNTs allow them to be widely applied in the field of energy storage, nanoelectronic, transistor, solar cell, and biological studies. The MWCNTs are applied for heavy metal ions' adsorption from wastewater by chemical interaction method, where the metal ions are adsorbed by functional groups present on the MWCNT surface. Based on experimental evidence, heavy metal ions like Cr^{6+}, Pb^{2+}, and Cd^{2+} are effectively adsorbed by oxidized MWCNTs from the effluent water and the adsorption affinity depends on the pH (Hiraoka et al. 2006). Nanocomposites having MWCNTs with a metal oxide, such as MWCNTs-Fe_2O_3, MWCNTs-Fe_3O_4, MWCNTs-Al_2O_3, and MWCNTs-MnO_2-Fe_2O_3, are applied for removal of Cr^{6+}, Hg^{2+}, Pb^{2+}, and Cu^{2+} ions, respectively from water (Hiraoka et al. 2006; Wu et al. 2019). The MWCNTs with surface functionalized have 20 times better adsorption capacity toward heavy metal ions in comparison to unoxidized MWCNTs (Fiyadh et al. 2019). The potential adsorption capacity of MWCNTs toward heavy metal ions further increases with acid treatment, mainly due to the generation of oxygen functional groups on the acidified MWCNT surface. The functionalized nanotube reacts with ions to form metal salt or complexes, which precipitate out. The nanocomposite formed by co-precipitation of MWCNT and metal oxides like Fe(III) oxide, zinc oxide, and aluminum oxide is used for the elimination of Cr, Ni, Pb, Cu, and As ions. The adsorption efficiency of nanocomposite depends on varying parameters of temperature and pH. The sorption volume for the nanocomposites increases from 10 to 31 $mg\,g^{-1}$ depending on the pH and temperature (Fiyadh et al. 2019).

8.4.6 Fullerenes

Fullerenes are made of a pentagonal and hexagonal closed ring structure, with a formula of C_{20+m} where m is an integer. Fullerenes are hydrophobic, with a good electron affinity to accept electrons, high surface area, high surface defects chemistry, long-term stability, and rich redox activity (Pan et al. 2020). The outstanding physicochemical properties of fullerene make them desirable materials for their applications in artificial photosynthesis, energy storage, sensors, cosmetic products, water treatment, and as semiconductor material (Baby et al. 2019). Fullerenes' molecules are ideal materials for energy storage due to their rich surface chemistry and high C—C bond energies, which allow their molecular cage structure to be readily reversed (Shen et al. 2011). Nanomaterial composite C_{60}/graphene shows a better specific capacitance value of 135.36 Fg^{-1} against pristine graphene. The synthesized composite shows a 1000 charge/discharge circle with a retaining duration of 92.35% (Ma et al. 2014). Fullerene has a narrow bandgap of 1.6–1.9 ev, which favors a rapid charge separation and electron transfer, hence resulting in a lower recombination rate. The 3D cage structure of fullerenes also provides structural stability. Consequently, owing to the desirable electronic properties of fullerene- (especially C_{60}) based nanomaterial, they exhibit good electron acceptors and fast electron transfer kinetics when applied as photocatalytic materials (Pan et al. 2020). The physicochemical properties of fullerenes, such as large surface area, lower accumulation affinity, and defect structure, fabricate a desirable candidate for heavy metal removal from polluted water. Fullerenes are considered adsorbed ionic species amid the defects present in the carbon nanoclusters. Alekseeva et al. reported the elimination of Cu^{2+} ions using nanocomposite fullerene with polystyrene, which follows the Langmuir model of adsorption. The adsorption efficiency of Cu^{2+} ions is superior in presence of fullerene against its bare counterpart. However, the high cost of fullerene is the main drawback to its wide practical applications (Alekseeva et al. 2016).

8.5 Adsorption Mechanisms

Heavy metal ions' adsorption on the carbon-based system is classified into five types as shown in Figure 8.3.

8.5.1 Physical Adsorption

The adsorption of heavy metal ions on the surface of the adsorbent material is controlled by the distribution of pore size and surface area. The surge in physical adsorption kinetics depends on an increase in the exposure of micropores and mesopores, which further cause an increase in the surface area of adsorbent material and promote metal ions' diffusion (Duan et al. 2020). Heavy metal adsorption by the physical method is triggered by van der Waals interactions between the surface of the AC materials and metal ions for removal. Physical adsorption is controlled by weak intermolecular forces, which show a probability for the reversible process (Mariana et al. 2021).

8.5.2 Electrostatic Interaction

The adsorption process is controlled by electrostatic attraction between an adsorbent and positively or negatively charged ions. Based on the reported literature, heavy metal ions' removal by carbon-based adsorbents like CNTs and AC occur through the processes of electrostatic interaction (Mariana et al. 2021). The oxidation states of metal ions, ionic radius, and the pH of the aqueous solution can directly affect the strength of electrostatic interaction. Metals with higher oxidation states give better interaction and offer improved adsorption capacity. The interaction between AC and charges on the metal surfaces depends on the charges of electrostatic attraction (Wang et al. 2014).

Figure 8.3 Different processes of heavy metal remediation.

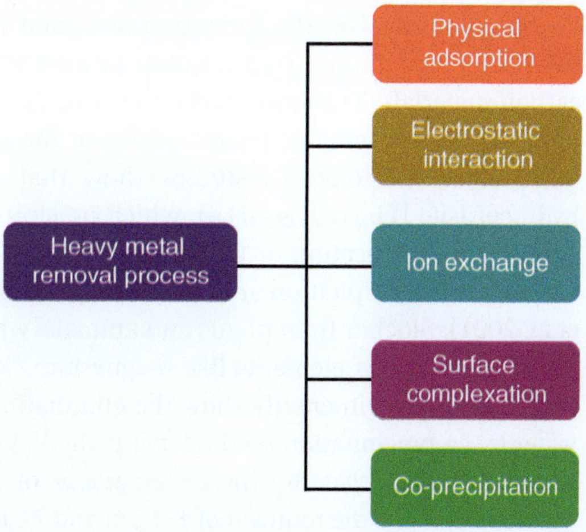

8.5.3 Ion Exchange

Ion exchange adsorption processes include the exchange of ions and Coulombic force taking place between metal ions with a positively charged and negatively charged surface of carbon nanomaterial. The processes of ionic exchange are based on the pH of the solution and take place rapidly. Ion exchange mainly occurs between the H^+ of oxygen—having functional groups (−COOH, −OH) and the positively charged metal cations (Mariana et al. 2021). The ionic exchange capacity, as reported by Wang et al. using TiO_2 nanotubes for the elimination of Pb^{2+}, Cd^{2+}, and Cr^{3+}, found that pH is the deciding factor for the adsorption processes, and pH 5 gives the best result (Wang et al. 2014). Another report by Cao et al. shows that the adsorption of Pb^{2+} by wheat straw is controlled by carbonization, which is further attributed to the precipitation of Pb^{2+} (Cao et al. 2019).

8.5.4 Surface Complexation

Surface complexation processes are another methodology for the adsorption of heavy metal ions in which the nature of adsorption depends upon the changes in ionic strength. Complex formation takes place between partly filled d-orbitals of transition metal ions with metal ligands to form a multiatom complex (Pintor et al. 2020). Adsorption is said to occur by outer-sphere complexation when the amount of adsorption decreases with the existence of other ions (Miao et al. 2014). However, if the ionic strength has a positive effect or no effect on the adsorption capacity, interaction is likely to be specific, which corresponds to inner-sphere complexes' formation (Antelo et al. 2005). On the carbon nanomaterial adsorbent surface, oxygen-containing functional groups react with metal with divalent (M^{2+}) or trivalent metal ions (M^{3+}) or hydroxide complexes (like Cd $(OH)^+$, Pb $(OH)^+$) for heavy metal removal from an aqueous medium (Duan et al. 2020). Pintor et al. (2020) reported As and phosphorus removal by the surface complexation method, in which lower ionic strength favored As(III) adsorption. Meanwhile, As(V) and P(V) adsorption improved with an increase in ionic strength as repulsive interactions decrease, and inner-sphere complexation is likely to take place in all three species (Pintor et al. 2020).

8.5.5 Precipitation/Coprecipitation

Precipitation involves the formation of a solid precipitate on the adsorbent surface or in solution. Precipitation is an effective process for the elimination of heavy metal ions by carbon materials. Transition metal ions like Zn, Co, Pb, Cr, etc., with ionization potentials of 2.5 to 9.5, can be precipitated easily on the surface of a nanomaterial as compared to other elements. Reported studies show that digested biochar with pH ~ 10.93 forms hydrocerusite [$Pb_3(CO_3)_2(OH)_2$], which precipitates Pb ions on the char surface (Inyanga et al. 2016). Adsorption of Cd ions on the biochar surface is reported by Mariana et al. (2021) by the precipitation or co-precipitation of $CdCO_3$, $Cd_3(PO_4)_2$, and $Cd(OH)_2$ (Mariana et al. 2021). Biochar from plants and animals, which are produced at high temperatures, is found to consist of elements like magnesium (Mg), Cu, Ca, Fe, and silicon (Si). Rahman et al. (2018) experimentally show the elimination of Fe, Pb, Cu, and Zn from mine wastewater by co-precipitation method using Mg-Al LDH. The elimination of Zn ions from sewage water takes place by the development of $Zn(NO_3)(OH).H_2O$ and $Zn_5(NO_3)_2(OH)_8$. Correspondingly, the removal of Fe, Cu, and Pb ions occurs by the formation of FeAl LDH, $Cu_2(OH)_3NO_3$, and $Pb_4(OH)_4(NO3)_4$, respectively (Rahman et al. 2018). Table 8.1

Table 8.1 Effect of various factors of certain heavy metal ions' adsorption.

Adsorbents	Adsorbates	Adsorption capacity	pH	Interaction mechanisms	Model	Reference
MWCNTs@ $mSiO_2$-MPTS	Hg^{2+}	349.7 $mg\,g^{-1}$	5–6	Surface complexation, chelation, surface reduction, precipitation, van der Waals', chemical, and cation-π interaction	Langmuir	Gupta and Gupta (2018)
SWCNTs	Hg^{2+}	41.66 $mg\,g^{-1}$	7.95	Physical or chemical	Freundlich	Alijani and Shariatinia (2018)
MWCNTs	Hg^{2+}	100 $mg\,g^{-1}$	5.26	Physical or chemical	Langmuir	Alijani and Shariatinia (2018)
MWCNTs/ CoS	Hg^{2+}	1666 $mg\,g^{-1}$	5.26	Physical or chemical	Langmuir	Alijani and Shariatinia (2018)
Char carbon (CC)	As^{3+}	89.2 $mg\,g^{-1}$	9	Complexation	Langmuir	Pintor et al. (2020)
Porous graphene	As^{3+}	90%	7	Surface complexation	Langmuir	Tabish et al. (2018)
MWCNTs-COOH	As^{3+}	133 $mg\,g^{-1}$	7.9	Electrostatic	Langmuir	Alimohammady et al. (2018)
rGO/MnO_2	Pb^{2+}	356.4 $mg\,g^{-1}$	5	Physisorption, hydrophobic or π-π interaction, and van der Waals interaction	Langmuir	Zeng et al. (2019)
MWCNTs	Pb^{2+}	93%	9	Chemical	NA	Chaudhari et al. (2016)
Zero-valent irons (NZVI)	Pb^{2+}	807.2 $mg\,g^{-1}$	6	NA	Langmuir	Dongsheng et al. (2019)
NZVI	Cd^{2+}	769.2 $mg\,g^{-1}$		Chemisorption	Langmuir	Boparai et al. (2011)
Biochar/ Fe_3O_4/ ZIF-67	Cd^{2+}	50.78 $mg\,g^{-1}$	6	NA	Langmuir	Foroutan et al. (2022)
DSAC/ZrO_2	Cd^{2+}	166.7 $mg\,g^{-1}$	6	NA	Langmuir	Sharma and Naushad (2020)
P(TA-TEPA)-PAM-RGO	Cr^{6+}	397.5 $mg\,g^{-1}$	2	Chemical	Langmuir	Zhang et al. (2018)
RGO/PEI/ Fe_3O_4	Cr^{6+}	266.6 $mg\,g^{-1}$	4	Electrostatic attraction	NA	Wang et al. (2017)
DSAC// MAX	Cr^{6+}	38.6 $mg\,g^{-1}$	6.5	NA	Experiment	Bharath et al. (2020)

summarizes the effect of various parameters for the adsorption of heavy metal ions by CNMs. The studies indicate the interaction mechanisms between CNMs and heavy metal ions, and the maximum adsorption capacity is performed by the Langmuir isotherm model.

8.6 Conclusion and Outlook

In conclusion, the present chapter highlights the practice of carbon-based nano-materials for heavy metal elimination. Carbon-based adsorbents like AC, graphene, SWCNTs, MWCNTs, biochar and fullerene are used for the removal of heavy metal ions such as Hg, Pb, As, Cr, and Cd from aqueous media. The main sources of heavy metal ions' contamination come from industrial effluent, biomedical waste, electronic waste, agricultural effluent, human activities, mining, and smelting. The parameters responsible for quantifying water concerning heavy metal removal are temperature, pH of the aqueous medium and ionic strength. The adsorption technique is considered the most suitable and economical method for heavy metal remediation. In conclusion, carbon-based nano-materials exhibit favorable physicochemical properties and a high potential for commercial wastewater remediation.

References

Agarwal, M. and Singh, K. (2017). Heavy metal removal from wastewater using various adsorbents: a review. *Journal of Water Reuse and Desalination* 7: 387–419. https://doi.org/10.2166/wrd.2016.104.

Alekseeva, O.V., Bagrovskaya, N.A., and Noskov, A.V. (2016). Sorption of heavy metal ions by fullerene and polystyrene/fullerene film compositions. *Protection of Metals and Physical Chemistry of Surfaces* 52: 443–447. https://doi.org/10.1134/S2070205116030035.

Ali, I., Basheer, A.A., Mbianda, X.Y. et al. (2019). Graphene based adsorbents for remediation of noxious pollutants from wastewater. *Environment International* 127: 160–180. https://doi.org/10.1016/j.envint.2019.03.029.

Alijani, H. and Shariatinia, Z. (2018). Synthesis of high growth rate SWCNTs and their magnetite cobalt sulfide nanohybrid as super-adsorbent for mercury removal. *Chemical Engineering Research and Design* 129: 132–149. https://doi.org/10.1016/j.cherd.2017.11.014.

Alimohammady, M., Jahangiri, M., Kiani, F., and Tahermansouri, H. (2018). Design and evaluation of functionalized multi-walled carbon nanotubes by 3-aminopyrazole for the removal of Hg(II) and As(III) ions from aqueous solution. *Research on Chemical Intermediates* 44: 69–92. https://doi.org/10.1007/s11164-017-3091-4.

Al-Khadi Ihsanullaha, F.A., Abusharkh, B., Khaled, M. et al. (2015). Adsorptive removal of cadmium (II) ions from liquid phase using acid modified carbon-based adsorbents. *Journal of Molecular Liquids* 204: 255–263. https://doi.org/10.1016/j.molliq.2015.01.033.

Anitha, K., Namsani, S., and Singh, J.K. (2015). Removal of heavy metal ions using a functionalized single-walled carbon nanotube: a molecular dynamics study. *The Journal of Physical Chemistry. A* 119: 8349–8358. https://doi.org/10.1021/acs.jpca.5b03352.

Antelo, J., Avena, M., Fiol, S. et al. (2005). Effects of pH and ionic strength on the adsorption of phosphate and arsenate at the goethite-water interface. *Journal of Colloid and Interface Science* 285: 476–486. https://doi.org/10.1016/j.jcis.2004.12.032.

Baby, R., Saifullah, B., and Hussein, M.Z. (2019). Carbon nanomaterials for the treatment of heavy metal-contaminated water and environmental remediation. *Nanoscale Research Letters* 14: 341. https://doi.org/10.1186/s11671-019-3167-8.

Bharath, G., Hai, A., Rambabu, K. et al. (2020). The fabrication of activated carbon and metal-carbide 2D framework-based asymmetric electrodes for the capacitive deionization of Cr (vi) ions toward industrial wastewater remediation. *Environmental Science: Water Research & Technology* 6: 351–361. https://doi.org/10.1039/C9EW00805E.

Boparai, H.K., Meera Joseph, M., and O'Carroll, D.M. (2011). Kinetics and thermodynamics of cadmium ion removal by adsorption onto nano zerovalent iron particles. *Journal of Hazardous Materials* 186: 458–465. https://doi.org/10.1016/j.jhazmat.2010.11.029.

Cao, Y., Xiao, W., Shen, G. et al. (2019). Carbonization and ball milling on the enhancement of Pb(II) adsorption by wheat straw: competitive effects of ion exchange and precipitation. *Bioresource Technology* 273: 70–76. https://doi.org/10.1016/j.biortech.2018.10.065.

Chai, W.S., Cheun, J.Y., Kumar, P.S. et al. (2021). A review on conventional and novel materials towards heavy metal adsorption in wastewater treatment application. *Journal of Cleaner Production* 296: 126589. https://doi.org/10.1016/j.jclepro.2021.126589.

Chaudhari, A.A., Ashmore, D., Nath, S.d. et al. (2016). A novel covalent approach to bio-conjugate silver coated single walled carbon nanotubes with antimicrobial peptide. *Journal of Nanbiotechnology* 14: 58–58. https://doi.org/10.1186/s12951-016-0211-z.

Chen, H., Zhao, J., Dai, G. et al. (2010). Adsorption characteristics of Pb(II) from aqueous solution onto a natural biosorbent, fallen cinnamomum camphora leaves. *Desalination* 262: 174–182. https://doi.org/10.1016/j.desal.2010.06.006.

Coleman, J.N., Khan, U., Blau, W.J., and Gun'ko, Y.K. (2006). Small but strong: a review of the mechanical properties of carbon nanotube–polymer composites. *Carbon* 44: 1624–1652. https://doi.org/10.1016/j.carbon.2006.02.038.

Dongsheng, Z., Wenqiang, G., Guozhang, C. et al. (2019). Removal of heavy metal lead(II) using nanoscale zero-valent iron with different preservation methods. *Advanced Powder Technology* 30: 581–589. https://doi.org/10.1016/j.apt.2018.12.013.

Duan, C., Ma, T., Wang, J., and Zhou, Y. (2020). Removal of heavy metals from aqueous solution using carbon-based adsorbents: a review. *The Journal of Water Process Engineering* 37: 101339. https://doi.org/10.1016/j.jwpe.2020.101339.

Elhafez, S.A., Hamad, H.A., Zaatout, A.A., and Malash, G.F. (2017). Management of agricultural waste for removal of heavy metals from aqueous solution: adsorption behaviors, adsorption mechanisms, environmental protection, and techno-economic analysis. *Environmental Science and Pollution Research* 24: 1397–1415. https://doi.org/10.1007/s11356-016-7891-7.

Ferraz, M.C.M.A. and Lourençlo, J.C.N. (2000). The influence of organic matter content of contaminated soils on the leaching rate of heavy metals. *Environmental Progress* 19: 53–58. https://doi.org/10.1002/ep.670190118.

Fiyadh, S.S., AlSaadi, M.A., Jaafar, W.Z. et al. (2019). Review on heavy metal adsorption processes by carbon nanotubes. *Journal of Cleaner Production* 230: 783–793. https://doi.org/10.1016/j.jclepro.2019.05.154.

Foroutan, R., Peighambardoust, S.J., Mohammadi, R. et al. (2022). Cadmium ion removal from aqueous media using banana peel biochar/Fe3O4/ZIF-67. *Environmental Research* 211: 113020. https://doi.org/10.1016/j.envres.2022.113020.

Fu, F. and Wang, Q. (2011). Removal of heavy metal ions from wastewaters: a review. *Journal of Environmental Management* 92: 407–418. https://doi.org/10.1016/j.jenvman.2010.11.011.

Ghorbani, M., Seyedin, O., and Aghamohammadhassan, M. (2020). Adsorptive removal of lead (II) ion from water and wastewater media using carbon-based nanomaterials as unique sorbents: a review. *Journal of Environmental Management* 254: 109814. https://doi.org/10.1016/j.jenvman.2019.109814.

Gu, S., Kang, X., Wang, L. et al. (2019). Clay mineral adsorbents for heavy metal removal from wastewater: a review. *Environmental Chemistry Letters* 17: 629–654. https://doi.org/10.1007/s10311-018-0813-9.

Gupta, N.K. and Gupta, A. (2018). 2D and 3D carbon-based adsorbents for an efficient removal of Hg II ions: a review. *FlatChem* 11: 1–14. https://doi.org/10.1016/j.flatc.2018.11.002.

Gupta, V.K. and Rastogi, A. (2008). Sorption and desorption studies of chromium (VI) from nonviable cyanobacterium nostoc muscorum biomass. *Journal of Hazardous Materials* 154: 347–354. https://doi.org/10.1016/j.jhazmat.2007.10.032.

Hadavifar, M., Bahramifar, N., Younesi, H., and Li, Q. (2014). Adsorption of mercury ions from synthetic and real wastewater aqueous solution by functionalized multi-walled carbon nanotube with both amino and thiolated groups. *Chemical Engineering Journal* 237: 217–228. https://doi.org/10.1016/j.cej.2013.10.014.

Hiraoka, T., Yamada, T., Hata, K. et al. (2006). Synthesis of single-and double-walled carbon nanotube forests on conducting metal foils. *Journal of the American Chemical Society* 128: 13338–13339. https://doi.org/10.1021/ja0643772.

Hoang, A.T., Nižetić, S., Cheng, C.K. et al. (2022). Heavy metal removal by biomass-derived carbon nanotubes as a greener environmental remediation: a comprehensive review. *Chemosphere* 287: 131959. https://doi.org/10.1016/j.chemosphere.2021.131959.

Hu, X., You, S., Li, F., and Liu, Y. (2022). Recent advances in antimony removal using carbon-based nanomaterials: a review. *Frontiers of Environmental Science & Engineering* 16: 487. https://doi.org/10.1007/s11783-021-1482-7.

Ihsanullah, Abbas, A., Al-Amer, A.M. et al. (2016). Heavy metal removal from aqueous solution by advanced carbon nanotubes: critical review of adsorption applications. *Separation and Purification Technology* 157: 141. https://doi.org/10.1016/j.seppur.2015.11.039.

Inyanga, M.I., Gao, B., Yao, Y. et al. (2016). A review of biochar as a low-cost adsorbent for aqueous heavy metal removal. *Critical Reviews in Environmental Science and Technology* 46: 406–433. https://doi.org/10.1080/10643389.2015.1096880.

Jha, M.K., Joshi, S., Sharma, R.K. et al. (2021). Surface modified activated carbons: sustainable bio-based materials for environmental remediation. *Nanomaterials* 11: 3140. https://doi.org/10.3390/nano11113140.

Jin, Z., Gao, H., and Hu, L. (2015). Removal of Pb (II) by nano-titanium oxide investigated by batch, XPS and model techniques. *RSC Advances* 5: 88520–88528. https://doi.org/10.1039/C5RA14004H.

Joseph, L., Jun, B.-M., Flora, J.R.V. et al. (2019). Removal of heavy metals from water sources in the developing world using low-cost materials: a review. *Chemosphere* 229: 142–159. https://doi.org/10.1016/j.chemosphere.2019.04.198.

Kabiri, S., Tran, D.N.H., Cole, M.A., and Losic, D. (2016). Functionalized three-dimensional (3D) graphene composite for high efficiency removal of mercury. *Environmental Science: Water Research & Technology* 2: 390–402. https://doi.org/10.1039/C5EW00254K.

Kumar, S., Nair, R.R., Pillai, P.B. et al. (2014). Graphene oxide–$MnFe_2$ O_4 magnetic nanohybrids for efficient removal of lead and arsenic from water. *ACS Applied Materials & Interfaces* 6: 17426–17436. https://doi.org/10.1021/am504826q.

Kumpiene, J., Lagerkvist, A., and Maurice, C. (2008). Stabilization of As, Cr, Cu, Pb and Zn in soil using amendments—a review. *Waste Management* 28: 215–225. https://doi.org/10.1016/j.wasman.2006.12.012.

Li, C., Yang, J., Zhang, L. et al. (2021). Carbon-based membrane materials and applications in water and wastewater treatment: a review. *Environmental Chemistry Letters* 19: 1457–1475. https://doi.org/10.1007/s10311-020-01112-8.

Ma, J., Guo, Q., Gao, H.L., and Qin, X. (2014). Synthesis of C_{60}/graphene composite as electrode in supercapacitors. *Fullerenes, Nanotubes, and Carbon Nanostructures* 23: 477–482. https://doi.org/10.1080/1536383X.2013.865604.

Malkoc, E. and Nuhoglu, Y. (2005). Investigations of nickel (II) removal from aqueous solutions using tea factory waste. *Journal of Hazardous Materials* 127: 120–128. https://doi.org/10.1016/j.jhazmat.2005.06.030.

Mariana, M., Abdul Khalil, H.P.S., Mistar, E.M. et al. (2021). Recent advances in activated carbon modification techniques for enhanced heavy metal adsorption. *The Journal of Water Process Engineering* 43: 102221. https://doi.org/10.1016/j.jwpe.2021.102221.

Miao, Y., Han, F., Pan, B. et al. (2014). Antimony(V) removal from water by hydrated ferric oxides supported by calcite sand and polymeric anion exchanger. *Journal of Environmental Sciences* 26: 307–314. https://doi.org/10.1016/S1001-0742(13)60418-0.

Mohan, D. and Pittman, C.U. (2007). Arsenic removal from water/wastewater using adsorbents—a critical review. *Journal of Hazardous Materials* 142: 1–53. https://doi.org/10.1016/j.jhazmat.2007.01.006.

Nadeem, M., Mahmood, A., Shahid, S.A. et al. (2006). Sorption of lead from aqueous solution by chemically modified carbon adsorbents. *Journal of Hazardous Materials* 138: 604–613. https://doi.org/10.1016/j.jhazmat.2006.05.098.

Pan, Y., Liu, X., Zhang, W. et al. (2020). Advances in photocatalysis based on fullerene C_{60} and its derivatives: properties, mechanism, synthesis, and applications. *Applied Catalysis. B, Environmental* 265: 118579. https://doi.org/10.1016/j.apcatb.2019.118579.

Pantsar-Kallio, M., Reinikainen, S., and Oksanen, M. (2001). Interactions of soil components and their effects on speciation of chromium in soils. *Analytica Chimica Acta* 439: 9–17. https://doi.org/10.1016/S0003-2670(01)00840-6.

Pintor, A.M.A., Vieira, B.R.C., Brandaõ, C.C. et al. (2020). Complexation mechanisms in arsenic and phosphorus adsorption onto iron-coated cork granulates. *Journal of Environmental Chemical Engineering* 8: 104184. https://doi.org/10.1016/j.jece.2020.104184.

Qasem, N.A.A., Mohammed, R.H., and Lawal, D.U. (2021). Removal of heavy metal ions from wastewater: a comprehensive and critical review. *npj Clean Water* 4: 36. https://doi.org/10.1038/s41545-021-00127-0.

Rahman, M.T., Kameda, T., Kumagai, S., and Yoshioka, T. (2018). Effectiveness of Mg–Al-layered double hydroxide for heavy metal removal from mine wastewater and sludge volume reduction. *International journal of Environmental Science and Technology* 15: 263–272. https://doi.org/10.1007/s13762-017-1385-0.

Sharma, G. and Naushad, M. (2020). Adsorptive removal of noxious cadmium ions from aqueous medium using activated carbon/zirconium oxide composite: isotherm and kinetic modelling. *Journal of Molecular Liquids* 310: 113025. https://doi.org/10.1016/j.molliq.2020.113025.

Sharma, M., Joshi, M., Nigam, S. et al. (2019). ZnO tetrapods and activated carbon-based hybrid composite: adsorbents for enhanced decontamination of hexavalent chromium from aqueous solution. *Chemical Engineering Journal* 358: 540–551. https://doi.org/10.1016/j.cej.2018.10.031.

Shen, Y.F., Reparaz, J.S., Wagner, M.R. et al. (2011). Assembly of carbon nanotubes and alkylated fullerenes: nanocarbon hybrid towards photovoltaic applications. *Chemical Science* 2: 2243–2250. https://doi.org/10.1039/c1sc00360g.

Smith, S.C. and Rodrigues, D.F. (2015). Carbon-based nanomaterials for removal of chemical and biological contaminants from water: a review of mechanisms and applications. *Carbon* 91: 122–143. https://doi.org/10.1016/j.carbon.2015.04.043.

Tabish, T.A., Memon, F.A., Gomez, D.E. et al. (2018). A facile synthesis of porous graphene for efficient water and wastewater treatment. *Scientific Reports* 8: 1817. https://doi.org/10.1038/s41598-018-19978-8.

Villaescusa, I., Fiol, N., Martínez, M. et al. (2004). Removal of copper and nickel ions from aqueous solutions by grape stalks wastes. *Water Research* 38: 992–1002. https://doi.org/10.1016/j.watres.2003.10.040.

Vimala, R. and Das, N. (2009). Biosorption of cadmium (II) and lead (II) from aqueous solutions using mushrooms: a comparative study. *Journal of Hazardous Materials* 168: 376–382. https://doi.org/10.1016/j.jhazmat.2009.02.062.

Wang, T., Liu, W., Xiong, L. et al. (2013). Influence of pH, ionic strength and humic acid on competitive adsorption of Pb(II), Cd(II) and Cr(III) onto titanate nanotubes. *Chemical Engineering Journal* 215: 366–374. https://doi.org/10.1016/j.cej.2012.11.029.

Wang, H., Liu, Y.G., Zeng, G.M. et al. (2014). Grafting of beta-cyclodextrin to magnetic graphene oxide via ethylenediamine and application for Cr(VI) removal. *Carbohydrate Polymers* 113: 166–173. https://doi.org/10.1016/j.carbpol.2014.07.014.

Wang, X.X., Chen, Z.S., and Yang, S.B. (2015). Application of graphene oxides for the removal of Pb(II) ions from aqueous solutions: experimental and DFT calculation. *Journal of Molecular Liquids* 211: 957–964. https://doi.org/10.1016/j.molliq.2015.08.020.

Wang, D., Zhang, G., Zhou, L. et al. (2017). Synthesis of a multifunctional graphene oxide-based magnetic nanocomposite for efficient removal of Cr(VI). *Langmuir* 33: 7007–7014. https://doi.org/10.1021/acs.langmuir.7b01293.

Wu, W., Jia, M., Zhang, Z. et al. (2019). Sensitive, selective and simultaneous electrochemical detection of multiple heavy metals in environment and food using a lowcost Fe_3O_4 nanoparticles/fluorinated multi-walled carbon nanotubes sensor. *Ecotoxicology and Environmental Safety* 175: 243–250. https://doi.org/10.1016/j.ecoenv.2019.03.037.

Xu, J., Cao, Z., Zhang, Y. et al. (2018). A review of functionalized carbon nanotubes and graphene for heavy metal adsorption from water: preparation, application, and mechanism. *Chemosphere* 195: 351–364. https://doi.org/10.1016/j.chemosphere.2017.12.061.

Yang, X., Wan, Y., Zheng, Y. et al. (2019). Surface functional groups of carbon-based adsorbents and their roles in the removal of heavy metals from aqueous solutions: a critical review (open access). *Chemical Engineering Journal* 366: 608–621. https://doi.org/10.1016/j.cej.2019.02.119.

Yu, G., Wang, X., Liu, J. et al. (2021). Applications of nanomaterials for heavy metal removal from water and soil: a review. *Sustainability* 13: 713. https://doi.org/10.3390/su13020713.

Zeng, T., Yu, Y., Li, Z. et al. (2019). 3D MnO_2 nanotubes@reduced graphene oxide hydrogel as reusable adsorbent for the removal of heavy metal ions. *Materials Chemistry and Physics* 231: 105–108. https://doi.org/10.1016/j.matchemphys.2019.04.019.

Zhang, Z., Gao, T., Si, S. et al. (2018). One-pot preparation of P(TA-TEPA)-PAM-RGO ternary composite for high efficient Cr(VI) removal from aqueous solution. *Chemical Engineering Journal* 343: 207–216. https://doi.org/10.1016/j.cej.2018.02.126.

9

Industrial Waste-Derived Materials for Adsorption of Heavy Metals from Polluted Water

Rahul Sharma[1], Pinki R. Agrawal[1], Chankit[2], Chanchal[2], Ittishree[3], Vinod Kashyap[4], Ashok K. Sharma[2], and V. Alagesan[5]

[1] *Environmental Sciences and Biomedical Metrology, CSIR-National Physical Laboratory, New Delhi, Delhi, India*
[2] *Department of Chemistry, Deenbandhu Chhotu Ram University of Science and Technology, Murthal (Sonipat), Haryana, India*
[3] *Department of Pharmacognosy, Hindu College of Pharmacy, Sonipat, Haryana, India*
[4] *Department of Chemistry, National Institute of Technology, Tiruchirappalli, Tamil Nadu, India*
[5] *School of Chemical and Biotechnology, SASTRA Deemed University, Thanjavur, Tamil Nadu, India*

9.1 Introduction

Due to rapid growth in industrialization and urbanization, the generation of waste has reached an unprecedented level in recent times. Waste has become the most serious and sensitive issue across the globe. Largely, most of the waste is untreated or inadequately treated and often unutilized and is incinerated, mishandled, and ends in landfills. This imposes severe threats on local environments, causing water pollution and imbalance in ecosystems. Therefore, recognition and inclusion of waste management strategies in decision making are fundamental for achieving sustainable development goals (Kılıç et al. 2008; Abdel-Shafy and Mansour 2018; Ferronato and Torretta 2019; Sharma et al. 2021b; Siddiqua et al. 2022).

Today, water pollution originating from solid waste, industrial effluents, domestic sewage, and agricultural runoffs is showing its catastrophic and devastating impacts worldwide. As per a report by the World Health Organization (WHO), water pollution causes 1.4 million deaths annually. Currently, the scarcity of freshwater sources is at its peak and with the current pace of water pollution, the scenario will become worst in the near future. Consequently, water is at the central core of waste management strategies as the unavailability of freshwater for agricultural, industrial, and domestic practices immediately exacerbates health, economic, and environmental issues (Thyagaraju 2016; Wang and Yang 2016; Lin et al. 2022).

Contaminated water generally has a cocktail mixture of deadly contaminants like synthetic dyes, cyanides, heavy metals, organic solvents, and aromatic pharmaceutical wastes. However, heavy metal contamination is widely accepted and recognized as the most threatening form of water pollution owing to the long persistence and extremely

Remediation of Heavy Metals: Sustainable Technologies and Recent Advances, First Edition. Edited by Rangabhashiyam Selvasembian, Binota Thokchom, Pardeep Singh, Ali H. Jawad, and Willis Gwenzi.

toxic nature of heavy metals. Moreover, bioaccumulation and biomagnification of heavy metals in aquatic environments translate their hazardous impacts to long-distance habitats via food chains and water tables (Ali et al. 2019; Huang et al. 2021). Anthropogenic sources are chiefly liable for periling levels of heavy metal pollution. In lack of proper treatment technologies and sometimes due to economic restrictions, compound organizations and ventures directly release defiled water in groundwater and surface water sources. In addition, it incorporates mining exercises, assembling and completing ventures, coal burning, composts, and drug squandering and reusing units (Fashola et al. 2016; Gafur et al. 2018). Moreover, the releases of heavy metals from research activities and exploratory squanders, which are straightforwardly tossed in the water, also fall under anthropogenic sources, which decay the water quality. Additionally, contemporary agriculture practices with unoptimized uses of fertilizers, herbicides, and pesticides for higher crop production contaminate soils with heavy metals. From such contaminated soils, heavy metals enter surface water and groundwater through agricultural runoffs. Increasing heavy metal pollution has significantly destroyed the ecological balance of rivers, lakes, and wetlands and can be prevented by the fact that almost 83% population of freshwater species (including animals and plants) have disappeared globally due to direct or indirect heavy metal contamination (Tchounwou et al. 2012; Sankhla et al. 2016; Alengebawy et al. 2021).

Notably, being neurotoxin species, heavy metals hinder the communication mechanism between aquatic species, particularly fishes and cause an interruption in communication between fish species and the surrounding environment (Djedjibegovic et al. 2020; Raissy et al. 2022). In heavy metal-contaminated waters, aquatic species suffer from several deformities like morphological, histopathological, cellular, organismic, or parasitic aspects of abnormalities. In fact, some of these deformities are considered excellent biomarkers of heavy metal contamination exposure in recent times (Sun et al. 2009; Sfakianakis et al. 2015). Moreover, the accumulation of heavy metals in conventional crops like wheat grains, brown rice, paddy plants, etc., through contaminated irrigation water is much more prevalent. Being the food for millions, aquatic species, including both animals and plants and conventional crops, have become the medium for the migration of toxic metals from polluted water to human life. This forces regular and stringent assessment of the adverse effects of heavy metal pollution in water bodies. This assessment also helps in establishing policies, regulatory measures, environmental legislation, and waste management strategies for industries and municipalities (Rai et al. 2019; Uddin et al. 2021).

Under the current scenario, it becomes indispensable to develop efficient heavy metal decontamination technologies, particularly those that can serve the purpose at the production site of effluents or contamination. Globally, engineered adsorption technology has emerged as a viable alternative for heavy metal decontamination due to its simplicity, lower footprints, and high-quality treatment. Sorbents have a unique ability to isolate contaminants into smaller spaces from large volumes of effluents (Hegazi 2013; Ali et al. 2019; Gupta et al. 2021; Yadav et al. 2021; Ekanayake et al. 2022; Segneanu et al. 2022). Moreover, adsorption technology is best suited for circular economy models, which are based upon the 4R principle, i.e. reduce, reuse, recycle (upcycle), and recover. In practice, it implies reducing waste to a minimum level and translating the same into value-added and ecologically viable products. In this regard, lots of efforts have been put forward by the research community to transform industrial waste into efficient adsorbents for the removal of heavy metals from aquatic environments. There are many interesting research reports for investigators to identify novel approaches to design industrial waste-derived adsorbents for heavy metal removal. It is noteworthy to know that most industrial wastes are not currently used in their pristine form but modified to reinforce the physicochemical characteristics through micro-/nanostructuring, activation, carbonization, and grafting (Mo et al. 2018; Park et al. 2020; Soliman and Moustafa 2020; Agasti 2021; Bayuo et al. 2022). This chapter explores a new aspect of the transformation of industrial waste into efficient adsorbents that have tunable surface functionalities, hierarchical porous architecture, ultra-low density, excellent recyclability, viable end-of-life utility, and disposal. Several case studies are also discussed to understand the criticality of the problem and reveal the hazardous effects of heavy metal contamination in water sources. It also reviews challenges and advancements in the field and provides cutting-edge knowledge for future possibilities.

9.2 Industrial Wastes: Origin, Amount, and Harmful Effects

Waste from industrial activity is any substance that is made useless during a manufacturing process, such as that of different kinds of factories, mining activities, fabric mills, food production, commercial products, occupational chemicals, publishing, and printing. Industrial solid waste, which can be either solid, liquid, or gas held in containers, is divided into two categories: hazardous and nonhazardous waste. Engineering and other industrial operations may produce hazardous waste. Hazardous waste also includes some commercial products that are thrown by businesses or individuals, such as cleaning agents, dyes, and insecticides. Nonhazardous industrial wastes are those that are not municipal waste and do not comply with the EPA's definition of toxic waste. Industrial waste has been causing concern ever since the dawn of civilization. Industrial waste may be poisonous, combustible, corrosive, or reactive. This garbage can have harmful effects on both the environment and human health if it is not properly managed. Most chemical components found in industrial wastewater are distinct and easy to identify. Concentrated toxic waste and organic pollutants are the principal sources of water contamination in a select few subsectors. This can be attributed in large part to the production of food goods and the processing of industrial chemicals. The majority of large enterprises have industrial wastewater treatment facilities, but this is not the case for small-scale industries, which have a very thin profit margin and cannot afford to make large expenditures in pollution control technology (Buczyńska et al. 1999). The many categories of industrial waste are described in the following several ways:

Solid waste: Industrial solid waste is one of the most common types of garbage, while the word "industrial waste" encompasses a wide range of items. Manufacturing processes that generate electricity, inorganic and organic chemicals used in agriculture, iron and steel production, plastic production, resin production for water treatment, and many more processes can result in the production of solid waste.

Liquid waste: Both domestically and industrially, liquid waste is created. In most industrial operations, large amounts of water are required, which may come into touch with hazardous items like radioactive materials, filthy water, organic liquids, rinse water, waste detergents, and even rainwater.

Chemical waste: Chemical waste, which encompasses all varieties of flammable, corrosive, poisonous, and explosive waste, is another byproduct of most industrial processes. Since chemical waste frequently contains dangerous chemical residue that is a threat to human, animal, and plant life, disposal of this material should be handled by experts. The most frequent places where chemical waste is generated are factories, processing centers, warehouses, and plants. Dangerous chemicals make up the majority of chemical waste.

Toxic industrial waste: In addition to industrial trash, toxic or hazardous waste is also present. This kind of industrial waste can contaminate rivers, lakes, and other bodies of water, which could harm people, animals, and the environment if it is not handled appropriately. Usually, in enterprises, hospitals, and manufacturing plants, this kind of industrial waste is produced as a byproduct of the manufacture of other materials (Tables 9.1 and 9.2).

Table 9.1 Examples of waste generated by industries (World Resources Institute 1994).

S. No.	Type of industries	Waste types
1.	Chemical manufacture	Acids and bases
		Spent solvents
		Reactive wastes
		Wastewater containing
		Organic constituents
2.	Printing industry	Heavy metal solution
		Waste inks
		Solvents
		Ink sludge containing
		Heavy metals
3.	Petroleum-refining industry	Wastewater containing
		Benzene and other
		Hydrocarbon
		Sludge from the refining process
		Solid waste
4.	Leather products manufacturing	Toluene and Benzene
		Salts, raw trimmings
		Hair waste, flashings
		Splitting wastes
		Chrome shavings
		Buffing dust
		Crust trimmings and
		Finished trimmings

Table 9.1 (Continued)

S. No.	Type of industries	Waste types
5.	Paper industry	Paint waste containing
		Heavy metals
		Ignitable solvents
		Wood-processing residuals
		Boiler and furnace ash
6.	Construction industry	Ignitable paint waste
		Spent solvents
		Strong acids and bases
		Concrete waste
		Wood waste
7.	Metal manufacturing	Sludge containing
		Heavy metals
		Cyanide waste
		Paint waste
		Iron ore fines

Source: Environmental protection agency, solving the hazardous waste problem: EPA's RCRA program (Washington, DC: EPA, November 1986), 8.

Table 9.2 Environmental and health risks due to waste.

Waste stream	Pollutants and hazards	Environmental and health risks
MSW open dumping	Leachates with high concentrations of heavy metals, BOD, COD, SO_4^{2-}, and NH_3	The leachate generated is released into the soil, polluting groundwater mainly used for drinking and household purposes. The risks concern the health of the population through direct and indirect (agriculture) intake
	Anaerobic digestion of organic fractions with the generation of landfill gases, mainly composed of methane, disease vectors living in the areas	The generation of methane and other GHGs increases global warming, the risk of local fires, and the pollution of the atmosphere surrounding the final disposal sites. The breeding of animals around the disposal sites and the presence of rodents and insects increase the risks of diseases transferring to the population through bites and direct contact with the animals
		The uncontrolled disposal causes the release of waste fractions, mainly plastics, into water bodies; contamination of rivers, lakes, and then the oceans; and then the oceans and the seas, causing the phenomena of the marine littering

(*Continued*)

Table 9.2 (Continued)

Waste stream	Pollutants and hazards	Environmental and health risks
MSW open burning	Generation of PCDD/F and cancer genic compounds, PM, BC, CO, CO_2, NO, and other GHG and hazardous compounds	The emissions due to uncontrolled waste fires produce significant amounts of contaminants that affect the health of the population. Respiratory illnesses, especially in children, are common in areas with open burning practices. The generation of BC, CO, CO_2, and other GHG, affects the GWP, more than the anaerobic degradation of organic waste
HW	Open dumping of sharp and infectious waste, Burned HW generates PCDD/F and other hazardous compounds	The presence of sharps and infectious waste in open dump sites increases risks to waste pickers that operate in the area. Indeed, recyclable materials are scavenged by informal recyclers, that are not aware of the issues due to HW. Moreover, these fractions can be targets for animal disease vectors. Finally, the open dumping of HW creates bacterial resistance that affects the performance of antibiotics for human use. Burned HW is a source of PCDD/F that directly affect the health of the population living near sites, increasing the risk of cancer and respiratory diseases
WEEE and used batteries	The open dumping of the waste generates leachates with high heavy metals concentrations. The open burning generates hazardous compounds like PCDD/F, BC and PM	Open WREE dumping generates leachates rich in heavy metals. Waste picking is also done on these fractions to collect precious metals. These practices affect the health of the pickers that operate in the dumping sites due to the presence of Hg, Pb, Cd, and Mn, all hazardous metals. Picking of WEEE leads to waste burning for the recovery of the metals. Uncontrolled WEEE burning affects the quality of the air and the atmosphere due to the generation of high amounts of PM, BC, and PCDD/F, affecting both the health of the populations surrounding the site and increasing the GWP
C&D waste	Landslides due to uncontrolled waste dumping, presence of hazardous materials within the waste, such as asbestos, lead, mercury and sharp waste	The main issue is the risk of landslides that can affect populated areas. The huge amounts of waste produced by C&D activities reduce the useful life of final disposal sites and their density, increasing the risk of land collapses. The presence of hazardous materials can be a source of pollution due to the leachates generated at the final disposal sites or directly for the waste pickers operating collecting recyclable materials. Pb and Hg waste can affect the health of the population due to respiratory, skin and other illnesses

Table 9.2 (Continued)

Waste stream	Pollutants and hazards	Environmental and health risks
Waste tires	The open dumping causes the presence of mosquitos and the risk of fire injections Open burning generates contaminants in the atmosphere, such as BC and So_x	Open dumping of waste tires is an area of mosquito growth, especially in tropical areas, where dengue, malaria, and yellow fever, among other diseases, are common. The presence of waste tires in open dump sites increases the risk of contraction of these illnesses. Moreover, the presence of this highly combustible waste fraction, can be a cause of fires Open burning of waste tires induces high generation of PM, BC and SO, increasing the GWP and the rain phenomenon due to the presence of sulfates in the atmosphere which generate H_2S in contact with water, increasing environmental pollution. Moreover, the generation of PM, containing heavy metals increases health issues in the populations that live near the areas, also affecting the air indoor
Industrial waste	Generation of hazardous leachates mainly composed of heavy metals	The presence of heavy metals affects the health of citizens, especially the children, which intake can be less than an adult in terms of concentrations assumed per day, boosting the health effects. The presence of heavy metals is persistent and affects the soil and groundwater quality, with possible intake by direct ingestion (e.g. food and water)

Source: Adapted from Ferronato and Torretta (2019).

9.3 Sources of Heavy Metal Contamination in Water Sources

The persistence of heavy metal contamination in water resources is fatal to all spheres of life and these include both anthropogenic and natural sources. However, anthropogenic sources are significantly liable for grave outcomes, taking everything into account. Nevertheless, the release of heavy metals from natural sources cannot be overlooked and is briefly discussed in the following section.

9.3.1 Natural Sources

Many phenomena like natural rock weathering, sea-salt sprays, forest fires, volcanic eruptions, etc., are some well-studied and recognized sources of heavy metal contamination in water. Besides, wind-borne soil particles and biogenic sources are also known to cause the emission of heavy metals. Moreover, the natural occurrence of heavy metals in the form of ores and minerals, like sulfides, oxides, hydroxides, silicates, phosphates, etc., also contributes to heavy metal contamination. However, they are much more rigid as far as their physical and chemical states are considered and are often typically less soluble in aquatic systems. In fact, they naturally occur in such states that are not promptly available for uptake by plants and aquatic species. The inherent affinity of naturally occurring heavy metals with soil is impregnable and hence they are relatively less toxic compared to anthropogenic counterparts (Starr et al. 2003; Zhou et al. 2020; Ahmad et al. 2021).

9.3.2 Anthropogenic Sources

Being highly soluble and reactive in nature, heavy metals originating from anthropogenic activities have outrageous bioavailability. Unsatisfactory treated effluents from manufacturing, e-waste recycling, printing, pharmaceutical, steel, textiles, mining, and electroplating industries are the major anthropogenic entry routes of heavy metals into water sources. Particularly, mining activities alter the geochemical cycle of heavy metals and open new pathways for their release in water sources. These practices generate effluents that contain huge quantities of heavy metals and are frequently discharged into natural water resources with adequate treatment (Morton-Bermea et al. 2009; Paul 2017; Ali et al. 2019). Owing to the nonbiodegradable and neurotoxin nature of heavy metals, they impose serious ecological threats.

Both organic and inorganic forms of Arsenic (As) are known to exist in an aquatic environment and are globally accepted as deadly poisons. Chemically, pentavalent arsenate ions [As(V)] and trivalent arsenite ions [As(III)] predominantly exist in surface and groundwater, respectively.

Groundwater represents the world's largest freshwater sources and hence its contamination with heavy metals is a major concern. Particularly, the coexistence of As and mercury in groundwater is extremely poisonous. Moreover, the less toxic inorganic form of mercury is readily transformed into highly toxic methyl mercury in aquatic environments. This further threatens ecological viability. Cadmium contamination in water arises from pigment, plastic and battery manufacturing, and automobile industries as well as from agrarian sources, for example, pesticides, manures, and insect poisons. Lead pollution in groundwater has additionally increased at an alarming pace, especially from manufacturing and recycling plants of lead-acid batteries. It is noteworthy that exposure to lead, even in trace amounts, is more fatal relative to other heavy metals. Tackling chromium (Cr) pollution in water is another daunting task as all its normally existing forms have high toxicity. The sources of Cr contamination include chrome plating, steel production, anti-corrosion coatings, and paints (Wang et al. 2004; Yadanaparthi et al. 2009; Prasad et al. 2021). In a nutshell, heavy metal contamination from anthropogenic sources has become a major global threat. In fact, it is marked as an environmental crime by experts as it antagonistically degrades biodiversity, disturbs the ecological balance, and causes irreversible climate changes. Heavy metal pollution in water has grave consequences as it adversely damages the environment and human health. Some of their ill impacts are summarized in the subsequent section.

9.3.2.1 Environmental and Health Hazards of Heavy Metal Contamination

Bioaccumulation, biomagnification, persistency, toxicity, and nonbiodegradability are the primary features that constitute the toxic nature of heavy metal contamination in aquatic environments. Overconcentration of As in drinking water is identified to cause cardiovascular, and peripheral vascular disease, developmental anomalies, behavioral disorders, portal fibrosis, hematologic disorders, and carcinoma. Moreover, As inhibits the production of adenosine triphosphate (ATP) during respiration by coagulating the proteins and interrupting the functionality of coenzymes (Mohammed Abdul et al. 2015; Fatoki and Badmus 2022).

The well-recognized pathway of As entry into human life is the food chain as it is readily bioaccumulated in traditional crops like rice from contaminated irrigation water. The WHO recognizes lead as the most toxic among heavy metals, particularly for young children. In young children, lead poisoning causes profound and permanent adverse health impacts and inhibits brain development and damages the nervous system. Furthermore, kidney damage, high blood pressures, premature birth, and miscarriage are other ill effects of lead consumption over prescribed limits. The aquatic species, particularly seafood, are found to contain significant amounts of lead through which it become an integral part of the human diet. Its overconsumption can cause death and permanent damage to the central nervous system (Vahter 2008; Bayuo et al. 2022). Although Nickel (Ni) is an essential trace element, however, its overconsumption has numerous side effects including allergy, baldness, lung fibrosis, and nasal cancer. In animals like dogs and rats, high doses of Ni cause serious weight loss. Being an immunotoxic and carcinogen agent, Cr can cause severe health issues including contact dermatitis, asthma, and respiratory tract cancer, depending upon its dose and exposure time. Water pollution from Ni and Cr is due to finishing, the use of liquid and solid fuels, and municipal and industrial waste. Importantly, the toxicity of Ni and Cr enhanced manifolds in the vicinity of other metals like copper (Cu), zinc (Zn), cobalt (Co), etc. (Yeganeh et al. 2013; Ghazzal et al. 2022).

Mercury has no physiological or biochemical role in humans. Its acute exposure has lethal health impacts. Moreover, its elemental as well as organic forms (methylated mercury) have been found to exhibit cytotoxicity and cause abortion, and complete blocking of the brain, and gastrointestinal and neurological disorders (Rice et al. 2014; Ha et al. 2017). The various ill impacts of heavy metals on human health and aquatic species are summarized in Table 9.3.

Undoubtedly, some heavy metals are essential for life as they act as micronutrients and have critical biological functions to perform in living organisms. However, their uptake above prescribed limits is always hazardous. After the release of heavy metals in aquatic environments, their potency to survive under diverse climate conditions makes them extremely chronic for biospheres. Particularly, in acidic conditions, ionic forms of heavy metals are predominant and this enhances the bioavailability of heavy metals manifolds and so does heavy metal toxicity to microorganisms, plants, and soils (10.4067/S0718-95162010000100005). Furthermore, contaminated soil limits plant biodiversity and results in ecological and nutritional problems. The magnitude of these problems varies with the type of heavy metal, soil composition, pH, and plant species. In contaminated zones, plant tissues absorb large amounts of heavy metals that kill plants or inhibit nutrient uptake, homoeostasis, and plant growth by causing oxidative stress, interruption in physiological, biochemical, and enzyme functions, destruction of cell organelles and biomass, delayed and imperfect germination, permanent genotoxicity, and inhibition of photosynthesis and respiration (Pichhode and Nikhil 2016;

Table 9.3 Hazardous effects of heavy metal toxicity on human health and aquatic species.

			Toxic effects of heavy metal contamination on			
S. No.	**Heavy metal**	**Source**	**Source human**	**Plants**	**Microorganisms**	**References**
1.	Antimony (Sb)	Natural phenomena like volcanic eruption, soil erosion, etc., coal-based thermal power plants. Mining. Smelting practices	Cancer cardiovascular diseases, conjunctivitis, dermatitis, liver diseases, nasal ulceration, respiratory diseases	Decreases synthesis of some metabolites, growth inhibition, inhibits chlorophyll synthesis	Inhibit enzyme activities, reduced growth rate	Boreiko and Rossman (2020), Bolan et al. (2022)
2.	Arsenic	Atmospheric deposition, mining, pesticides, rock sedimentation, smelting	Brain damage, cardiovascular and respiratory disorder, conjunctivitis, dermatitis, skin cancer	Damage cell membrane, inhibition of growth, inhibits roots extension and proliferation, interferes with critical metabolic processes, loss of fertility, yield and fruit production, physiological disorders, oxidative stress	Deactivation of enzymes	Bissen and Frimmel (2003), Finnegan and Chen (2012), Ali et al. (2019)
3.	Beryllium (Be)	Coal and oil combustion, volcanic dust	Allergic reactions, berylliosis, cancer, heart diseases, lung diseases	Inhibits seed germination	Chromosomal aberration, mutation	Gordon (2003), Blais et al. (2008)
4.	Chromium (Cr)	Dyeing. Electroplating, paints production, steel fabrication, tanning, textile	Bronchopneumonia. Chronic bronchitis, diarrhea, emphysema, headache, irritation of the skin, itching of respiratory tract, liver diseases, lung cancer, nausea, renal failure, reproductive toxicity, vomiting	Chlorosis, delayed. Senescence, wilting, biochemical oxidative stress lesions, reduced biosynthesis germination, stunted growth, oxidative stress	Elongation of lag phase, growth inhibition, inhibition of oxygen uptake	Barakat (2011), Mohanty et al. (2012)

5.	Cadmium (Cd)	Fertilizer, mining, pesticide, plastic, refining, welding	Bone disease, coughing, emphysema, headache, hypertension, itaiitak, kidney diseases, lung and prostate cancer, lymphocytosis, microcytic hypochromic anemia, testicular atrophy, vomiting	Chlorosis, decrease in plant nutrient content, growth inhibition, reduced seed germination	Damage nucleic acid, denature protein, inhibit cell division and transcription, inhibits carbon and nitrogen mineralization	Chibuike and Obiora (2014), Fashola et al. (2016), Singh et al. (2016)
6.	Mercury (Hg)	Batteries, coal combustion, geothermal activities. Mining. paint industries, paper industry, volcanic eruption, weathering of rocks	Ataxia, attention deficit, blindness, deafness, decrease rate of fertility, dementia, dizziness, dysphasia, gastrointestinal irritation, gingivitis, kidney problem, loss of memory, pulmonary edema, reduced immunity, sclerosis	Affects antioxidative system, affects photosynthesis. Enhance lipid peroxidation, induced genotoxic effect, inhibit plant growth, yield, nutrient uptake and homeostasis, oxidative stress	Decrease population size, denature protein, disrupt cell membrane, inhibits enzyme function	Wang et al. (2012), Ali et al. (2013), Yeganeh et al. (2013)
7.	Lead	Coal combustion. Electroplating, manufacturing of batteries, mining, paint, pigments	Anorexia. Chronic nephropathy, damage to neurons, high blood pressure, hyperactivity, insomnia, learning deficits, reduced fertility, renal system damage risk factor for Alzheimer's disease, shortened attention span	Affects photosynthesis and growth, chlorosis, inhibit enzyme activities and seed germination, oxidative stress	Denatures nucleic acid and protein, inhibits enzymes activities and transcription	Nagajyoti et al. (2010), Wuana and Okieimen (2011)
8.	Nickel (Ni)	Electroplating, nonferrous metal, paint, porcelain enameling	Cardiovascular diseases, chest pain, dermatitis, dizziness, dry cough and shortness of breath, headache, kidney diseases, lung and nasal cancer, nausea	Decrease chlorophyll content, inhibit enzyme activities and growth, reduced nutrient uptake	Disrupt cell membrane, inhibit enzyme activities, oxidative stress	Malik (2004), Chibuike and Obiora (2014), Fashola et al. (2016)
9.	Selenium (Se)	Coal combustion, mining	Dysfunction of the endocrine system, gastrointestinal disturbances, impairment of natural killer cells activity, liver damage	Alteration of protein properties, reduction of plant biomass	Inhibits growth rate	Germ et al. (2007), Dixit et al. (2015)

(*Continued*)

Table 9.3 (Continued)

S. No.	Heavy metal	Source	Toxic effects of heavy metal contamination on			References
			Source human	**Plants**	**Microorganisms**	
10.	Thallium	Cement production, combustion of fossil fuels, metal smelting, oil refining	Alopecia, ataxia, burning feet syndrome, coma, convulsions, delirium, fatigue, gastroenteritis, hair fall, hallucinations, headache, hypotension, insomnia, nausea, tachycardia, vomiting	Inhibits enzyme activities, reduced growth	Damages DNA, inhibits enzyme activities growth	Babula et al. (2008), Blais et al. (2008)
11.	Zinc (Zn)	Brass manufacturing, mining, oil refinery, plumbing	Ataxia, depression, gastrointestinal irritation, hematuria, icterus, impotence, kidney and liver failure, lethargy, macular degeneration, metal fume fever, prostate cancer, seizures, vomiting	Affects photosynthesis, inhibits growth rate, germination rate and plant biomass, reduced chlorophyll content	Death, decrease in biomass, inhibits growth	Chibuike and Obiora (2014), Gumpu et al. (2015)

Singh et al. 2016). Overall, heavy metal pollution disturbs the naturally established equilibrium between different spheres of life, such as the atmosphere, water, air, and living creatures. The changes that heavy metal pollution effectuates are generally irreversible in nature, and therefore, this issue needs to be addressed immediately. Integrated strategies that involve the imposition of stringent legislation on the discharge of industrial effluents and the development of efficient heavy metal decontamination technologies need to be coined, evolved, and realized based on the nature of effluents and the performance of treatment and control technologies. The existing regulatory measures for heavy metal discharge and permissible limits in water sources are frequently reviewed and modified by national and international agencies and local governments. The permissible limits of different heavy metal ions in water sources, viz. surface water, drinking water, and marine water, are listed in the next section.

9.3.2.2 Regulatory Legislations and Permissible Limits of Heavy Metals in Water

Present-day review studies show that heavy metal contamination at high as well as low concentrations begets trouble. International and national agencies like the WHO, European Union (EU), and Bureau of Indian Standards (BIS) have specified permissible limits for various heavy metals in water sources including drinking and irrigation water. The permissible limits of heavy metals in water are presented in Table 9.4. The WHO explicitly recognized that freshwater resources are essential for the realization of fundamental human rights. The purpose of regulatory legislation is to force individuals, organizations, industrial units, and other stakeholders that discharge effluents shall treat their waste in accordance with the requirements of established legislation.

Table 9.4 Permissible limits of the discharge of some heavy metals in water.

	Permissible limits (mg l^{-1})				
Heavy metal	**Drinking water**	**Inland surface water**	**Public sewers**	**Irrigation water**	**Marine coastal water**
Mercury (Hg)	0.001	0.01	0.01	0.01	0.01
Arsenic (As)	0.05	0.2	0.2	0.2	0.2
Lead (Pb)	0.1	0.1	1.0	—	2.0
Cadmium (Cd)	0.01	2.0	1.0	—	2.0
Zinc (Zn)	5.0	5.0	15	—	15
Nickel (Ni)	0.1	3.0	3.0	—	5.0
Copper (Cu)	1.3	3.0	3.0	—	3.0
Selenium (Se)	0.01	0.05	0.05	—	0.05
Chromium (Cr)	0.05	0.1	2	—	1

Source: Sharma et al. (2021a).

9.3.2.3 Remediation Technologies for Water Pollution

Not far from today, access to clean water will be a daunting task for the entire world. The deployment of toxic metals in water sources through industrial effluents has depleted both the quality and quantity of potable water. Providing foolproof solutions to such scenarios is a tedious job. The public authority likewise propounded numerous establishments for attention to individuals to clean the water. Importantly, they ought to make a suitable move on the synthetic ventures that straightforwardly release their loss into water assets. The industrial waste ought to be dealt with properly prior to release into the water. In developing economies, a substandard administrative and technological framework is additionally liable for water defilement. Conventional bioremediation processes are generally used to eliminate metal contaminants from water on account of minimal expense and eco-accommodating. Many survey articles have examined the potential of different remediation techniques like oxidation/precipitation, coagulation, electrolysis, membrane filtration, and adsorption technique for the elimination of toxic metals from tainted water (Mukherjee et al. 2015; Sikosana et al. 2019; Obotey Ezugbe and Rathilal 2020; Rashid et al. 2021; Ighalo et al. 2022). The above-mentioned heavy metal removal techniques are discussed in the subsequent section.

a) *Oxidation/precipitation*

This is, perhaps, the most facile and widely used technique for the removal of water contaminants. For the last few years, the oxidation strategy has been intended to eliminate heavy metals from tainted water. This methodology was presented during the 1980s for a water disinfecting process using very strong oxidizing agents like hydroxyl and sulfate ions. Presently, oxidizing/precipitation strategy is generally utilized for treating industrial effluents. In this technique, the reactive oxygen-based oxidizing reagents such as ozone, hydrogen peroxide, and dioxygen, energy sources like ultraviolet radiations and photocatalysts such as TiO_2 are delivered for oxidizing the water contaminants. In provincial and semi-metropolitan regions, homegrown water providers are taking advantage of chlorine as an oxidizing agent for water cleaning. Moreover, this strategy is utilized for the inactivation of microorganisms and microbe pointers. Additionally, the oxidation/precipitation method can also be utilized for removing the natural and inorganic compounds from wastewater. The role of various types of oxidizing agents can be assessed for the disposal of tainted water deeply probed by Deng et al. A compound oxidation strategy is investigated by Hodaifa et al. for the treatment wastewater streams of olive oil factory (Burke et al. 2013; Deng and Zhao 2015; Garrido-Cardenas et al. 2019; Hodaifa et al. 2019).

b) *Ion exchange*

The ion-exchange strategy is generally employed for making wastewater consumable for certain practical applications. In 1850, it emerged as a result of a trial made by Thomas. Thomas passed the aqueous solution of ammonium sulfate through the dirt. Generally, the particle trade technique is shown to eliminate ionic debasements including harmful metals. Owing to this, the ion exchange strategy is cast off in turn around the assimilation (RO) strategy. Particle trade is a kind of synthetic interaction which assists exchange between the ionic pollutions from water with comparative charged counterparts (ionic species). It is one of the best-suited techniques for dealkalization, deionization, and in general, sanitization of wastewater. Additionally, the ion exchange technique is utilized for eliminating the hardness of water as magnesium (Mg) and calcium (Ca) ions are readily replicable with hydronium ions in ion exchange columns while the anionic counterparts are exchanged with hydroxide ions. Recently, new novel engineered ion-exchange resins have been proposed that claim to be more efficient in serving the purpose. Zeolites are among the marathon materials of such resins (Shi et al. 2018; Bashir et al. 2019).

c) *Electrokinetic process*
This technique is best known for the sequestration of heavy metals from defiled soils and water. The electrokinetic remediation (ER) technique is comprised of three-stage processes, namely electroosmosis (EO), electrorelocation (EM), and electrophoresis (EP). Herein, a low-intensity current acts as a cleansing agent where heavy metal ions are either electrolyzed or electrodeposited via EO, EM, and EP. It can be further finetuned through in situ and ex situ mechanisms. In the EO method, the liquid stream flows from the anode to the cathode. The ER process is known as the electromovement process, where at the cathode, the decrease happens due to the reduction of cations, and at the anode, the oxidation process occurs inferable from loss of electrons by anions in the presence of an external electric field. Along with heavy metals, other organic pollutants like pesticides, pharmaceutical solvents, biphenyls, and polychlorinates can also be targeted in electrochemical strategies (Cherifi et al. 2011; Fuwad et al. 2019).

d) *Membrane filtration/reverse osmosis*
Recently, membrane filtration such as nanofiltration (NF) and reverse osmosis (RO) have emerged as the most efficient techniques for the elimination of heavy metals from wastewater streams. Besides, heavy metals have been utilized for various sorts of contaminations like natural, suspended solids and organic and inorganic pollutants. This technique primarily depends upon the size exclusion principle. For instance, ultrafiltration (UF) presents as a penetrable layer that has a pore size of 5–20 nm to macromolecules and suspended solids from tainted water, and it just permits the isolation of particles that have 1000–100,000 Da atomic weight. It is generally permeable for low sub-atomic weight solutes and nano-size particles. Particularly, for heavy metals, NF and RO are more productive and have gained immense attention worldwide. Strictly speaking, RO is the most effective strategy for heavy metal decontamination among all types of membrane filtrations. However, its commercialization at every level is limited due to high setup and operational costs, poor durability of membranes, and rapid membrane fouling with acidic feedstocks (Homaeigohar and Elbahri 2017; Siddique et al. 2020; Zunita 2021).

e) *Bioremediation*
The bioremediation process is also known as phytoremediation. It involves utilizing bio-origin and living organisms like microbes and bacteria to decontaminate heavy metals from polluted water. It includes processes such as phytoextraction, phytomining, phytostabilization, rhizofiltration, and phytovolatilization. Interestingly, in a few research reports *Pteris vittata* and Vetiver grass have been shown to be promising in the remediation of heavy metals through the phytoremediation process. The phytoremediation interaction is truly usable with minimal expense and has ecofriendly and ecologically amicable nature (Kapahi and Sachdeva 2019; Sharma et al. 2022).

f) *Adsorption method*
Engineered adsorption technology has been widely recognized as a promising alternative to the above-discussed technologies for heavy metal decontamination due to its simplicity, low cost, lower footprints, high-quality treatment, and availability of various types of adsorbents, such as macro-/micro- and nano-adsorbents. For the last many years, carbon materials, specifically activated carbon (AC), have been the key players in commercially available water filters for heavy metal decontamination owing to their low cost, high surface area, green nature, and environmental benignity. In general, adsorbents are utilized either in powdered form or in monolithic form. However, both forms have their own pros and cons.

Powder materials have been used from antiquated times for the treatment of industrial water. As zeolite, iron center, resin, and AC are exhibited to adsorb the water's foreign substances to further develop the properties of the adsorbents, the examination is continuing and setting up the most reasonable adsorbent which can dispense with the toxic metals from water to being recyclable after use. Carbon materials like graphene, carbon nanotubes, AC, and carbon fiber are additionally used materials to take out the various kinds of poisonous metals from water as a result of permeable construction, enormous surface region, and accessibility of additional practical gatherings, which assists with drawing in the harmful metal on their site, and thermally steady (Mittal et al. 2021; Ighalo et al. 2022). Then again, the nanomaterials with their novel properties are likewise getting more consideration for toxic metal expulsion from water. Even though powder materials have excellent sorption properties to wipe out the poisonous metals from water, notwithstanding these benefits, scientists are dealing with a few challenges, particularly with the isolation of powdered materials after the adsorption process. There is always a chance of secondary contamination as powdered adsorbents tend to remain in the solution phase. To address this, magnetic functionalities are imparted into the adsorbent material. However, sometimes, magnetic nanoparticles leach out into the aquatic systems and cause secondary contamination (Makar et al. 1975; De Gisi et al. 2016; Schieweck 2020; Kumar et al. 2021). In keeping perspective on this, the powder materials are supplanted by the monolithic and self-assembled materials to avoid secondary contamination in water sources.

Carbon materials owing to the hierarchical porous architecture, desirable surface functionalities, and large surface areas have been exhibited as a layout material for sequestration of toxic metals from industrial wastewater. Furthermore, the designing of monolithic structures with interconnected pores using carbon nanostructures is a facile and rapid process. Based on literature investigations, Burke et al. and Lee et al. have synthesized 3D carbon foams (CFs) with mesoporous structures. The as-prepared foams are shown to have good adsorption capacity for Cu and Pb. Parallelly, 3D graphene oxide foam, graphene oxide/$CuFe_2O_4$, and 3D graphene oxide/Fe_3O_4 foam have been presented for toxic metals, As, and Cr elimination, respectively. Agrawal et al. have presented some 3D-assembled adsorbents for the removal of As from standard aqueous solutions such as resin-based CF, multiwall carbon nanotube-based carbon froth MWCNTs-CF, graphite intercalation compounds functionalized carbon foam (GICCF) and a mixture of zirconia oxide decorated rGO functionalized carbon foam (ZRCF) for As elimination from the contaminated water (Burke et al. 2013; Agrawal et al. 2016, 2018, 2021). The 3D carbon porous material is creating more interest in water disinfecting on the grounds that there is a compelling reason needed to partition strategies to isolate the material after the adsorption process. Also, they can be reused, and have improved effectiveness, and usefulness toward weighty metal adsorption.

9.4 Sequestration of Heavy Metals Using Industrial Waste-Derived Adsorbents

As discussed earlier, adsorption is a universal method for wastewater treatment and recycling. The demand for sustainable adsorbents has led to the rapid growth of research interests in this field. In the current era, circular economy models are obligatory for environmental sustainability. These models are mainly based upon the 4R principle, i.e. reduce, reuse, recycle (upcycle), and recover (Kazerooni Sadi et al. 2012; Stefanakis et al. 2021). In practice, it implies creating climate-resilient, enduring, self-reliant ecosystems that close the loop and reduce waste to a minimum level and translate the same into value-added and ecologically viable products. Principally, adsorption technology is best suited for such models. Particularly, with the advancement of nanoscience and nanotechnology, it is now viable to design sorbent materials from waste materials like agricultural, domestic, and industrial wastes (Puyol et al. 2017; Spiridon 2020; Solangi et al. 2021). On this ground, numerous research reports have been published showing the potential of industrial waste-derived sorbents in the decontamination of heavy metals from aquatic environments. It represents an innovative approach providing dual advantages of waste upcycling and removal of toxic metals thereafter. For example, lignin – a natural polymer (found in the cell mass of woody trees) – is the second most abundant natural polymer after cellulose and is extensively studied and utilized for developing efficient adsorbents. Brazil's sugarcane industry is the largest source of lignin in the world, followed by southeast Asia and Malaysia's oil palm refinery industries (Spiridon 2020). Luo et al. synthesized the modified biochar (MBC) by the waste raw corncobs that contain lignin in significant amounts. The cyanide group was implanted in the lignin of MBC350 successfully under the charring temperature conditions of 350 °C. As per the experimental data, MBC350 have exhibited excellent adsorption capacity (>90%) and acid resistance for Cd(II). The Cd(II) removal mechanism on the surface of the synthesized adsorbent follows ion exchange and adsorption-complexation (Vardhan et al. 2019). In another study, Mohan et al. extracted lignin from black liquor waste. The extracted black liquor lignin has a higher sorption capacity than most adsorbents/biosorbents/carbonaceous adsorbents, which was later employed for the elimination of Cu and Cd ions from wastewater. The maximum adsorption capacities at 25 °C were found to be 1.37 and 1.22 $mmol\,g^{-1}$ for Cu(II) and Cd(II), respectively (Lim et al. 2021). Zou and Han (2001) synthesized the ACs from Chinese hydrolysis lignin by using chemical activation. The report reveals the role carbonization step and activation temperature on the physicochemical characterization of the prepared adsorbent. The chemical activation introduced new adsorption sites on the adsorbent surface and resulted in a large BET surface area of 2263–2753 $m^2\,g^{-1}$. The activation temperature range was varied from 750 °C to 850 °C. Moreover, it is shown that the carbonization step produces a material with low volatility, and this can be attributed to the fact that new cross-linkings are developed within the lignin. Moreover, it firmly favors the formation of

Table 9.5 Adsorption capacities of some industrial (food, cement, sugarcane industries, etc.) waste-derived sorbents.

S. No.	Adsorbent	Heavy metal ions	Adsorbent capacity ($mg\,g^{-1}$)	Reference
1	Orange peel	Cd(II)	128	Akinhanmi et al. (2020)
2	Maize cope and husk	Pb(II) Cd(II) Zn(II)	456 493.7 495.9	Igwe and Abia (2003)
3	Activated sugarcane bagasse with Zn chloride at an optimum pH of 8.58	Cr(VI)	>87% ($C_o = 77.5\,mg\,l^{-1}$)	Cronje et al. (2011)
4	Rice husk	Cd(II)	67.917% ($C_o = 0.48\,mg\,l^{-1}$)	Hegazi (2013)
5	Fly ash	Cd(II)	73.542% ($C_o = 0.48\,mg\,l^{-1}$)	Hegazi (2013)
6	Activated carbon from coconut shell	Cd(II)	98%	Deng et al. (2021)
7	Apple pomace activated with Xanthate moiety	Cd(II)	112.35	Chand et al. (2014)
8	Raw almond shell Activated almond shell activated with Ortho-phosphoric acid and H_3PO_4, respectively	Hg(II)	3.7 37.1	Taha et al. (2018)
9	Activated onion skin with thioglycolic acid	Pb(II)	6.173	Olasehinde et al. (2018)
10	Maize husk modified with tartaric acid/phenol/ methonic acid	Cu	35.714	Duru et al. (2019)

numerous aromatics in the cross-linking reaction that was carried out with activating agent KOH (Luo et al. 2018). The heavy metal adsorption capacities of various industrial waste-derived sorbents are given in Table 9.5.

Apart from agro-industries, coal combustion-based thermal plants produce a large amount of fly ash which is a rich source of AC. It primarily consists of many inorganic substances like silica, Mg oxide, alumina, ferric oxide, and carbon. Daci et al. showed the role of Kosova lignite fly as an adsorbent for the decontamination of heavy metals from industrial wastewater streams, deliberately prepared standard solutions, and natural water bodies. Their accomplished results were based on the contact time of adsorbents with a contaminated solution and the removal efficiency of Kosova lignite fly ash adsorbent material. As per their observations, 10 minutes of contact were appropriate for the removal of ions with efficiency up to 99.89 wt.% from standard solutions and natural water body samples and 60 minutes' contact time for the maximum removal of ions up to 98.12% from an industrial wastewater sample.

The percentage removal efficiency for Pb and Cd from batteries' industrial wastewater using fly ash as adsorbent is 90.75% and 86.66%, while for fertilizers, industrial wastewater is 65% and 50%, respectively (Daci et al. 2011). Attari et al. contemplated a cost-effective removal of hazardous metal mercury by the approach based on the synthesis of the zeolite Linde type A (LTA) from coal fly ash (CFA-ZA). In all examined adsorbent/solution ratios, the synthesized zeolite has strong efficiency toward Hg(II). The Hg(II) ion adsorption capacity of CFA-ZA FOR Hg(II) was observed as 94% with $10\,mg\,l^{-1}$ initial concentration. The as-revealed adsorption capacity of CFA-ZA is quite comparable with some ACs (Ma et al. 2018). Ma et al. have investigated the removal of various heavy metals, viz. Cu(II), Ni(I3+I), Co(II), and Zn(II) ions from the standard aqueous solutions by using Ca silicate powder (CSP). This waste material is obtained from alumina that is a byproduct of coal ash. The maximum adsorption capacities for Cu(II), Ni(II), Zn(II) and Co(II) were found to be 680.93, 420.17, 251.89, and $235.29\,mg\,g^{-1}$, respectively. The removal efficiency for Ni(II) was found to be 100% when the feed solution of Ni(II) has a concentration of $100\,mg\,l^{-1}$, and it suggests that CSP is eminently effective even with low adsorbent doses (Ma et al. 2018).

Steel-processing units are also major sources of coke and coal-based wastes. Blast furnace slag from steel and sludge from the electroplating industries are effective adsorbents for heavy metals. Owing to the presence of alumina and iron oxides, blast furnace slag has a higher affinity for heavy metal uptake relative to sludge. However, the efficiency of such adsorbents governed by many factors including nature of impurity, dose of adsorbent, temperature, and pH conditions. In general, the optimum pH for the removal of Hg(II) and Pb(II) ions by pristine biomass is found to be 5.8 and 3.5, respectively (Kılıç et al. 2008). A significant enhancement in adsorption capacity was observed for activated biomass sludge. The activation process involves dewatering of raw biomass followed by drying at room temperature, conditioning in the 0.5 N NaOH solution for 15 minutes and then autoclaving for 30 minutes (Naiya et al. 2009). Oxygen furnace waste from steel plants has been used as an economical adsorbent for the capturing of Pb(II) ions from aqueous solutions. The batch adsorption studies were carried out with variable experimental conditions. The maximum adsorption capacity was observed at pH 5 and higher pH conditions. However, the precipitation of hydroxide ions onto the adsorbent surface occurs in strongly acidic conditions within the pH range of 3–7. Furthermore, the optimum adsorbent dose corresponding to maximum uptake was found to be $5\,g\,l^{-1}$. The adsorption of Pb(II) from aqueous solution was achieved at the equilibrium contact time of 30 minutes (Oh et al. 2012).

Oh. C et al. studied the removal of As(III) and As(V) in an aqueous system using slag obtained from steel industries. Slag consists of two main elements with Fe (36%) and Ca (35%). Ca become a prevalent substance in the leachate as it can be easily solubilized out of the slag. When the initial pH was lower than 3, Ca concentration in leachates showed a sharp increase. It is noteworthy that some of Ca transformed into amorphous $CaCO_3$ in the alkaline condition due to the high basicity of slag. The adsorption of As occurs on the surface of amorphous $CaCO_3$ by binding with iron oxides of the slag. The binding affinity of as proposed slag for As was too strong and the regeneration step can only be executed with complexing or reducing agents. Under strong acidic and alkaline conditions, a small amount of heavy metals is found to be leached out from the slag. However, owing to the high removal efficiency of slag under both acidic and basic conditions make slag as the pertinent adsorbent for the removal of As in aqueous solution (Oh et al. 2012).

The sugarcane industry is another primary source of valuable waste materials. The raw sugar waste, chemically modified sugar waste, can be utilized for the fabrication of adsorbents for heavy metal removal (Ezeonuegbu et al. 2021). For instance, Sun et al. demonstrated the amide-modified cellulose from sugarcane bagasse (SCB) for mercury ion removal from water. Mir has synthesized the green nanocomposite from sugarcane waste for the removal of Cr ions from water (Sun et al. 2018). Yong et al. have proposed SCB cellulose-capped silver nanoparticles for the adsorption of heavy metals. They have shown in-depth characterization of the proposed material and also revealed the role of ultrasonic irradiation in the synthesis of nanoparticles. SCB and beet pulp (BP) have been investigated for water treatment and desalination (Yong et al. 2020). Recently, Arthi and co-workers have reported agricultural waste for the elimination of heavy metals from water. In the study, they reported the adsorption efficiency of a modified form of rice husk and coconut husk-based AC toward Zn^{+2}, Co^{+2}, and Cu^{+2} (Arthi et al. 2021). Agri-food industry waste is investigated by Sandowal et al. for the removal of heavy metals from water. The valorized broccoli stalks, cauliflower cores, and coconut shell wastes have been utilized for removal of heavy metal from defiled water (Landin-Sandoval et al. 2020). Similarly, Ince et al. explored Cu(II) ion adsorption onto several dried biomasses such as tea leaves, banana peels, and chestnut shells. These adsorbents were functional in mineral water and industrial wastewater samples and the adsorption performance was investigated for both mineral and industrial wastewater. The adsorption capacities of banana peel, chestnut shell, and tea waste for Cu(II) ions were reported as 1.94 $mg\,g^{-1}$, 2.25 $mg\,g^{-1}$, and 3.36 $mg\,g^{-1}$, respectively (Ince 2017). Recently, electronic waste has drawn more attention due to its rapidly increasing quantity day by day. By utilizing e-waste, Xu et al. and Hadi et al. have developed AC-based adsorbent from printed track and circuit boards for the elimination of heavy metals from water (Hadi et al. 2013; Xu et al. 2014). Though the future of industrial waste-derived adsorbents looks promising, nevertheless, to achieve higher adsorption capacity and cost-effectiveness, a more rigorous research is a requisite.

Kantarli et al. developed AC by different types of leather-shaving wastes using chemical and physical activation and utilized for evacuation of toxic materials. Prepared AC has been studied toward methylene blue, phenol, and Cr (Kantarli and Yanik 2010). He et al. demonstrated zeolite from coal fly ash for the removal of Ni^{+2} from wastewater. The efficiency has been found to be up to 94% and the adsorption capacity (47 $mg\,g^{-1}$ with 100 $mg\,l^{-1}$ Ni^{+2}) (He et al. 2020). Visa and his team have derived hydrothermally modified fly ash for the evacuation of heavy metals and dyes (Visa and Chelaru 2014). Buema et al. studied modified fly ash for the removal of Cu ions from aqueous solution (Buema et al. 2021). He found the maximum adsorption capacity 27.32–57.48 $mg\,g^{-1}$ toward Cu ion. Singh and his team reported a review on coal fly ash as an emerging material for water remediation (Singh et al. 2022).

9.5 Conclusion

This chapter reflects on the status of heavy metal contamination in natural water sources, its environmental hazards and various natural and anthropogenic sources of contamination. Both developing and developed countries are suffering from heavy metal pollution and this stimulates the policy makers to design efficient remediation strategies and protocols. Thereafter, the tolerable limits of heavy metal concentration in water resources are now established across the world. This chapter summarized the tolerable limits of heavy metals. The harmful effects of heavy metal contamination on human health and the aquatic environment are progressively demonstrated. A clear prospect on the toxicity of heavy metals is presented to understand the heavy metal pollution in different water assets. In our opinion, more stringent guidelines and restrictions need to be imposed on the discharge of industrial effluents into freshwater bodies. Further, remediation techniques need to design as per the accessibility and efficiency of local resources and demands. A centralized database with detailed information on the mapping of anthropogenic sources, quality, and volume of industrial effluents will be a key as it will help in the execution of remediation plans in a more effective way. The recent progression in adsorption technology seems to be effective for the remediation of heavy metal pollution. In the recent past, nanotechnology revolutionized this field and new novel adsorbent materials in the form of hybrids and composites of carbon nanomaterials, conducting polymers, metal/metal oxide nanoparticles, AC, etc. Moreover, presently lots of efforts are paid to transform industrial waste into adsorbent materials as it can contribute to *waste-to-wealth approach*, reduce the load of natural resources, and help in a circular economy. The potential of industrial waste-derived adsorbents has been shown in many research reports. This chapter reviewed the key findings of these research reports. It can be concluded that industrial waste-derived adsorbents exhibited excellent heavy metal decontamination efficiency under diverse conditions of pH, temperature, and nature of pollutants. The performance of such adsorbents is comparable with commercially available adsorbents in many cases. But most of these adsorbents are investigated in lab-scale experiments only, and the results cannot be extrapolated to real-time situations. Therefore, pilot scale experiments and studies need to be carried out by involving real-time parameters like volume and nature of effluents, treatment time, operational cost, probability of secondary contamination, disposal, and end-life utility of adsorbents. The results of only such studies decide the fate of technology and help in the realization of commercially and ecologically viable products.

References

Abdel-Shafy, H.I. and Mansour, M.S.M. (2018). Solid waste issue: sources, composition, disposal, recycling, and valorization. *Egyptian Journal of Petroleum* 27 (4): 1275–1290. https://doi.org/10.1016/j.ejpe.2018.07.003.

Agasti, N. (2021). Decontamination of heavy metal ions from water by composites prepared from waste. *Current Research in Green and Sustainable Chemistry* 4: 100088. https://doi.org/10.1016/j.crgsc.2021.100088.

Agrawal, P.R. et al. (2016). Novel 3D lightweight carbon foam as an effective adsorbent for arsenic(v) removal from contaminated water. *RSC Advances* 6 (36): 29899–29908. https://doi.org/10.1039/C6RA02208A.

Agrawal, P.R. et al. (2018). Multiwall carbon nanotube embedded phenolic resin-based carbon foam for the removal of As (V) from contaminated water. *Materials Research Express* 5 (3): 035601. https://doi.org/10.1088/2053-1591/aaaf7c.

Agrawal, P.R. et al. (2021). Rapid adsorption of arsenate from water on a novel hybrid of zirconia oxide anchored rGO functionalised carbon foam. *Colloid and Interface Science Communications* 40: 100350. https://doi.org/10.1016/j.colcom.2020.100350.

Ahmad, W. et al. (2021). Toxic and heavy metals contamination assessment in soil and water to evaluate human health risk. *Scientific Reports* 11 (1): 17006. https://doi.org/10.1038/s41598-021-94616-4.

Akinhanmi, T.F. et al. (2020). Orange peel as low-cost adsorbent in the elimination of Cd(II) ion: kinetics, isotherm, thermodynamic and optimization evaluations. *Bioresources and Bioprocessing* 7 (1): 34. https://doi.org/10.1186/s40643-020-00320-y.

Alengebawy, A. et al. (2021). Heavy metals and pesticides toxicity in agricultural soil and plants: ecological risks and human health implications. *Toxics* 9 (3): 42. https://doi.org/10.3390/toxics9030042.

Ali, H., Khan, E., and Sajad, M.A. (2013). Phytoremediation of heavy metals—concepts and applications. *Chemosphere* 91 (7): 869–881. https://doi.org/10.1016/j.chemosphere.2013.01.075.

Ali, H., Khan, E., and Ilahi, I. (2019). Environmental chemistry and ecotoxicology of hazardous heavy metals: environmental persistence, toxicity, and bioaccumulation. *Journal of Chemistry* 2019: 1–14. https://doi.org/10.1155/2019/6730305.

Ali, I. et al. (2019). Graphene based adsorbents for remediation of noxious pollutants from wastewater. *Environment International* 127: 160–180. https://doi.org/10.1016/j.envint.2019.03.029.

Arthi, D. et al. (2021). Removal of heavy metal ions from water using adsorbents from agro waste materials. *Materials Today: Proceedings* 45: 1794–1798. https://doi.org/10.1016/j.matpr.2020.08.738.

Babula, P. et al. (2008). Uncommon heavy metals, metalloids and their plant toxicity: a review. *Environmental Chemistry Letters* 6 (4): 189–213. https://doi.org/10.1007/s10311-008-0159-9.

Barakat, M.A. (2011). New trends in removing heavy metals from industrial wastewater. *Arabian Journal of Chemistry* 4 (4): 361–377. https://doi.org/10.1016/j.arabjc.2010.07.019.

Bashir, A. et al. (2019). Removal of heavy metal ions from aqueous system by ion-exchange and biosorption methods. *Environmental Chemistry Letters* 17 (2): 729–754. https://doi.org/10.1007/s10311-018-00828-y.

Bayuo, J., Rwiza, M., and Mtei, K. (2022). A comprehensive review on the decontamination of lead(ii) from water and wastewater by low-cost biosorbents. *RSC Advances* 12 (18): 11233–11254. https://doi.org/10.1039/D2RA00796G.

Bissen, M. and Frimmel, F.H. (2003). Arsenic—a review. Part I: occurrence, toxicity, speciation, mobility. *Acta Hydrochimica et Hydrobiologica* 31 (1): 9–18. https://doi.org/10.1002/aheh.200390025.

Blais, J.F. et al. (2008). Metals precipitation from effluents: review. *Practice Periodical of Hazardous, Toxic, Waste Management* 12 (3): 135–149. https://doi.org/10.1061/(ASCE)1090-025X(2008)12:3(135).

Bolan, N. et al. (2022). Antimony contamination and its risk management in complex environmental settings: a review. *Environment International* 158: 106908. https://doi.org/10.1016/j.envint.2021.106908.

Boreiko, C.J. and Rossman, T.G. (2020). Antimony and its compounds: health impacts related to pulmonary toxicity, cancer, and genotoxicity. *Toxicology and Applied Pharmacology* 403: 115156. https://doi.org/10.1016/j.taap.2020.115156.

Buczyńska, A., Rolecki, R., and Tarkowski, S. (1999). Industrial wastes and health hazards. *Medycyna Pracy* 50: 179–190.

Buema, G., Harja, M., Lupu, N. et al. (2021). Adsorption performance of modified fly ash for copper ion removal from aqueous solution. *Water* 13: 207.

Burke, D.M., Morris, M.A., and Holmes, J.D. (2013). Chemical oxidation of mesoporous carbon foams for lead ion adsorption. *Separation and Purification Technology* 104: 150–159. https://doi.org/10.1016/j.seppur.2012.10.049.

Chand, P. et al. (2014). Improved adsorption of cadmium ions from aqueous solution using chemically modified apple pomace: mechanism, kinetics, and thermodynamics. *International Biodeterioration & Biodegradation* 90: 8–16. https://doi.org/10.1016/j.ibiod.2013.10.028.

Cherifi, M. et al. (2011). Electrokinetic removal of aluminum from water potabilization treatment sludge. *Desalination* 281: 263–270. https://doi.org/10.1016/j.desal.2011.07.071.

Chibuike, G.U. and Obiora, S.C. (2014). Heavy metal polluted soils: effect on plants and bioremediation methods. *Applied and Environmental Soil Science* 2014: 1–12. https://doi.org/10.1155/2014/752708.

Cronje, K.J. et al. (2011). Optimization of chromium(VI) sorption potential using developed activated carbon from sugarcane bagasse with chemical activation by zinc chloride. *Desalination* 275 (1–3): 276–284. https://doi.org/10.1016/j.desal.2011.03.019.

Daci, M.N. et al. (2011). Coal ash as adsorbent for heavy metal ions in standard solutions, industrial wastewater and streams. *Ecohydrology & Hydrobiology* 11 (1–2): 129–132. https://doi.org/10.2478/v10104-011-0037-0.

De Gisi, S. et al. (2016). Characteristics and adsorption capacities of low-cost sorbents for wastewater treatment: a review. *Sustainable Materials and Technologies* 9: 10–40. https://doi.org/10.1016/j.susmat.2016.06.002.

Deng, Y. and Zhao, R. (2015). Advanced oxidation processes (AOPs) in wastewater treatment. *Current Pollution Reports* 1 (3): 167–176. https://doi.org/10.1007/s40726-015-0015-z.

Deng, Z. et al. (2021). Modification of coconut shell-based activated carbon and purification of wastewater. *Advanced Composites and Hybrid Materials* 4 (1): 65–73. https://doi.org/10.1007/s42114-021-00205-4.

Dixit, R. et al. (2015). Bioremediation of heavy metals from soil and aquatic environment: an overview of principles and criteria of fundamental processes. *Sustainability* 7 (2): 2189–2212. https://doi.org/10.3390/su7022189.

Djedjibegovic, J. et al. (2020). Heavy metals in commercial fish and seafood products and risk assessment in adult population in Bosnia and Herzegovina. *Scientific Reports* 10 (1): 13238. https://doi.org/10.1038/s41598-020-70205-9.

Duru, C., Nnabuchi, M., and Duru, I. (2019). Adsorption of cu onto maize husk lignocellulose in single and binary Cu-Zn solution systems: equilibrium, isotherm, kinetic, thermodynamic and mechanistic studies. *Egyptian Journal of Chemistry* 62: 1295–1305. https://doi.org/10.21608/ejchem.2019.6377.1539.

Ekanayake, A. et al. (2022). Amino-functionalized biochars for the detoxification and removal of hexavalent chromium in aqueous media. *Environmental Research* 211: 113073. https://doi.org/10.1016/j.envres.2022.113073.

Ezeonuegbu, B.A. et al. (2021). Agricultural waste of sugarcane bagasse as efficient adsorbent for lead and nickel removal from untreated wastewater: biosorption, equilibrium isotherms, kinetics and desorption studies. *Biotechnology Reports* 30: e00614. https://doi.org/10.1016/j.btre.2021.e00614.

Fashola, M., Ngole-Jeme, V., and Babalola, O. (2016). Heavy metal pollution from gold mines: environmental effects and bacterial strategies for resistance. *International Journal of Environmental Research and Public Health* 13 (11): 1047. https://doi.org/10.3390/ijerph13111047.

Fatoki, J.O. and Badmus, J.A. (2022). Arsenic as an environmental and human health antagonist: a review of its toxicity and disease initiation. *Journal of Hazardous Materials Advances* 5: 100052. https://doi.org/10.1016/j.hazadv.2022.100052.

Ferronato, N. and Torretta, V. (2019). Waste mismanagement in developing countries: a review of global issues. *International Journal of Environmental Research and Public Health* 16 (6): 1060. https://doi.org/10.3390/ijerph16061060.

Finnegan, P.M. and Chen, W. (2012). Arsenic toxicity: the effects on plant metabolism. *Frontiers in Physiology* 3: 182. https://doi.org/10.3389/fphys.2012.00182.

Fuwad, A. et al. (2019). An electrokinetic approach to fabricating aquaporin biomimetic membranes for water purification. *Desalination* 452: 9–16. https://doi.org/10.1016/j.desal.2018.10.010.

Gafur, N. et al. (2018). A case study of heavy metal pollution in water of bone river by artisanal small-scale gold mine activities in eastern part of Gorontalo, Indonesia. *Water* 10 (11): 1507. https://doi.org/10.3390/w10111507.

Garrido-Cardenas, J.A. et al. (2019). Wastewater treatment by advanced oxidation process and their worldwide research trends. *International Journal of Environmental Research and Public Health* 17 (1): 170. https://doi.org/10.3390/ijerph17010170.

Germ, M. et al. (2007). Combined effects of selenium and drought on photosynthesis and mitochondrial respiration in potato. *Plant Physiology and Biochemistry* 45 (2): 162–167. https://doi.org/10.1016/j.plaphy.2007.01.009.

Ghazzal, M. et al. (2022). Bubalus bubalis blood as biological tool to track impacts from cobalt: bioaccumulation and health risks perspectives from a water-soil-forage-livestock ecosystem. *Biological Trace Element Research* 201: 706–719. [Preprint]. https://doi.org/10.1007/s12011-022-03206-6.

Gordon, T. (2003). Beryllium: genotoxicity and carcinogenicity. *Mutation Research, Fundamental and Molecular Mechanisms of Mutagenesis* 533 (1–2): 99–105. https://doi.org/10.1016/j.mrfmmm.2003.08.022.

Gumpu, M.B. et al. (2015). A review on detection of heavy metal ions in water—an electrochemical approach. *Sensors and Actuators B: Chemical* 213: 515–533. https://doi.org/10.1016/j.snb.2015.02.122.

Gupta, A. et al. (2021). A review of adsorbents for heavy metal decontamination: growing approach to wastewater treatment. *Materials* 14 (16): 4702. https://doi.org/10.3390/ma14164702.

Ha, E. et al. (2017). Current progress on understanding the impact of mercury on human health. *Environmental Research* 152: 419–433. https://doi.org/10.1016/j.envres.2016.06.042.

Hadi, P., Barford, J., and McKay, G. (2013). Synergistic effect in the simultaneous removal of binary cobalt–nickel heavy metals from effluents by a novel e-waste-derived material. *Chemical Engineering Journal* 228: 140–146. https://doi.org/10.1016/j.cej.2013.04.086.

He, X., Yao, B., Xia, Y. et al. (2020). Coal fly ash derived zeolite for highly efficient removal of Ni2+ inwaste water. *Powder Technology* 367: 40–46.

Hegazi, H.A. (2013). Removal of heavy metals from wastewater using agricultural and industrial wastes as adsorbents. *HBRC Journal* 9 (3): 276–282. https://doi.org/10.1016/j.hbrcj.2013.08.004.

Hodaifa, G. et al. (2019). Chemical oxidation methods for treatment of real industrial olive oil mill wastewater. *Journal of the Taiwan Institute of Chemical Engineers* 97: 247–254. https://doi.org/10.1016/j.jtice.2019.02.001.

Homaeigohar, S. and Elbahri, M. (2017). Graphene membranes for water desalination. *NPG Asia Materials* 9 (8): e427. https://doi.org/10.1038/am.2017.135.

Huang, Z. et al. (2021). Distribution, toxicity load, and risk assessment of dissolved metal in surface and overlying water at the Xiangjiang River in southern China. *Scientific Reports* 11 (1): 109. https://doi.org/10.1038/s41598-020-80403-0.

Ighalo, J.O., Yap, P.-S. et al. (2022). Adsorption of persistent organic pollutants (POPs) from the aqueous environment by nano-adsorbents: a review. *Environmental Research* 212: 113123. https://doi.org/10.1016/j.envres.2022.113123.

Ighalo, J.O., Rangabhashiyam, S. et al. (2022). Zeolitic Imidazolate frameworks (ZIFs) for aqueous phase adsorption—a review. *Journal of Industrial and Engineering Chemistry* 105: 34–48. https://doi.org/10.1016/j.jiec.2021.09.029.

Igwe, J. and Abia, A. (2003). Maize cob and husk as adsorbents for removal of Cd, Pb and Zn ions from wastewater. *The Physical Science* 2: 83–94.

Ince, M. (2017). Using food waste biomass as effective adsorbents in water and wastewater treatment for Cu(II) removal. *Atomic Spectroscopy* 38 (5): 142–148. https://doi.org/10.46770/AS.2017.05.004.

Kantarli, I.C. and Yanik, J. (2010). Activated carbon from leather shaving wastes and its application in removal of toxic materials. *Journal of Hazardous Materials* 179: 348–356.

Kapahi, M. and Sachdeva, S. (2019). Bioremediation options for heavy metal pollution. *Journal of Health and Pollution* 9 (24): 191203. https://doi.org/10.5696/2156-9614-9.24.191203.

Kazerooni Sadi, M.A. et al. (2012). Reduce, reuse, recycle and recovery in sustainable construction waste management. *Advanced Materials Research* 446–449: 937–944. https://doi.org/10.4028/www.scientific.net/AMR.446-449.937.

Kılıç, M. et al. (2008). Effect of conditioning for Pb(II) and Hg(II) biosorption on waste activated sludge. *Chemical Engineering and Processing: Process Intensification* 47 (1): 31–40. https://doi.org/10.1016/j.cep.2007.07.019.

Kumar, R., Bhattacharya, S., and Sharma, P. (2021). Novel insights into adsorption of heavy metal ions using magnetic graphene composites. *Journal of Environmental Chemical Engineering* 9 (5): 106212. https://doi.org/10.1016/j.jece.2021.106212.

Landin-Sandoval, V.J. et al. (2020). Valorization of agri-food industry wastes to prepare adsorbents for heavy metal removal from water. *Journal of Environmental Chemical Engineering* 8 (5): 104067. https://doi.org/10.1016/j.jece.2020.104067.

Lim, H.Y. et al. (2021). Review on conversion of lignin waste into value-added resources in tropical countries. *Waste and Biomass Valorization* 12 (10): 5285–5302. https://doi.org/10.1007/s12649-020-01307-8.

Lin, L., Yang, H., and Xu, X. (2022). Effects of water pollution on human health and disease heterogeneity: a review. *Frontiers in Environmental Science* 10: 880246. https://doi.org/10.3389/fenvs.2022.880246.

Luo, M. et al. (2018). A novel modification of lignin on corncob-based biochar to enhance removal of cadmium from water. *Bioresource Technology* 259: 312–318. https://doi.org/10.1016/j.biortech.2018.03.075.

Ma, J. et al. (2018). Heavy metal removal from aqueous solutions by calcium silicate powder from waste coal fly-ash. *Journal of Cleaner Production* 182: 776–782. https://doi.org/10.1016/j.jclepro.2018.02.115.

Makar, A.B. et al. (1975). Formate assay in body fluids: application in methanol poisoning. *Biochemical Medicine* 13 (2): 117–126. https://doi.org/10.1016/0006-2944(75)90147-7.

Malik, A. (2004). Metal bioremediation through growing cells. *Environment International* 30 (2): 261–278. https://doi.org/10.1016/j.envint.2003.08.001.

Mittal, J. et al. (2021). Batch and bulk adsorptive removal of anionic dye using metal/halide-free ordered mesoporous carbon as adsorbent. *Journal of Cleaner Production* 321: 129060. https://doi.org/10.1016/j.jclepro.2021.129060.

Mo, J. et al. (2018). A review on agro-industrial waste (AIW) derived adsorbents for water and wastewater treatment. *Journal of Environmental Management* 227: 395–405. https://doi.org/10.1016/j.jenvman.2018.08.069.

Mohammed Abdul, K.S. et al. (2015). Arsenic and human health effects: a review. *Environmental Toxicology and Pharmacology* 40 (3): 828–846. https://doi.org/10.1016/j.etap.2015.09.016.

Mohanty, M. et al. (2012). Bio-concentration of chromium—an in situ phytoremediation study at South Kaliapani chromite mining area of Orissa, India. *Environmental Monitoring and Assessment* 184 (2): 1015–1024. https://doi.org/10.1007/s10661-011-2017-7.

Morton-Bermea, O. et al. (2009). Assessment of heavy metal pollution in urban topsoils from the metropolitan area of Mexico City. *Journal of Geochemical Exploration* 101 (3): 218–224. https://doi.org/10.1016/j.gexplo.2008.07.002.

Mukherjee, R. et al. (2015). Nanostructured polyaniline incorporated ultrafiltration membrane for desalination of brackish water. *Environmental Science: Water Research & Technology* 1 (6): 893–904. https://doi.org/10.1039/C5EW00163C.

Nagajyoti, P.C., Lee, K.D., and Sreekanth, T.V.M. (2010). Heavy metals, occurrence and toxicity for plants: a review. *Environmental Chemistry Letters* 8 (3): 199–216. https://doi.org/10.1007/s10311-010-0297-8.

Naiya, T.K., Bhattacharya, A.K., and Das, S.K. (2009). Clarified sludge (basic oxygen furnace sludge)—an adsorbent for removal of Pb(II) from aqueous solutions—kinetics, thermodynamics and desorption studies. *Journal of Hazardous Materials* 170 (1): 252–262. https://doi.org/10.1016/j.jhazmat.2009.04.103.

Obotey Ezugbe, E. and Rathilal, S. (2020). Membrane technologies in wastewater treatment: a review. *Membranes* 10 (5): 89. https://doi.org/10.3390/membranes10050089.

Oh, C. et al. (2012). Removal characteristics of As(III) and As(V) from acidic aqueous solution by steel making slag. *Journal of Hazardous Materials* 213–214: 147–155. https://doi.org/10.1016/j.jhazmat.2012.01.074.

Olasehinde, E.F. et al. (2018). Sequestration of aqueous lead (II) using modified and unmodified red onion skin. *Analytical Letters* 51 (17): 2710–2732. https://doi.org/10.1080/00032719.2018.1448989.

Park, J.-H. et al. (2020). Exploration of the potential capacity of fly ash and bottom ash derived from wood pellet-based thermal power plant for heavy metal removal. *Science of the Total Environment* 740: 140205. https://doi.org/10.1016/j.scitotenv.2020.140205.

Paul, D. (2017). Research on heavy metal pollution of river Ganga: a review. *Annals of Agrarian Science* 15 (2): 278–286. https://doi.org/10.1016/j.aasci.2017.04.001.

Pichhode, M. and Nikhil, K. (2016). Effect of heavy metals on plants: an overview. *International Journal of Application or Innovation in Engineering & Management* 5: 56–66. https://doi.org/10.13140/RG.2.2.27583.87204.

Prasad, S. et al. (2021). Chromium contamination and effect on environmental health and its remediation: a sustainable approaches. *Journal of Environmental Management* 285: 112174. https://doi.org/10.1016/j.jenvman.2021.112174.

Puyol, D. et al. (2017). Resource recovery from wastewater by biological technologies: opportunities, challenges, and prospects. *Frontiers in Microbiology* 7: 2106. https://doi.org/10.3389/fmicb.2016.02106.

Rai, P.K. et al. (2019). Heavy metals in food crops: health risks, fate, mechanisms, and management. *Environment International* 125: 365–385. https://doi.org/10.1016/j.envint.2019.01.067.

Raissy, M. et al. (2022). A systematic review of the concentration of potentially toxic elements in fish from the Persian Gulf: a health risk assessment study. *Food and Chemical Toxicology* 163: 112968. https://doi.org/10.1016/j.fct.2022.112968.

Rashid, R. et al. (2021). A state-of-the-art review on wastewater treatment techniques: the effectiveness of adsorption method. *Environmental Science and Pollution Research* 28 (8): 9050–9066. https://doi.org/10.1007/s11356-021-12395-x.

Rice, K.M. et al. (2014). Environmental mercury and its toxic effects. *Journal of Preventive Medicine and Public Health* 47 (2): 74–83. https://doi.org/10.3961/jpmph.2014.47.2.74.

Sankhla, M.S. et al. (2016). Heavy metals contamination in water and their hazardous effect on human health-a review. *SSRN Electronic Journal* 5 (10): 759–766. https://doi.org/10.2139/ssrn.3428216.

Schieweck, A. (2020). Adsorbent media for the sustainable removal of organic air pollutants from museum display cases. *Heritage Science* 8 (1): 12. https://doi.org/10.1186/s40494-020-0357-8.

Segneanu, A.-E. et al. (2022). Highly efficient engineered waste eggshell-fly ash for cadmium removal from aqueous solution. *Scientific Reports* 12 (1): 9676. https://doi.org/10.1038/s41598-022-13664-6.

Sfakianakis, D.G. et al. (2015). Effect of heavy metals on fish larvae deformities: a review. *Environmental Research* 137: 246–255. https://doi.org/10.1016/j.envres.2014.12.014.

Sharma, R., Agrawal, P.R. et al. (2021a). Current scenario of heavy metal contamination in water. In: *Contamination of Water*, 49–64. Elsevier 10.1016/B978-0-12-824058-8.00010-4.

Sharma, R., Kumar, R. et al. (2021b). Groundwater extractions and climate change. In: *Water Conservation in the Era of Global Climate Change*, 23–45. Elsevier 10.1016/B978-0-12-820200-5.00016-6.

Sharma, R. et al. (2022). Biosorption for eliminating inorganic contaminants (IOCs) from wastewater. In: *Biosorption for Wastewater Contaminants*, 1ste (ed. R. Selvasembian and P. Singh), 42–62. Wiley 10.1002/9781119737629.ch3.

Shi, J. et al. (2018). Preparation and application of modified zeolites as adsorbents in wastewater treatment. *Water Science and Technology* 2017 (3): 621–635. https://doi.org/10.2166/wst.2018.249.

Siddiqua, A., Hahladakis, J.N., and Al-Attiya, W.A.K.A. (2022). An overview of the environmental pollution and health effects associated with waste landfilling and open dumping. *Environmental Science and Pollution Research* 29: 58514–58536. [Preprint]. https://doi.org/10.1007/s11356-022-21578-z.

Siddique, T., Dutta, N.K., and Roy Choudhury, N. (2020). Nanofiltration for arsenic removal: challenges, recent developments, and perspectives. *Nanomaterials* 10 (7): 1323. https://doi.org/10.3390/nano10071323.

Sikosana, M.L. et al. (2019). Municipal wastewater treatment technologies: a review. *Procedia Manufacturing* 35: 1018–1024. https://doi.org/10.1016/j.promfg.2019.06.051.

Singh, S. et al. (2016). Heavy metal tolerance in plants: role of transcriptomics, proteomics, metabolomics, and Ionomics. *Frontiers in Plant Science* 6: 1143. https://doi.org/10.3389/fpls.2015.01143.

Singh, N., Agarwal, A., De, A., and Singh, P. (2022). Coal fly ash: an emerging material for water remediation. *International Journal of Coal Science & Technology* 9: 44. https://doi.org/10.1007/S40789-022-00512-1.

Solangi, N.H. et al. (2021). Development of fruit waste derived bio-adsorbents for wastewater treatment: a review. *Journal of Hazardous Materials* 416: 125848. https://doi.org/10.1016/j.jhazmat.2021.125848.

Soliman, N.K. and Moustafa, A.F. (2020). Industrial solid waste for heavy metals adsorption features and challenges; a review. *Journal of Materials Research and Technology* 9 (5): 10235–10253. https://doi.org/10.1016/j.jmrt.2020.07.045.

Spiridon, I. (2020). Extraction of lignin and therapeutic applications of lignin-derived compounds. A review. *Environmental Chemistry Letters* 18 (3): 771–785. https://doi.org/10.1007/s10311-020-00981-3.

Starr, M. et al. (2003). Weathering release of heavy metals from soil in comparison to deposition, litterfall and leaching fluxes in a remote, boreal coniferous forest. *Applied Geochemistry* 18 (4): 607–613. https://doi.org/10.1016/S0883-2927(02)00157-9.

Stefanakis, A.I., Calheiros, C.S.C., and Nikolaou, I. (2021). Nature-based solutions as a tool in the new circular economic model for climate change adaptation. *Circular Economy and Sustainability* 1 (1): 303–318. https://doi.org/10.1007/s43615-021-00022-3.

Sun, P. et al. (2009). Morphological deformities as biomarkers in fish from contaminated Rivers in Taiwan. *International Journal of Environmental Research and Public Health* 6 (8): 2307–2331. https://doi.org/10.3390/ijerph6082307.

Sun, N., Wen, X., and Yan, C. (2018). Adsorption of mercury ions from wastewater aqueous solution by amide functionalized cellulose from sugarcane bagasse. *International Journal of Biological Macromolecules* 108: 1199–1206. https://doi.org/10.1016/j.ijbiomac.2017.11.027.

Taha, A. et al. (2018). Comparative biosorption study of Hg (II) using raw and chemically activated almond shell. *Adsorption Science & Technology* 36 (1–2): 521–548. https://doi.org/10.1177/0263617417705473.

Tchounwou, P.B. et al. (2012). Heavy metal toxicity and the environment. In: *Molecular, Clinical and Environmental Toxicology*, Experientia Supplementum (ed. A. Luch), 133–164. Basel: Springer Basel 10.1007/978-3-7643-8340-4_6.

Thyagaraju, N. (2016). Water pollution and its impact on environment of society. *International Research Journal of Management, IT and Social Sciences* 3 (5): 1–7. https://doi.org/10.21744/irjmis.v3i5.10.

Uddin, M.M. et al. (2021). Heavy metal accumulation in rice and aquatic plants used as human food: a general review. *Toxics* 9 (12): 360. https://doi.org/10.3390/toxics9120360.

Vahter, M. (2008). Health effects of early life exposure to arsenic: early life arsenic exposure. *Basic & Clinical Pharmacology & Toxicology* 102 (2): 204–211. https://doi.org/10.1111/j.1742-7843.2007.00168.x.

Vardhan, K.H., Kumar, P.S., and Panda, R.C. (2019). A review on heavy metal pollution, toxicity and remedial measures: current trends and future perspectives. *Journal of Molecular Liquids* 290: 111197. https://doi.org/10.1016/j.molliq.2019.111197.

Visa, M. and Chelaru, A.-M. (2014). Hydrothermally modified fly ash for heavy metals and dyes removal in advanced wastewater treatment. *Applied Surface Science* 303: 14–22.

Wang, Q. and Yang, Z. (2016). Industrial water pollution, water environment treatment, and health risks in China. *Environmental Pollution* 218: 358–365. https://doi.org/10.1016/j.envpol.2016.07.011.

Wang, Q. et al. (2004). Sources and remediation for mercury contamination in aquatic systems—a literature review. *Environmental Pollution* 131 (2): 323–336. https://doi.org/10.1016/j.envpol.2004.01.010.

Wang, J. et al. (2012). Remediation of mercury contaminated sites—a review. *Journal of Hazardous Materials* 221–222: 1–18. https://doi.org/10.1016/j.jhazmat.2012.04.035.

World Resources Institute (1994). *The… Information Please Environmental Almanac.* Houghton Mifflin Company.

Wuana, R.A. and Okieimen, F.E. (2011). Heavy metals in contaminated soils: a review of sources, chemistry, risks and best available strategies for remediation. *ISRN Ecology* 2011: 1–20. https://doi.org/10.5402/2011/402647.

Xu, M. et al. (2014). Removal of cadmium ions from wastewater using innovative electronic waste-derived material. *Journal of Hazardous Materials* 273: 118–123. https://doi.org/10.1016/j.jhazmat.2014.03.037.

Yadanaparthi, S.K.R., Graybill, D., and von Wandruszka, R. (2009). Adsorbents for the removal of arsenic, cadmium, and lead from contaminated waters. *Journal of Hazardous Materials* 171 (1–3): 1–15. https://doi.org/10.1016/j.jhazmat.2009.05.103.

Yadav, A. et al. (2021). Advances in decontamination of wastewater using biomass-basedcomposites: a critical review. *Science of the Total Environment* 784: 147108. https://doi.org/10.1016/j.scitotenv.2021.147108.

Yeganeh, M. et al. (2013). Mapping of human health risks arising from soil nickel and mercury contamination. *Journal of Hazardous Materials* 244–245: 225–239. https://doi.org/10.1016/j.jhazmat.2012.11.040.

Yong, B.C. et al. (2020). Synthesis and characterization of sugarcane bagasse cellulose-capped silver nanoparticle using ultrasonic irradiation for the adsorption of heavy metal. *Asia-Pacific Journal of Chemical Engineering* 15 (3): e2433. https://doi.org/10.1002/apj.2433.

Zhou, Q. et al. (2020). Total concentrations and sources of heavy metal pollution in global river and lake water bodies from 1972 to 2017. *Global Ecology and Conservation* 22: e00925. https://doi.org/10.1016/j.gecco.2020.e00925.

Zou, Y. and Han, B.-X. (2001). Preparation of activated carbons from Chinese coal and hydrolysis lignin. *Adsorption Science and Technology* 19 (1): 59–72. https://doi.org/10.1260/0263617011493971.

Zunita, M. (2021). Graphene oxide-based nanofiltration for Hg removal from wastewater: a mini review. *Membranes* 11 (4): 269. https://doi.org/10.3390/membranes11040269.

10

Biological Remediation of Heavy Metals from Acid Mine Drainage—Recent Advancements

Suparna Datta[1]*, Keisham Radhapyari*[2]*, Snigdha Dutta*[2]*, Rinkumoni Barman*[2]*, and Anadi Gayen*[1]

[1] *Central Ground Water Board, Eastern Region, Department of Water Resources, River Development and Ganga Rejuvenation, Ministry of Jal Shakti, Kolkata, West Bengal, India*
[2] *Central Ground Water Board, North Eastern Region, Department of Water Resources, River Development and Ganga Rejuvenation, Ministry of Jal Shakti, Guwahati, Assam, India*

10.1 Introduction

Over the last century, the mining industry has taken a lead role in the expansion of the global economy, but, at the same time, this is severely affecting environmental health (Bachtlin 2019; Braga et al. 2020; Yuan et al. 2021). The pollutants released from the active as well as abandoned mines have endangered the aquatic and terrestrial ecosystem, making it life-threatening (Rambabu et al. 2020; Tabelin et al. 2020). In a mining operation, during the generation and consumption of different mineral resources, a significant amount of sulfur compounds, viz. pyrite (FeS_2), pyrrhotite, etc., are exposed to rainwater, air and other microbes (Kaksonen et al. 2006). Thus, through different physical, chemical, and biological changes in the system, AMD is produced, which is highly acidic ($pH < 2$; $Sulfate > 20\,g\,l^{-1}$) (Abinandan et al. 2020; Lukhele et al. 2020; Daraz et al. 2021). In such an acidic environment, minerals get dissolved, including toxic metals (e.g. silver [Ag], iron [Fe], manganese [Mn], lead [Pb], copper [Cu], cobalt [Co], zinc [Zn], cadmium [Cd], and chromium [Cr]) and metalloids (e.g. arsenic [As]) and contaminate nearby water bodies, rivers, aquifers, soil, etc. (Torregrosa et al. 2019; Vieira et al. 2020). It is presumed that a single mine can generate hundreds to thousands of cubic meters of AMD, and in global statistics, annually, 20,000–25,000 million tonnes of solid mine wastes are produced (Lottermoser 2010; Buzzi et al. 2013).

High concentrations of dissolved metals and metalloids, acidic pH and higher sulfate concentration may pose a serious health risk to the human being, aquatic life, microbial community, and biodiversity surrounding the mining activity and abandoned mines (Konhauser et al. 2011; Sun et al. 2019a, b; Xu et al. 2020). To prevent the catastrophe of AMD and sustainable management of the mine effluents, convenient technology is the need of the hour.

Remediation of Heavy Metals: Sustainable Technologies and Recent Advances, First Edition. Edited by Rangabhashiyam Selvasembian, Binota Thokchom, Pardeep Singh, Ali H. Jawad, and Willis Gwenzi.

In addition to the most used active and passive treatment methods, such as chemical neutralization, filtration, Precipitation, adsorption, construction of wetland, biochemical reactors, and reactive barriers (Cunha et al. 2018; Yigit et al. 2020), researchers are in quest of a novel, eco-friendly, sustainable, and economically viable means. However, due to the involvement of expensive machinery, a steady supply of chemicals and resources, routine maintenance and operation workforce, conventional methods are ineffective for long-run treatment, particularly for abandoned mining sites (Bolan et al. 2014; Giordani et al. 2019). Furthermore, chemical treatment methods also generate a massive quantity of secondary sludges, whose management is an additional concern (Yang et al. 2016; Bruneel et al. 2019).

In the present era, applying biosorption is considered a lucrative approach to heavy metal removal. The toxic metals are not degradable; only they can be transformed into a lesser toxic or more stable form, having reduced activity to affect the living cells (Raja Sathendra et al. 2018). Microbes play a crucial role in the biogeochemical cycles of metals and metalloids through mineral dissolution, redox transformation, selective chelation, formation of organometallic complexes, co-precipitation, and transportation (Lounate et al. 2020; Ruehl and Hiibel 2020). They can survive in a diverse physicochemical environment and extreme conditions and have developed resistance and metabolic contraption against higher concentrations of metals in their surroundings. Thus, they are helpful in the remediation of mining sites (Abinandan et al. 2018; Gruzdev et al. 2020). In recent times, tremendous progress has been achieved in biotechnology and genetic engineering to use bioremediation as an effective alternative to environmental remediation in a sustainable manner. The use of mixed culture, low-cost carbon source, biowastes in the combination of resistant organisms, nanobioremedial options, biofilm, etc., both in situ and laboratory conditions, came out satisfactorily in the remediation of AMD. Bioremediation is also effective in biomining to "reuse and recover" expensive metals for various uses from abandoned ores and tailings (Ryu et al. 2020).

The present chapter will show an insight into the recent development in the treatment methods of AMD using biological means. Sincere efforts are made to discuss the use of different prokaryotes and eukaryotes in the bioremediation of heavy metals and metalloids present extensively in mining sites and abandoned mines. The chapter describes a chronological account of the onset of AMD and the role of sulfate-reducing bacteria (SRB) and other organisms in the formation and treatment of AMD. The chapter also deals with various available treatment options and highlights the latest advancement in the application of biological technologies and emerging trends for treating AMD. Finally, the chapter also provides detailed information on the various applications of bioremediation techniques in removing toxic metals from AMD study areas and essential findings and assessments.

10.2 Acid Mine Drainage

10.2.1 Overview of Acid Mine Drainage

AMD, produced due to diverse mining operations and related industries, is a significant environmental problem (Xuwen and Fuqin 2010) due to its detrimental effects. The polluted water caused by AMD originating from operating and inactive, exposed, and underground mines contains elevated concentrations of metals, metalloids, and sulfuric acid, leading to severe contamination (Akcil and Koldas 2006; Gutiérrez et al. 2016). AMD is linked to oilfield production processes, such as drilling, acid fracturing, and well-testing (Chen et al. 2014), as well as secondary sources like mine waste and tailing dumps, Fe ore stockpiles, haul roads, pit lakes, etc. (Shiyi and Qiang 2018). The hydrogeochemical weathering of sulfide-bearing rocks such as arsenopyrite, FeS_2, and marcasite occurs when they interact with water and oxygen, leading to the formation of AMD. Moreover, the catalysis of the reaction is facilitated in the presence of sulfur-oxidizing microorganisms and iron. Nevertheless, the reaction rate and the ensuing acidogenesis primarily rely on mineralogical variability, water and oxygen availability, and microbial kinetics (Lapakko 2002). Despite the potential for this process to occur naturally, diverse anthropogenic mining pursuits are recognized to accelerate the emergence of AMD (Akcil and Koldas 2006).

10.2.2 Environmental Effects of Acid Mine Drainage

Improperly treated discharge of AMD leads to alarming water contamination, posing immediate threats to humans, animals, plants, and surrounding ecosystems within the proximity of AMD-generating sites. AMD's highly acidic metal stream inflicts substantial harm on water, sediments, and soil, causing contamination (Anawar 2015; Coudert et al. 2019; Zhang et al. 2019). Additionally, AMD corrodes human skin and pollutes the water source once it enters surface water, leading to soil hardening, structural breakdown, and hindered crop growth, severely compromising crop production. Furthermore, the environmental repercussions of AMD can be broadly categorized into four domains: biological, physical, ecological, and chemical, as highlighted in Table 10.1 (Kumari et al. 2010; Han et al. 2017; Chen et al. 2021; Tong et al. 2021).

Table 10.1 Environmental effects of acid mine drainage.

Biological effects	Physical effects	Ecological effects	Chemical effects
Acute and chronic toxicity	Adsorption of metal ions into sediment	Alteration in food chain	pH reduction
Reproduction	Decrease in light penetration	Loss of food source or prey	Acidity increased
Respiratory	Modification of the substrate	Alteration in habitat	Increase in soluble metal concentration
Behavioral	Stream velocity increased	Niche loss	Impairment of bicarbonate buffering system
Disturbance in acid–base balance in living beings	Sedimentation affected	Removal of vulnerable/susceptible species	Elevation of particulate metals
Extinction of vulnerable/sensitive species	Reduction in turbulence due to the process of sedimentation	Minimization of primary productivity	—
Migration and avoidance	Color changes in streams	Bioaccumulation within the food chain	—
Imbalance in osmoregulation	Turbidity affected	—	—

10.2.3 Remediation Options/Technologies

The remediation of AMD is challenging due to its distinctive physicochemical attributes, encompassing high pH levels and elevated heavy metal concentrations, which exacerbate the harmful effects on the environment and human well-being; indiscriminate discharge of untreated AMD results in a cascade of environmental issues, engendering prolonged ecological contamination. Park et al. (2019) delineated various alternate strategies to redress AMD. The optimal AMD treatment methodologies are succinctly encapsulated in Figure 10.1.

10.3 Role of Microorganisms in the Formation and Remediation of AMD

10.3.1 Role of Microorganisms in the Formation of Acid Mine Drainage

The genesis of AMD transpires when rocks comprising sulfide minerals, predominantly FeS_2, the most abundant and chief gangue sulfide mineral in metal sulfide ores and coal, are laid bare to the ambient surroundings (Figure 10.2). The key steps for AMD generation involved the slow chemical oxidation of the FeS_2 near neutral pH as per the production of acid (Hallberg 2010; Tabelin et al. 2020). The genesis of acidic mine drainage has been acknowledged as an intricate triphasic progression of chemical reactions (Kleinmann et al. 1981). The reactions involve several oxidation–reduction, complexation, hydrolysis, and exothermic processes. In nature, FeS_2 is fairly defiant to acidic and alkaline reactions and has profuse permanence over pH = 2–10 (Da Costa et al. 2017). Therefore, the dissolution primarily depends on the existentiality of oxidants, viz. dissolved oxygen (DO) and ferric ions (Fe^{3+}) (Lehner et al. 2007; Kumari et al. 2010; Tabelin et al. 2018).

$$FeS_2(S) + 3.5O_2(aq) + H_2O \rightarrow Fe^{2+} + 2H^+ + 2SO_4^{2-} \tag{10.1}$$

$$Fe^{2+} + 0.25O_2(aq) + H^+ \rightarrow Fe^{3+} + 0.5H_2O \tag{10.2}$$

$$FeS_2(S) + 14Fe^{3+} + 8H_2O \rightarrow 15Fe^{2+} + 16H^+ + 2SO_4^{2-} \tag{10.3}$$

The initial stage (Eq. (10.1)) of FeS_2 oxidation is a kinetically sluggish/slow chemical process (Park et al. 2019) and at pH values less than 4, chemical oxidation of ferrous iron is insignificant (Chandra and Gerson 2010; Stumm and Morgan 2012; Tabelin et al. 2017; Nicol 2020). At this stage, the acidophilic iron-oxidizing microorganisms (*Thiobacillus ferrooxidans, Leptospirillum ferrooxidans*, etc.) take the foremost part in the genesis of AMD by catalyzing the process via oxidization of ferrous to ferric iron.

In general, the rate of acid production is amplified by various orders of magnitude through the regeneration of Fe(III) ions (Eq. (10.2)) (Johnson and Hallberg 2003; RoyChowdhury et al. 2015). Singer and Stumm (1970) demonstrated this reaction as the rate-determining step in the oxidative dissolution of FeS_2. According to a research, Edwards et al. (2000) reported that the biological oxidation of FeS_2 is solely responsible for ~75% of AMD occurrences. Further, the generated Fe^{3+} oxidizes FeS_2 to form Fe^{2+}, $SO_4{}^{2-}$, and H^+ (Eq. (10.3)). Eqs. (10.1) and (10.3) coalesce to create a fast-paced and recurring sequence

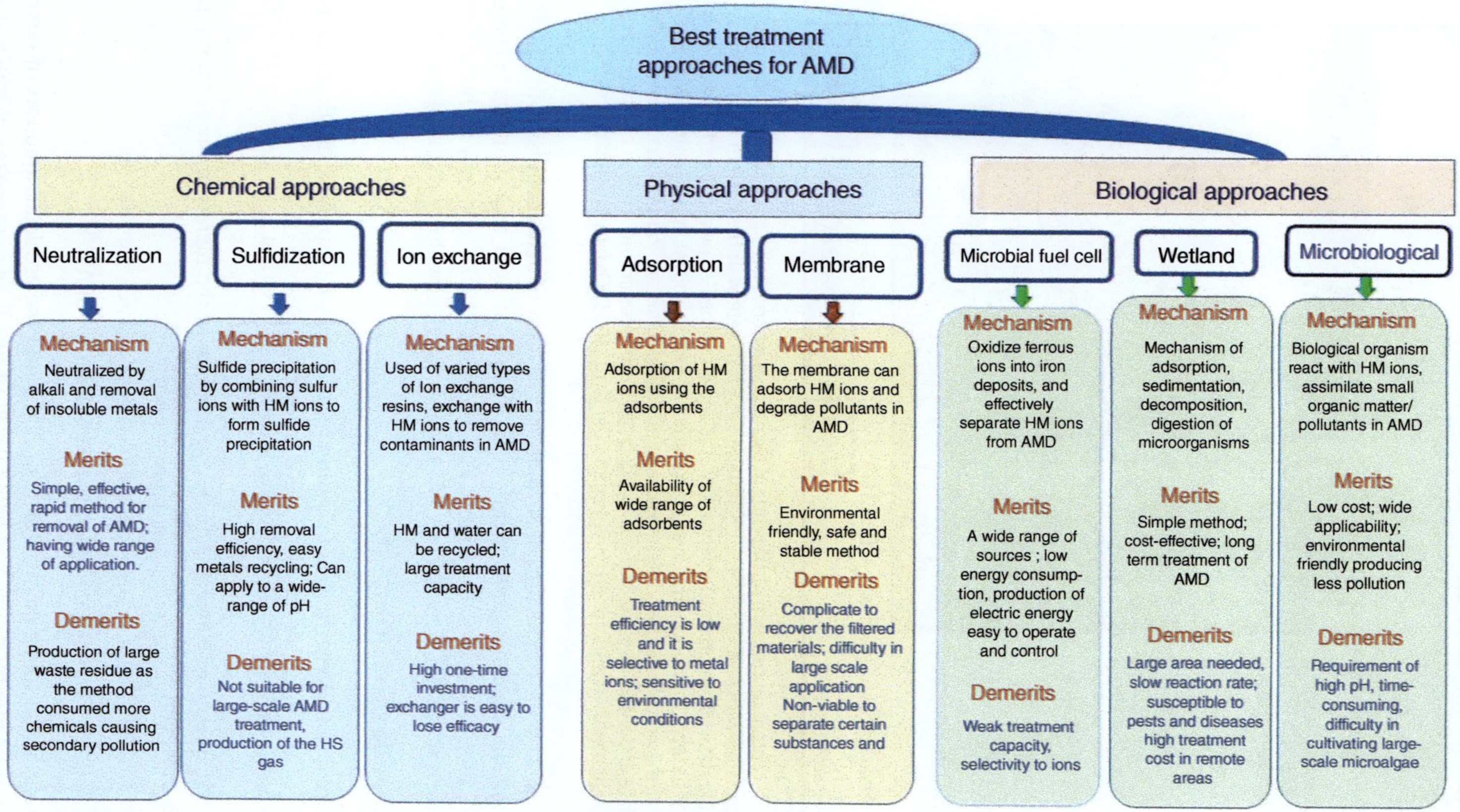

Figure 10.1 Illustrating the mechanism, merits, and demerits of best treatment approaches for acid mine drainage.

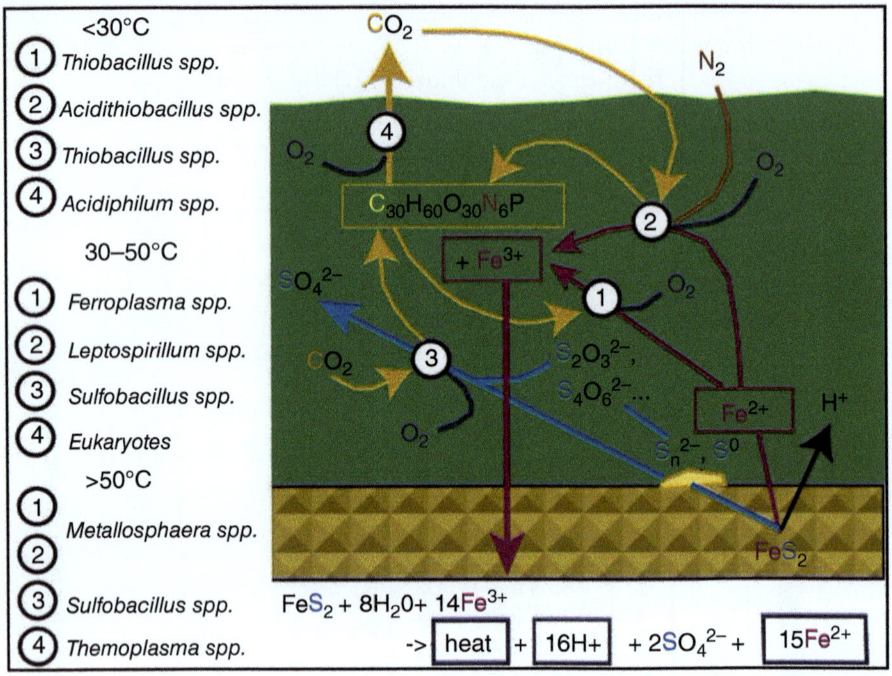

Figure 10.2 Potential iron, sulfur, and carbon cycling based on known metabolic capabilities (1, 2, 3, and 4) associated with AMD members. Crystalline FeS_2 is in yellow at the bottom and green is representing the AMD solution. Elemental sulfur is shown at the FeS_2-water interface as a possible inhibitor of surface dissolution. The overall oxidation of FeS_2 is shown at the bottom, with Fe^{3+} indicated as the primary oxidant. Intermediate sulfur compounds are indicated as follows: $S_2O_3{}^{2-}$ is thiosulfate and $S_2O_6{}^{2-}$ is tetrathionate. $C_{30}H_{60}O_{30}N_6P$ indicates organic carbon compounds. *Source:* Adapted with permission from Baker and Banfield (2003)/Oxford University Press.

characterized by the production of substantial amounts of acids coupled with the liberation of metallic elements into a solution. At this phase, acid rock drainage becomes an issue. Fe^{3+} precipitates as yellow-orange $Fe(OH)_3$ (Eq. (10.4)), commonly known as "yellow boy," if the system's pH is more than 3.5–4.0 units (RoyChowdhury et al. 2015).

$$Fe^{+3} + 3H_2O \rightarrow Fe(OH)_3 + 3H^+ \tag{10.4}$$

The overall stoichiometric FeS_2 oxidation reaction may be expressed as (Eq. (10.5)):

$$FeS_2 + 3.75O_2 + 3.5H_2O \rightarrow Fe(OH)_3 + 2SO_4^{2-} + 4H^+ + \text{Heat} \tag{10.5}$$

As a consequence of elevated acidity, the mobility of the metals in the surroundings is markedly augmented.

10.3.2 Role of Microorganisms in the Remediation of AMD

On account of the acidic characteristic of AMD and the escalated amounts of dissolved metals, it is frequently deemed as being bereft of any living organisms. While this is, in general, true for higher life forms, it has been acknowledged for over half a century that microorganisms thrive in AMD (Colmer and Hinkle 1947). For example, the earliest microbe isolated from the AMD (Colmer et al. 1950) was an iron-oxidizing microbe called *Thiobacillus ferrooxidans* (now known as *Acidithiobacillus ferrooxidans*).

The foundation of AMD bioremediation lies in the capability of certain microorganisms to generate alkalinity and immobilize metals, effectively reversing the chemical reactions that are accountable for the formation of AMD. This process occurs by either modifying the redox state, such as the oxidation of ferrous to ferric, leading to precipitation, or by creating insoluble minerals, for example, the precipitation of chalcophilic metals in the form of insoluble metal sulfides. In AMD, oxidation reactions are typically conducted by autotrophic iron-oxidizing acidophiles that utilize inorganic carbon. On the other hand, the reductive reactions are facilitated by organic carbon as the electron donor and source of carbon.

Due to the prevalence of ferric iron and sulfate in AMD, the process of alkali genesis resulting from the reduction of these two species can have a significant impact on water impacted by AMD. This bioremediation process typically depends on the presence of iron-oxidizing microorganisms in aerobic conditions or SRB in anaerobic conditions (Johnson and Hallberg 2005; Hallberg 2010). In the absence of oxygen, SRB facilitates the reduction of sulfate to hydrogen sulfide (H_2S, represented by Eq. (10.6)), which can then react with dissolved metals and metalloids (represented by M in Eq. (10.7)) to create insoluble metal sulfides. This phenomenon can also be utilized for the remediation of AMD (Johnson and Hallberg 2005; Zhang et al. 2016; Li et al. 2018a, b).

Reactions involved (Eqs. (10.6) and (10.7)):

Anaerobic Condition: SRB

$$SO_4^{2-} + 2CH_2O + 2H^+ \rightarrow Fe(OH)_3 + 2H_2CO_3 + H_2S \quad (10.6)$$

$$M^{2+}(Aq) + H_2S \rightarrow MS + 2H^+ \quad (10.7)$$

Moreover, the metabolism of SRB also facilitates the neutralization of acidity in AMD by generating alkalinity through its biochemical reactions. This aids in the effortless removal of metal sulfides from the liquid phase of the reaction system through filtration. As AMD usually contains elevated concentrations of soluble Fe(II) and is rich in oxygen, the natural remediation phenomenon predominantly relies on Fe oxidation and precipitation (as per Eq. (10.7)), succeeded by the adsorption of soluble metals and metalloids onto the formed ferric minerals comprising a diverse range of Fe (oxyhydr)oxides and hydroxy sulfates, such as jarosite, schwertmannite, ferrihydrite, or goethite (Eq. (10.8)) (Ferris et al. 2000; Morin et al. 2003; Johnson and Hallberg 2005; Rowe and Johnson 2008; Hallberg 2010).

Aerobic Condition:

$$Fe^{+3} + 3H_2O \rightarrow Fe(OH)_3 + 3H^+ \quad (10.8)$$

In general, Fe(III) is insoluble, above pH 2.5 (Johnson and Hallberg 2005). The natural dwindling of transition metals in AMD is acknowledged to happen (Sánchez-Andrea et al. 2016), primarily due to Fe oxidation and precipitation and other metals' adsorption and metalloids to the ferric minerals formed.

10.4 Bioremediation of Heavy Metals in AMD

10.4.1 Arsenic

As is a noxious trace metalloid that is generally distributed in natural environments. Arsenite [As(III)] and arsenate [As(V)] are the most abundant As species in aqueous systems. In addition, it is well known that As(III) exhibits much greater mobility and injuriousness than As(V) (Straif et al. 2009). As released due to AMD that originates from mining metallic sulfide ore containing arsenopyrite is a global ecological concern (Jang and Dempsey 2008; Stafilov et al. 2010; Migaszewski et al. 2019). AMD favors the release of As unswervingly through the oxidation of As-bearing sulfide or by the dissolution of As-rich iron-oxyhydroxides) (Majzlan et al. 2014).

The production of As-rich waters is prominent in arid and semi-arid climates due to elevated evaporation rates resulting in concentrating surface runoff (Dold and Fontboté 2001). The removal of As from AMD poses a great challenge due to its acidic conditions that promote the dissolution of As species and hinder its removal. Various conventional treatment technologies had been reported for As removal, viz. chemical neutralization, adsorption, co-precipitation, or ion exchange processes, have several limits, both economic and operational.

Furthermore, many researchers reported on the use of microbial-oxidation of Fe(II) and/or As(III) accompanied by co-precipitation of Fe(III) oxyhydroxides/hydroxysulfates with As(V) and/or As(III) for the removal of As-containing AMD (Morin et al. 2003; Fernandez-Rojo et al. 2019). As can be scavenged through the formation of ferric arsenite sulfate mineral, ferric arsenate, and/or As-bearing Fe(III) minerals such as ferrihydrite, goethite, schwertmannite via the As-oxidizing or Fe-oxidizing bacteria mediated resulting up to 97% As removal efficiency (Fernandez-Rojo et al. 2019) and 87% removal, respectively (Morin et al. 2003).

However, the efficiency and duration of microbial-mediated As removal are negatively affected by time-consuming microbial incubation and inhibiting microbial growth in As-rich AMD (Paikaray 2015). Another technique to limit the mobility of As is the bio xidation of Fe(II) and/or As(III) in AMD followed by Fe(III)-As(III/V) co-precipitates (Yuan et al. 2019). The various bioremediation techniques for As removal from AMD are as below:

10.4.1.1 Sulfate-Reducing Bacteria in Arsenic Removal

One of the effective methods for As removal from polluted water, AMD, metallurgy industry effluents, etc., is the microbial sulfate reduction based on sulfidogenic process (Battaglia-Brunet et al. 2012; Le Pape et al. 2017).

Serrano and Leiva (2017) in their work had investigated a sulfur-mining site located on the slopes of the volcano (Figure 10.3), where an accumulation of elemental sulfur (S0) tailings was observed (Figure 10.1(II)). Furthermore, the generation of AMD is associated with S0 tailings, categorized by pH in the range of 1.9–4.0. As shown in Figure 10.3, S0 tailings are disposed of near a natural wetland, where plant organic matter decomposes and creates anaerobic environments that facilitate sulfate reduction processes to occur continuously. Figure 10.3(II) depicts that the removal of As and Fe was most pronounced with the increase of pH, signifying the initial pH-dependent mechanism in association with the sulfate reduction.

During this process, SRB are utilized to generate hydrogen sulfide using organics as electron (Eq. (10.9)).

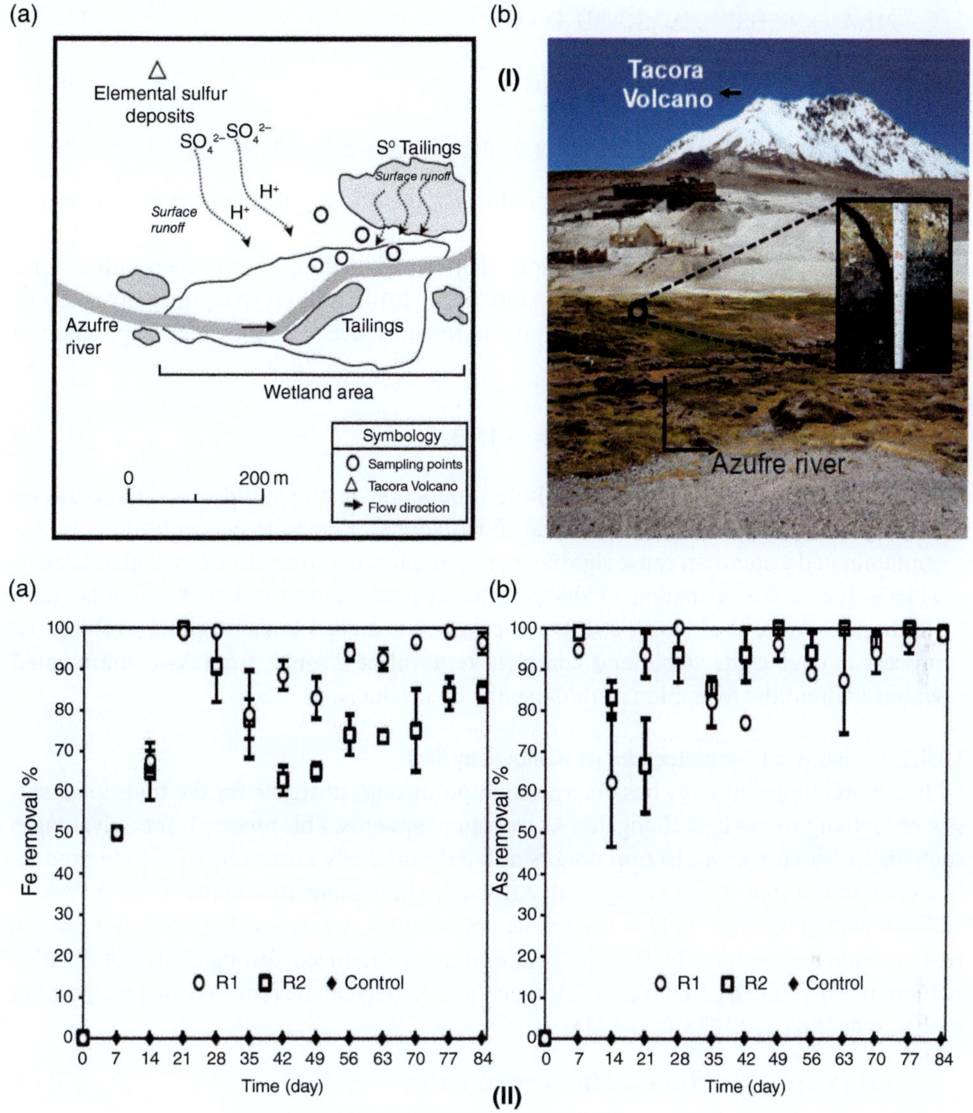

Figure 10.3 (I) (a and b) Schematic representation of the study area; Photographs of the wetland area (II) Fe and As removal during batch reactor operation. *Source:* Serrano and Leiva (2017)/ Reproduced with permission from MDPI/CC BY 4.0.

In addition, As sulfide precipitates (As_2S_3, AsS) when As(III) reacts with sulfide (Eqs. (10.10) and (10.11)) (Rittle et al. 1995; Le Pape et al. 2017; de Matos et al. 2018). Rodriguez-Freire et al. (2014) explained that SRB-based sulfidogenic process can effectively remove arsenate through the microbial reduction of As(V) to As(III) by SRB and subsequent As(III) precipitation with sulfide.

$$CH_3COOH + SO_4^{2-} \xrightarrow{SRB} HS^- + 2CO_2 + OH^- + H_2O \tag{10.9}$$

$$2H_3AsO_3 + 3HS^- \rightarrow As_2S_3 + 3H_2O + OH^- \tag{10.10}$$

$$H_3AsO_3 + HS^- + 2H^+ \rightarrow AsS + 3H_2O \tag{10.11}$$

However, there are some limitations of SRB in As removal, such as:

- Formation of soluble thioarsenite by-products in the process, making the process unstable (Battaglia-Brunet et al. 2012).
- Equation (10.1) depicts the generation of a huge amount of hydrogen sulfide and hydroxyl ions, elevates the pH level (Altun et al. 2014), and favoring the conversion of insoluble arsenite sulfide into thioarsenite in the presence of sulfide and at near-neutral pHs (Eq. (10.12)) (Bostick et al. 2005).

$$As_2S_3 + HS^- + 3OH^- \rightarrow 2As(OH)S_2^{2-} + H_2O \tag{10.12}$$

- Although it is possible to control sulfide production in the sulfidogenic structure by varying the supply of organic material, fluctuations in arsenite concentrations in As-contaminated waters can cause significant fluctuations in the arsenite removal efficiency. This is due to the formation of thioarsenite by-products at the low As/S molar ratio (Rodriguez-Freire et al. 2014, 2016). Hence, there is a need to develop the sulfidogenic process to enable the stable and complete removal of arsenite from As-contaminated waters without the formation of thioarsenite by-products.

10.4.1.2 Improved Technology for As Removal by SRB

Of late, novel biotechnology has emerged as a promising substitute for the traditional sulfate reduction process in sulfidogenic As treatment systems. This process offers advantages such as decreased use of electron donors and a significantly faster rate of sulfide production (Florentino et al. 2015; Zhang et al. 2018a, b). These innovative approaches, driven by sulfur-reducing bacteria, hold considerable potential for the removal of arsenite from the As-contaminated waters. In this method, sulfide is produced through microbial sulfur reduction (Eq. (10.13)) (Zhang et al. 2018a, b), and As(III) can be removed via precipitation with sulfide (Eqs. (10.10) and (10.11)).

$$CH_3COOH + S^0 + 2H_2O \rightarrow 4HS^- + 2CO_2 + 4H^+ \tag{10.13}$$

Microbial sulfur reduction has been established to use 66–80% less organic carbon than microbial sulfate reduction (Qiu et al. 2017; Sun et al. 2018; Zhang et al. 2018a, b).

Furthermore, the sulfur reduction produces protons, as shown in Eq. (10.13), which reduce the pH rise in the As-contaminated waters. This, in turn, moves the formation of thioarsenite by-products. However, the vital challenge is proficiently producing sulfide by S0RB for arsenite removal at the pH levels relevant to the acidic As-contaminated waters (2–5.0).

10.4.1.3 Novel Sulfur-Reducing Bacteria for Arsenic Removal

In one study, Sun et al. (2019a, b) investigated the sulfide production by sulfur-reducing bacteria under acidic environments to remove (As III) removal by the proposed sulfidogenic process. In the method, a remarkably high rate of sulfide production via microbial sulfur reduction by Desulfurella (an acid-tolerant sulfur-reducing genus) was ascertained in a sulfur-packed bed bioreactor (Figure 10.4) at around pH 4.3. This method could remove As(III) more than 99%, even in the presence of extreme sulfide.

Figure 10.2 depicts the performance of sulfide production by sulfur-reducing bacteria and the subsequent removal of As(III) in the sulfidogenic process. Figure 10.5(iA)

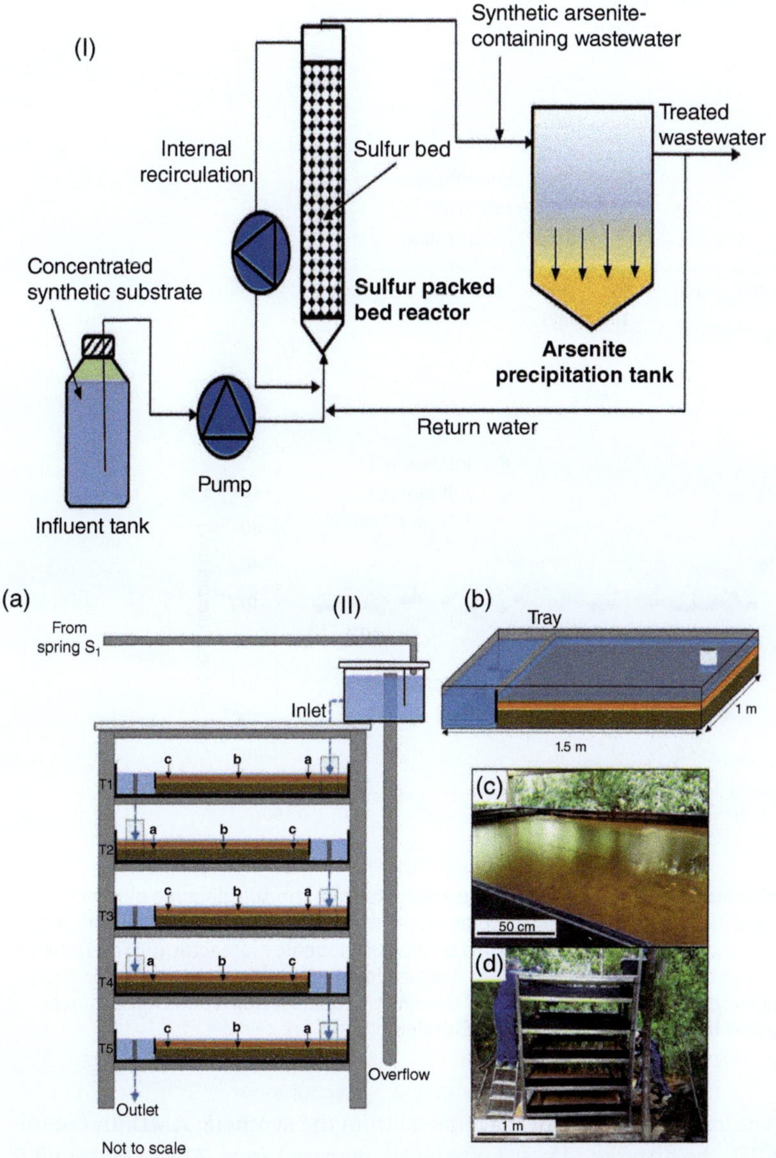

Figure 10.4 Schematic presentation of (I) the sulfidogenic system driven by SORB for the removal of arsenite from the synthetic wastewater, (II) the field bioreactor with the five trays (T = tray) (A) and detail of each tray (B). Pictures of one of the trays showing the orange biogenic precipitate below the thin water layer (C), and of the whole field bioreactor (D) from June 2016. Letters a, b, and c in (A) indicate the biogenic precipitates sampling locations. *Source:* Adapted with permission from (I) Sun et al. (2019a, b)/Elsevier; (II) L. Fernandez-Rojo et al. 2019/Reproduced with permission from Elsevier.

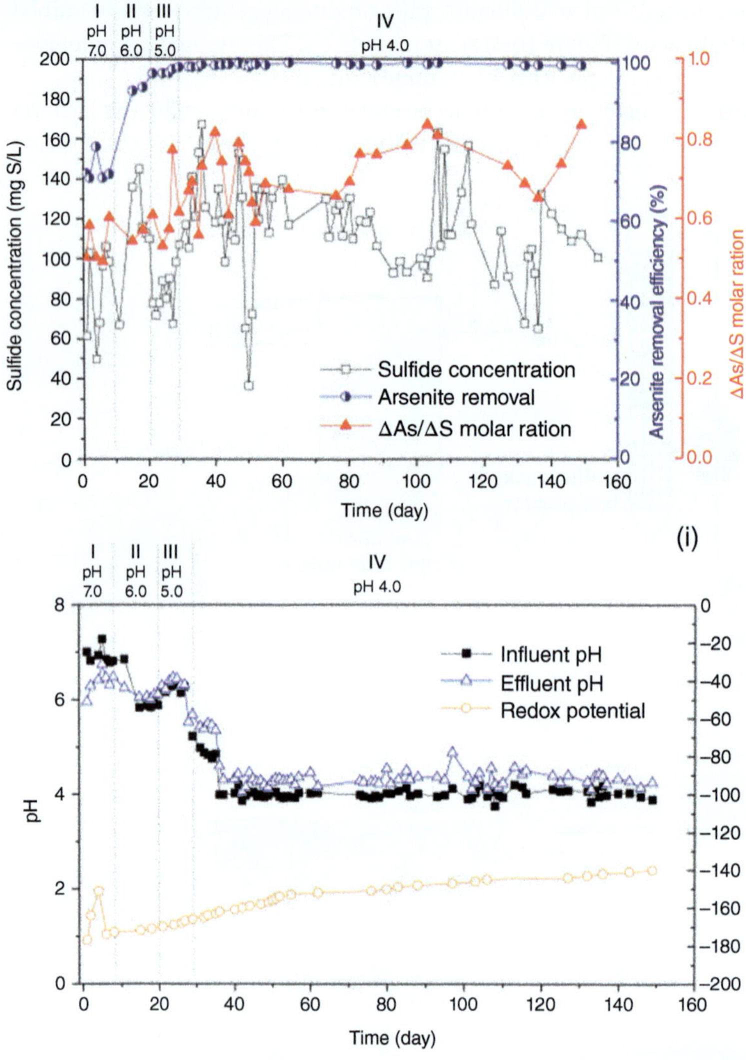

Figure 10.5 (i–iii), (iA) Sulfide concentrations in the effluents from the sulfidogenic bioreactor, and the As(III) removal efficiencies and DAs/DS molar ratios in the precipitation tank. (B) Influent and effluent pH values and effluent redox potentials of the sulfidogenic bioreactor. (ii) The rates of sulfide production and TOC consumption and C/S mass ratios in the sulfur-reducing bioreactor, (iii) the proposed mechanisms of sulfur reduction by *Desulfurella* under acidic conditions. *Source:* Adapted with permission from Sun et al. (2019a, b)/Elsevier.

illustrates the high efficiencies of the As(III) removal from the synthetic AMD. In the initial three Phases I–III, the arsenite removal efficiency increased from 70% to 97%, and it remained above 99% removal efficiency in Phase IV. According to the findings, under acidic conditions, the complete conversion of arsenite into the insoluble orpiment precipitates (Eq. (10.10)) was accomplished without the formation of soluble As by-products (e.g. thioarsenic complexes). The redox potentials of the effluents from the bioreactor were

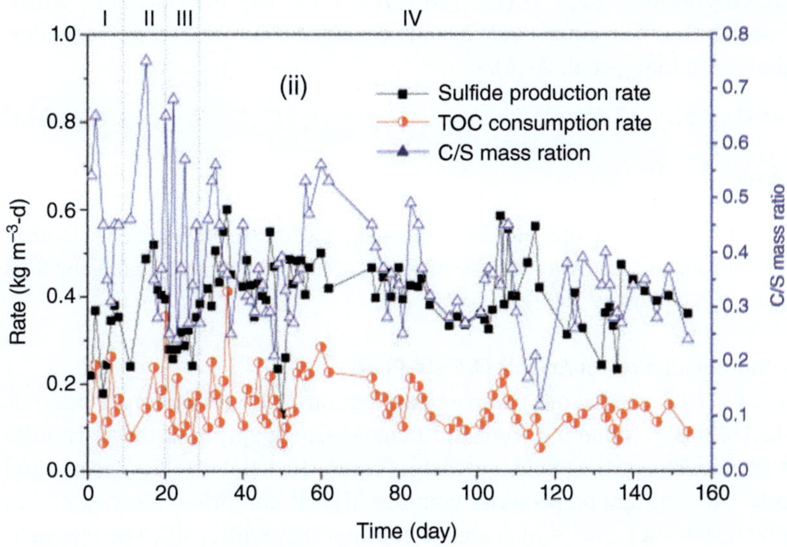

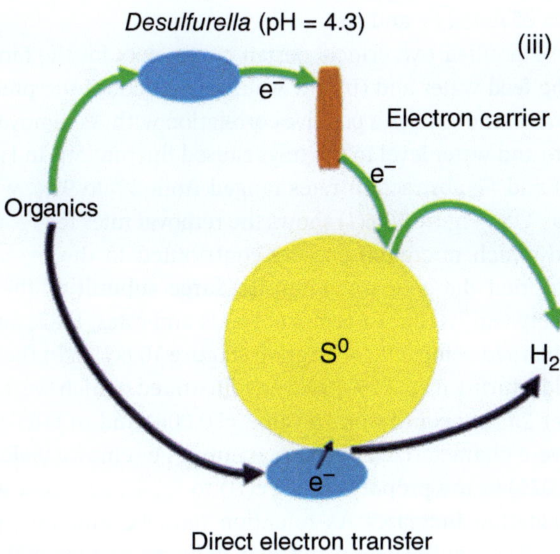

Figure 10.5 (Continued)

measured in the range of −141 to −178 mV in all four phases (Figure 10.5(ii)), favoring the formation of As sulfide precipitates (Kumar et al. 2016).

During the microbial sulfur reduction process, protons and bisulfate ions are produced (Eq. (10.13)), and the reactions of bisulfide with protons at pH lower than 7 (Eq. (10.14)) can offset the pH decrease caused by the proton production. Therefore, mechanisms of microbial sulfur reduction under acidic conditions may differ from those under near-neutral conditions (Figure 10.5(iii)). In near-neutral conditions, the reaction between sulfide and sulfur polysulfides ($S_{n+1}{}^{2-}$) (Eq. (10.15)) are the key intermediates which increase the bio-availability of insoluble sulfur and accelerate microbial sulfur reduction (Boyd and Druschel 2013; Liang et al. 2016).

$$HS^- + H^+ \leftrightarrow H_2S_{aq} pK_a = 7.05 \quad (10.14)$$

$$HS^- + nS^o \leftrightarrow S_{n+1}{}^{2-} + H^+ \quad (10.15)$$

Desulfurella bacteria present in the biofilm, summary sulfur through direct electron transfer which is an important pathway for the microbial reduction of iron oxides by SRB and iron-reducing bacteria (Lovley 1997).

10.4.1.4 Passive Bioremediation for Arsenic (A Field-Pilot)

Fernandez-Rojo et al. (2019) carried out passive treatment on AMD containing dangerous levels of As, up to $152\,mg\,l^{-1}$, obtained from the Reigous Creek spring at the Carnoulès mine in southern France. The passive field bioreactor (Figure 10.4(II)) employs the method of iron oxidation by autochthonous bacterial communities. It comprises a series of five shallow trays, each measuring $1.5\,m^2$, which are fed continuously with AMD. The researchers have diligently observed the flow rate, physicochemical properties of the inlet and outlet water, and the speciation of redox Fe and As.

Fernandez-Rojo et al. (2019) identified two critical design parameters for the bioreactor: (a) adequate oxygenation of the feed water and (b) low water height above the precipitate. The hydraulic retention time was found to have a positive correlation with As removal yields. However, variations in flow rate and water level in the trays caused fluctuations in HRT during the monitoring period. As and Fe abstraction rates ranged from 3% to 98%, while the maximum Fe removal yield was 11%. Figure 10.6(I) shows the removal rates for Fe and As.

To determine the degree to which microbial activity contributed to the oxidation of arsenite, the researchers quantified the gene encoding the large subunit of the As(III) oxidase (aioA). The relations between As and Fe removal yields and rates, HRT, and physicochemical variables were explained using a PCA diagram (Figure 10.6(II)). In their study, the maximum As removal yields during days 115–194 were illustrated, which were considerably correlated to inlet water DO concentration (p-value = 0.006) and to HRT (p-value <0.001). Days 11, 27, and 32 were characterized by the maximum Fe removal yields, negatively correlated (p-value = 0.025) to the proportion of Fe(II) to total Fe in inlet water by natural flow. Since As(III) oxidation increases As retention onto Fe minerals, bacteria with the ability to oxidize As(III) were measured through their marker gene aioA. The As-bearing phases producing amorphous ferric arsenate are beneficial in the perception of long-term storage of this treatment sludge. This solid phase has a high As load and a low solubility compared to As(III)-Fe(III) phases (Langmuir et al. 2006; Maillot et al. 2013). However, the present field bioreactor needs further improvements, which are listed in Table 10.2. Furthermore, the effect of storage environments on sludge stability must be fully estimated. Hence, future determinations ought to emphasize the improvement in the As removal rate by testing higher oxygenation in addition to the introduction in the trays of physical support with high pore volume, such as pozzolana, to avoid speedy clogging and to maximize the surface area of Fe- and As-oxidizing biofilm per unit of water volume (Battaglia-Brunet et al. 2006).

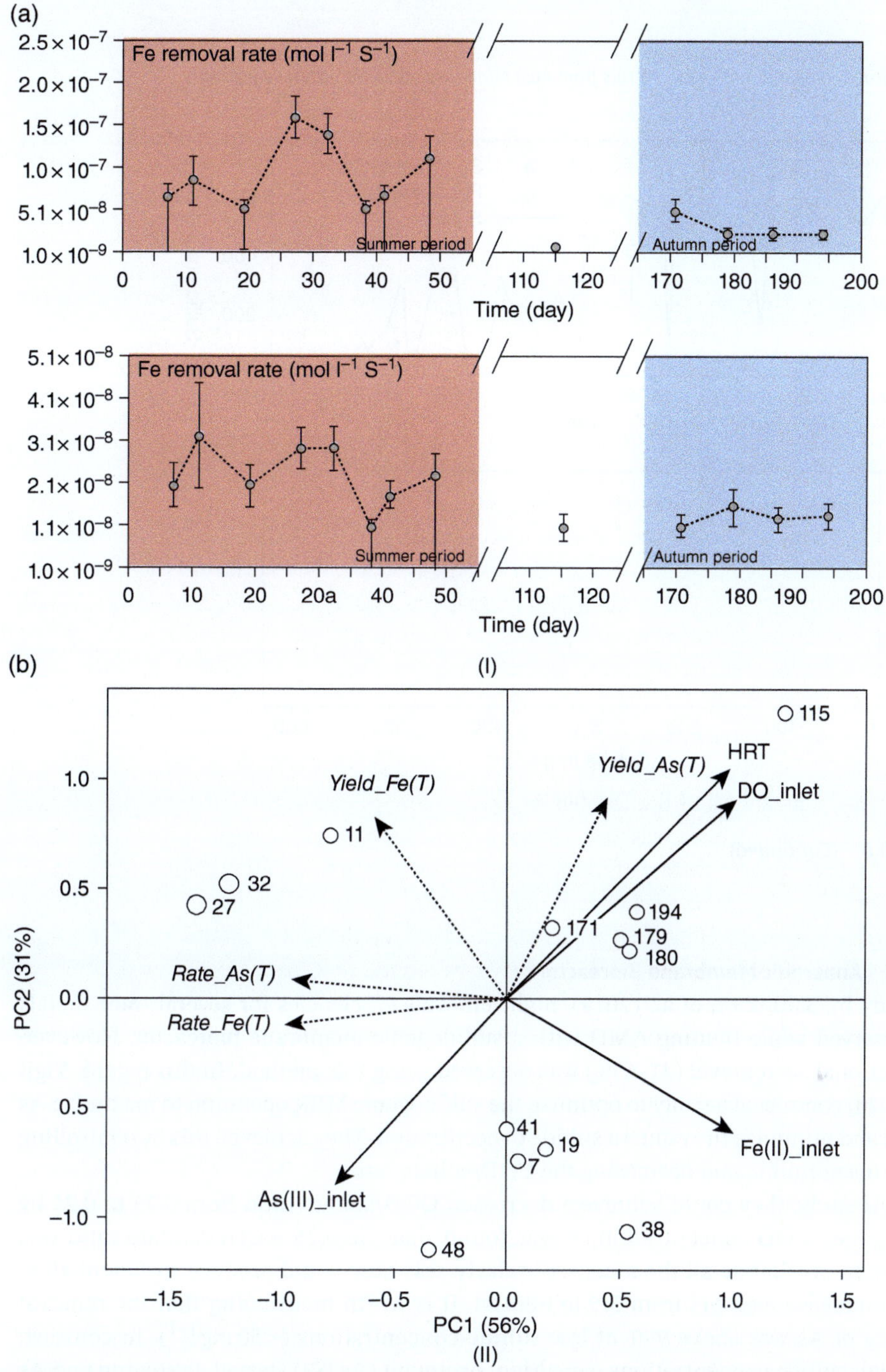

Figure 10.6 (I) Monitoring of the Fe (a) and As removal rates (b) in the field bioreactor. Rate values represented by a high negative SD correspond to maximum removal rates estimated from the analytical detection limit (no significant total Fe or total As concentration decrease between inlet and outlet). (II) Principal Component Analysis (PC1 × PC2) based on functional parameters (total As and total Fe removal yields and rates) of the bioreactor across time. Circles with numbers represent the sampling days after the bioreactor setup. Vectors in the bi-plot overlay were constructed from a data frame containing HRT and water physicochemistry parameters (inlet and outlet temperature, DO, pH, Eh, conductivity, Fe, and As concentrations and proportions of Fe(II) to total Fe and As(III) to total As). Only correlations with p-values ⩽0.05 were included. The angle and length of the vector indicate the direction and strength of the variable, (III) Relationship between permeate sulfide and As concentrations. *Source:* Adapted with permission from (II) Fernandez-Rojo et al. (2019)/Elsevier (III) Adapted with permission from Yigit et al. (2020).

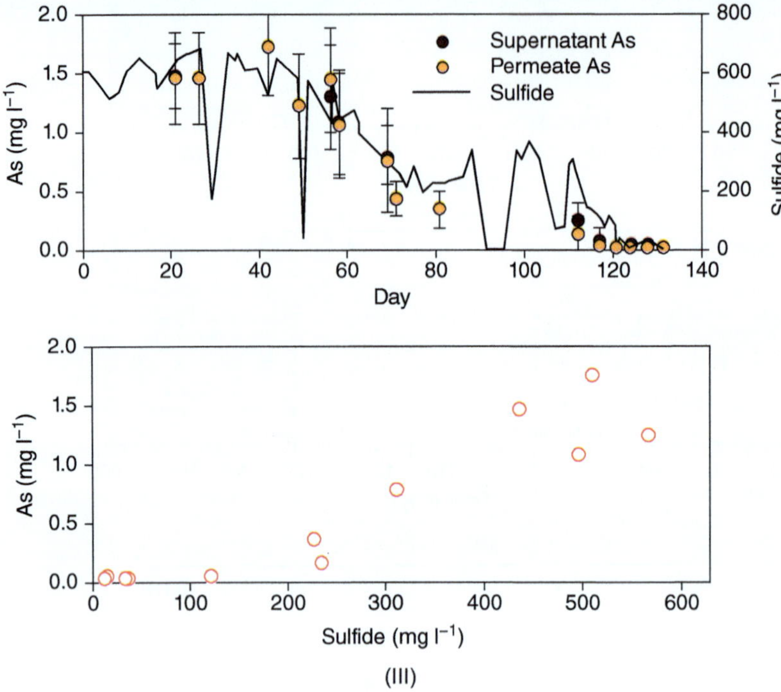

Figure 10.6 (Continued)

10.4.1.5 Anaerobic Membrane Bioreactor

In a study by Sahinkaya et al. (2019), high removals efficiencies for several trace metals were observed while treating AMD with a sulfidogenic membrane bioreactor. However, only fractional As removal (41–67%) was observed using this method. In this regard, Yigit et al. (2020) conducted a study to optimize the sulfidogenic MBR operation to maximize As removal and minimize the effluent sulfide concentration. They achieved this by controlling sulfide concentration and decreasing the COD/sulfate ratio.

In their study, they could achieve a decreased COD/sulfate ratio from 0.75 to 0.25 by reducing feed COD concentration. It was found that the 0.25 – COD/sulfate ratio was sufficient to precipitate all the metals. Similarly, this ratio could produce sufficient alkalinity to increase feed pH from 3.5 to neutral. It is worth mentioning that the removal efficiency of As was above 99% at low sulfide concentrations (<50 mg l^{-1}). In contrast, when high sulfide concentrations were high, orpiment (As2S3) started dissolving and As concentration in permeate increased almost linearly in the bioreactor ((Figure 10.6(III) (reaction (10.6)).

SEM-EDS analyses of the cake layer show that the inorganic foulants – Fe, S, Cu, and Zn concentrations – in the sludge were relatively high (Figure 10.7a). The developed cake layer of 75 μm thickness was scrubbed, and extracted metals were measured using ICP (Figure 10.7b). It was observed that Fe concentration reached around 1.6 g m^{-2} and Ca^{2+} and Mg^{2+} concentrations reached 600 mg m^{-2}.

Table 10.2 Application of bioremediation technique in removal of trace metals from acid mine drainage.

Microbial type	Metals	Reaction condition	Removal efficiency (%), concentration in mg l^{-1}	Bioremediation technique	Advantages/limitation	Reference
Acidithiobacillus ferrooxidans along with Biochar	Copper (Cu)	pH 2.0	Reduced the biodissolution rate of chalcopyrite from 57.1% to 39.4%	Formation of passivation layer on Cu containing minerals and adsorbing bacteria with the biochar	Biochar significantly inhibits chalcopyrite biodissolution was significantly reduced with the use of Biochar along with the Fe-oxidizing bacteria	Yang et al. (2020)
SRB *Citrobacter* sp.	Copper (Cu)	pH 6.6	Bioremoval of 99% copper and 12.8 mM of sulfate	Copper precipitation and sulfate reduction	Citrobacter sp. strain DBM is capable of sulfate reduction and can survive in sulfate-rich AMD. Sulfate reduction activity ceased at high Cu concentration of 0.6 mM.	Qiu et al. (2009)
Sulfate-reducing bacteria with Fe^0	Copper (Cu)	pH 4.0	Above 95%	Bench scale continuous run	Double efficiency of SRB + Fe0 system in metal removal in compared to the SRB system	Bai et al. (2013)
Desulfovibrio sp. sulfate-reducing bacterial (SRB) with microalgae	Copper (Cu)	pH 5.5	72.4–74.4% sulfate and over 91.7% of Cu(II)	Batch experiments	Immobilized SRB beads with abundant low-cost microalgae as a nutrient source	Li et al. (2018a, b)
Sulfate-reducing bacteria and cellulolytic bacteria	Copper (Cu)	pH 4.2	0.33 mol m^3 day^1 (99% metal removal)	Sulfidogenic diffusive exchange system (SDES) reactors through reactive bed	Indirect diffusive transport system protected the microorganisms in the reactor bed from direct exposure to AMD Recovery of valuable minerals	Pérez et al. (2017)
Sulfate-reducing bacteria and cellulolytic bacteria	Copper (Cu) and Zinc (Zn)	pH < 5.0	100% at high sulfate removal rates (0.44–0.47 mol m^3 day^1)	SDES permeable reactive barrier (PRB) bench-scale reactors	Thinner layer reactor failed in the effluent after 150 days	Torregrosa et al. (2019)
Acidophilic sulfate-reducing bacteria (SRB)	Copper (Cu)	pH 4.5	90%	Sulfidogenic bioreactor	Recovery of valuable minerals	Silva et al. (2019)

(*Continued*)

Table 10.2 (Continued)

Microbial type	Metals	Reaction condition	Removal efficiency (%), concentration in mg l^{-1}	Bioremediation technique	Advantages/limitation	Reference
Microalgae *Nannochloropsis oculate*	Copper (Cu)	pH 8.5	99.92 ± 0.04%	Laboratory condition through algal remediation	Microalga can metabolize large amounts of Cu can be utilized widely in AMD to eliminate the high concentration of Cu. Inhibition of algal growth and lipid accumulation at higher Cu concentration.	Martínez-Macias et al. (2019)
Fungi *Trichoderma lixii* CR700	Copper (Cu)	pH 7.0	84.6% at 10 mg l^{-1} of Cu^{2+} concentration	Batch experiment	Tolerance to high Cu concentration The fungi are bio-pesticide and plant growth promoter	Kumar and Dwivedi (2021)
SRB with the genera *Clostridiaceae, Eubacterium, Pseudobutyrivibrio*, and *Clostridium*	Copper (Cu) and Zinc (Zn)	pH 2.0	98% and 85% for Cu and Zn respectively	Column experiment	Resist high acidic conditions in the media and good metal removal capacity.	Lu et al. (2011)
Sulfur reducing bacteria and municipal compost (C Source)	Fe, Al, Zn and Cu	pH 3.0	60% of the influent Zn >99% for Fe, Al and Cu	Batch experiment	Saturation of sorption sites for composts.	Gibert et al. (2005)
SRB with multiple substrate (spent mushroom compost [SMC], limestone, activated sludge [AS], and woodchips [WC])	Mn, Cu, Pb, and Zn	pH < 3.1	89–100% Mn, Cu, Pb, and Zn	Batch experiment	Fe- reduction not sufficient in the bioreactor and further treatment phase is required for removal of Fe	Muhammad et al. (2017)

Engineered *Escherichia coli* having cadmium binding peptides.	Cadmium (Cd)	pH 3–9	10–41% and 47–89% at pH 7 and 9	Bioengineered beads	Cost-effectively, can be produced in large scale, biodegradable and safe	Marques et al. (2020)
Fe–Mn oxidizing bacteria *Flavobacterium, Brevundimonas, Stenotrophomonans* and *Thermomonas*	Iron (Fe) and Manganese (Mn)	pH 4.2	Fe (99.8%) and Mn (87.5%)	Batch experiment	Increase in pH from 4.2–7.2 during the treatment.	Hou et al. (2020)
Mn(II)-oxidizing microorganisms	Al, Mn and Fe	pH 3.66	Majority of Al, Mn and Fe removed	Field pilot passive bioremediation technique	The system was operative for long run (13 years) with efficient metal removal.	Crafton et al. (2019)
Acidiphilium multivorum JZ-6 and *cidithiobacillus ferrooxidans* LX5	Fe	pH 2	Fe 100% and 40% sulfate	Sequence batch reactor (SBR)	Recovery of Iron bearing mineral Schwertmannite, useful for absorption of Arsenic and Chromium.	Jin et al. (2020)
A. ferrooxidans	Fe	pH 2.5	100% removal of Iron	Biomineralization	Recovery of valuable minerals	Wang et al. (2020)
Sulfate-reducing bacteria (SRB)	Iron (Fe) and Manganese (Mn)	pH 3–5.5	Fe^{2+}, Mn^{2+} removed 98.70% and 79.97% respectively.	Column experiment	Stable columns without any secondary release at the end of the experiment.	Dong et al. (2020)
Yeast species *Rhodotorula mucilaginosa*	Manganese (Mn)	pH 6.5	0.91 mM	Laboratory experiment	Efficient in Mn removal.	Ruas et al. (2020)
Sulfur reducing bacteria (Desulfurella)	As(III)	pH (pH = 4.3),	99%, $10\,mg\,l^{-1}$	Batch reactor	The residual sulfide can serve as a good precipitant for the removal of other heavy metals (such as Zn, Cu and Cd) co-present in the arsenic contaminated waters, such as AMD and metallurgy industry effluents.	Sun et al. (2019a, b)

(*Continued*)

Table 10.2 (Continued)

Microbial type	Metals	Reaction condition	Removal efficiency (%), concentration in $mg\,l^{-1}$	Bioremediation technique	Advantages/limitation	Reference
Passive treatment of As-rich acid mine drainage (AMD); gene aioA; auto- chthonous bacterial communities	As(III) and Fe	pH 2.6–3.5; 84–152; As/Fe molar ratio; 0.29–0.86	As(III) ranged 3% and 97% Fe 11%	Field pilot passive bioremediation technique 5 aerobic ponds stacked on top of each other connected by cascades. Carnoulès mining site (southern France)	*Advantage*. The biotreatment system we designed proved to be suitable for high As DMA. The formation of sludge highly enriched into As(V) rather than As(III) is advantageous in the perspective of long term storage *Limitations:* • Water circulation in the system must be improved to stabilize the HRT in future work. • Fe oxidation and precipitation yield appeared to be limited by the large water height above the precipitate • The low amount of iron, maximum 11% precipitated, compared to other systems where the AMD is neutralized, enables the production of low amounts of sludge, highly concentrated in arsenic (up to 20% dry weight), thus limiting the cost of As-rich sludge disposal in engineered hazardous wastes landfill facilities.	Fernandez-Rojo et al. (2019)

Sulfidogenic anaerobic membrane bioreactor	AS (III)	PH = 3.5, sulfide concentrations (<50 mg l^{-1}), COD/sulfate of 0.75 and 0.25, Metal concentrations 75 or 150 mg l^{-1} Fe, 25 mg l^{-1} Cu, 5 mg l^{-} Zn, 5 mg l^{-} Co, 5 mg l^{-1} Mn, 2.5 mg l^{-1} Ni and 2.5 mg l^{-1} As.	99%	Batch reactor	• COD limiting conditions, COD oxidation efficiency was higher than 95% • High sulfide concentration led to dissolution of orpiment (As2S3) and low As removal efficiency. • Whereas sulfate reduction efficiency decreased from around 95% to 40% at COD/sulfate of 0.75 and 0.25. • High metal in the feed increased sludge filterability in the MBR, which may be due to behaving of metal precipitates as filtration aids and decreasing colloidal particles and generating a more permeable cake layer.	Yigit et al. (2020)
Passive treatment of As-rich acid mine drainage (AMD); aioA genes	As (III)	As-rich AMD (~30–40 mg L-1); water heights (4 and 25 mm) and with/without forced aeration	A maximum of 80% As removal was achieved within 500 min at the lowest water height.	bench-scale continuous flow channel bioreactor	Higher water height slowed down Fe(II) oxidation, Fe precipitation and As removal, in relation with limited oxygen transfer through the water column. • This study proved the ability of this bioreactor-system to treat extreme As concentrations and may serve in the design of future in situ bioremediation system able to treat As-rich AMD	Fernandez-Rojo et al. (2019)

(Continued)

Table 10.2 (Continued)

Microbial type	Metals	Reaction condition	Removal efficiency (%), concentration in $mg\,l^{-1}$	Bioremediation technique	Advantages/limitation	Reference
Sulfate-reducing bacteria	As(III)	(pH 6–8). Our results show that SRB tolerate low pH (up to 3.5) and high concentrations of As (~3.6 $mg\,l^{-1}$).	As removal of up to 73%, Iron (Fe) removal higher than 78% and a neutralization of pH from acidic to circum-neutral conditions	Batch reactor	• Additional efforts are needed to describe the mechanisms that are involved in the tolerance to As, Fe and low pH, and the tolerance to other metals. Future research should include the study of these processes and the interaction of acid/metal-tolerant SRB with other acidophilic or acid-tolerant microorganisms living in AMD waters.	Serrano and Leiva (2017)

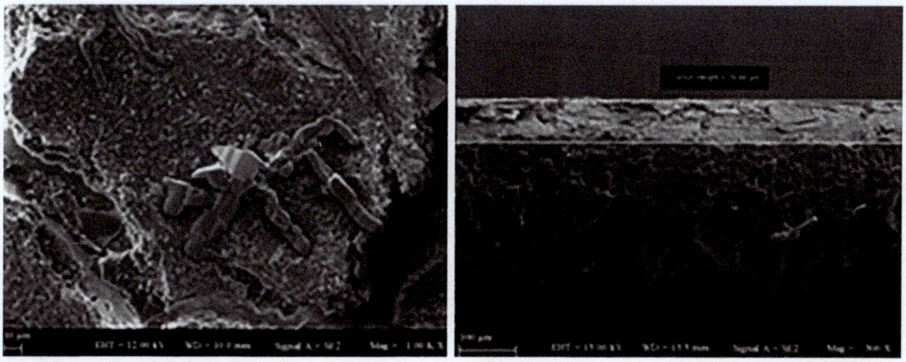

(a)

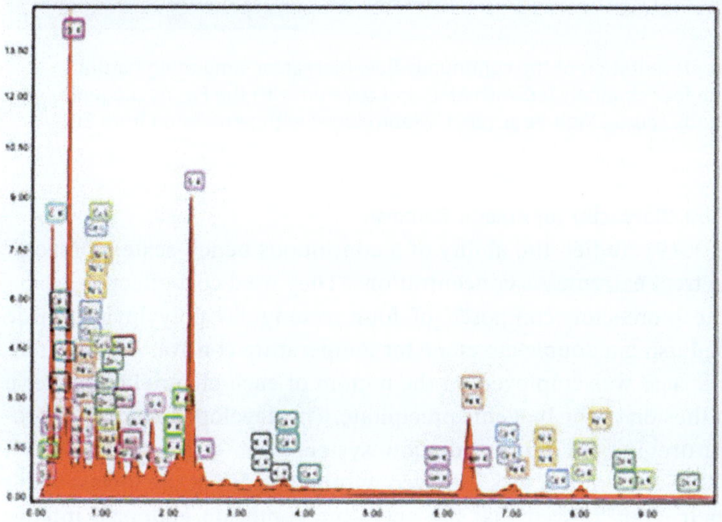

Element	Weight %
C	20.8
N	51
O	22
Mg	0.7
As	0.4
Al	1.2
Si	1.7
P	0.9
S	10.9
K	0.6
Ca	0.7
Mn	0.9
Fe	19.9
Co	2.2
Ni	1.5
Cu	6.5
Zn	2.2

(b)

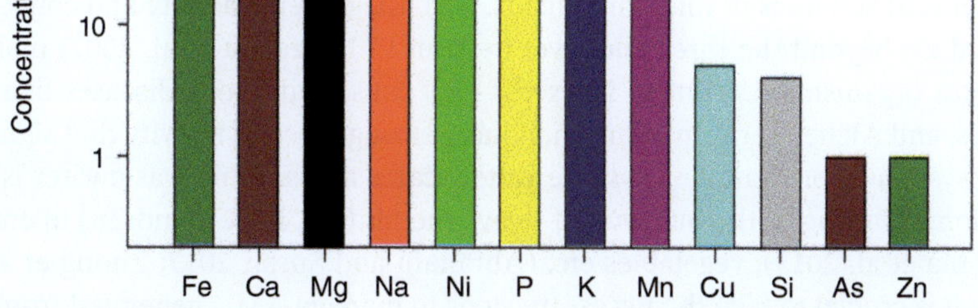

Figure 10.7 (a) SEM images and SEM-EDS analyses of the developed cake layer on the membrane surface, (b) concentration of inorganics in the cake layer. *Source:* Yigit et al. (2020)/Reproduced with permission from Elsevier.

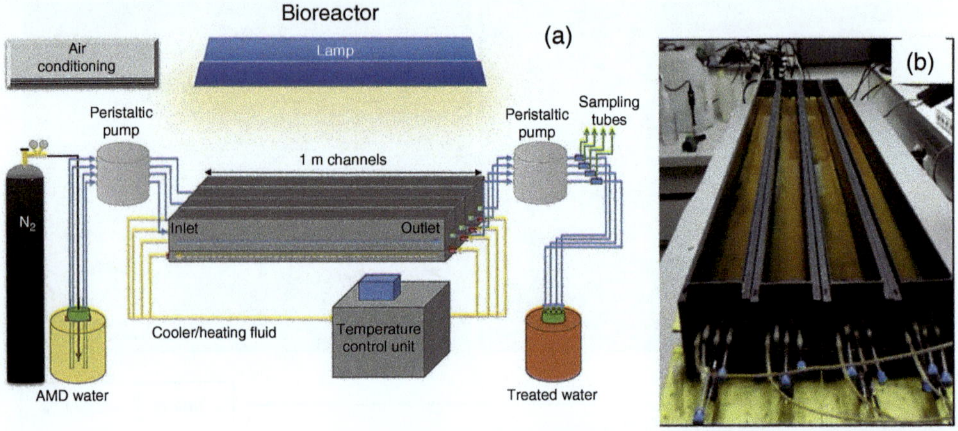

Figure 10.8 Schematic representation of the continuous flow bioreactor simulating natural attenuation of AMD (a). The four channels fed with AMD and covered with the Fe–As biogenic precipitates at the bottom (b). *Source:* Yigit et al. (2020)/Reproduced with permission from Elsevier.

10.4.1.6 Continuous Flow Bioreactor for Arsenic Removal

Fernandez-Rojo et al. (2019) studied the ability of a continuous bench-scale bioreactor system (Figure 10.8) to treat extreme As concentrations. They used cost-effective biodegradable canvas for the bioreactor, composed of four rectangular polyvinyl chloride channels, each equipped using a double envelope for temperature control. A biodegradable canvas of polylactic acid was employed on the bottom of each channel to deliver a rough surface for the adhesion of the biogenic precipitate. The developed bioreactor provides a compelling future in situ bioremediation system that could treat As-rich AMD. Approximately, 80% As removal was achieved within 500 min at the lowest water height. In addition, their study proved that the operating condition endorsed intense Fe(II) microbial oxidation and consequent precipitation of amorphous ferric arsenate and As-bearing schwertmannite. The chemical composition of the feed waters of AMD is shown in Table 10.3.

10.4.2 Copper

Another abundant element found in AMD is Cu (Siew et al. 2020). Cu is essential for the physicochemical activities of microorganisms. Still, the extreme acidity and concentration of dissolved Cu beyond the threshold level ($>10\,\text{mg}\,\text{l}^{-1}$) (Azabou et al. 2007) make AMD toxic to most organisms. Cu intake in excess may cause numerous diseases like Wilson, Parkinson's, and Alzheimer's. In addition, it has a strong biocidal activity that significantly affects ecosystems. For example, in some water-scarce areas, mine wastewater is utilized for irrigation, resulting in the build-up of heavy metals (Cu, Pb, Cr, and Zn) in crops such as wheat (Ma et al. 2015), vegetables etc. (Abraham and Susan 2017; Zhong et al. 2018). Therefore, it is crucial to take the necessary steps to manage AMD generated from chalcopyrite dissolution so that release of Cu ions can be controlled. Most Cu bioremediation research is done on a laboratory scale and pilot project basis.

Table 10.3 Inlet water chemistry for the whole duration of the experiments.

Code	Experimental conditions	Date of water collection	Experiment duration (days)		pH	T (°C)	EC ($mS\,cm^{-1}$)	Eh (mV)	Fe(II) ($mg\,l^{-1}$)	Total Fe ($mg\,l^{-1}$)	As(III) ($mg\,l^{-1}$)	As(V) ($mg\,l^{-1}$)	Total As ($mg\,l^{-1}$)	SO_4^{2-} ($g\,l^{-1}$)
	Water height: 4 mm			Average	3	20	3.21	535	439	452	13	17	30	1.8
Exp 1(FF)	Floating film: Yes	20/01/2015	17	SD	0.1	1	0.06	39	30	26	4	6	9	0.1
	Air bubbling: No			SD (%)	2	3	2	7	7	6	32	36	31	5
				n	20	19	8	8	20	8	8	8	8	8
	Water height: 4 mm			Average	3.1	19.7	3	547	432	441	19	12	31	1.8
Exp 1(OA)	Floating film: No	20/01/2015	34	SD	0.2	5	0.2	34	29	23	8	7	7	0.1
	Air bubbling: No	12/02/2015		SD (%)	5	3	7	6	7	5	40	54	22	4
				n	27	25	15	15	27	15	15	15	15	15
	Water height: 4 mm			Average	3.4	20	2.89	499	480	484	30	6	36	1.9
Exp 2(4 mm)	Floating film: No	23/03/2015	35	SD	0.3	0.8	0.06	45	25	19	10	2	10	0.1
	Air bubbling: No			SD (%)	9	4	2	9	5	4	34	42	29	6
				n	35	35	22	23	35	15	18	18	18	15
	Water height: 4 mm (22 days) + 25 mm (13 days)			Average	3.4	19.8	2.89	494	476	482	33	6	39	1.9
Exp 2(25 mm)	Floating film: No	23/03/2015	35	SD	0.3	0.8	0.06	46	29	16	12	2	11	0.1
	Air bubbling: No			SD (%)	10	4	2	9	6	3	37	40	30	3
				n	28	28	16	16	28	8	11	11	11	8
	Water height: 4 mm (22 days) + 25 mm (13 days)			Average	3.4	19.8	2.87	492	472	479	33	5	38	1.9
Exp 2(25 mm. Air)	Floating film: No	23/03/2015	35	SD	0.3	0.8	0.07	48	29	13	12	2	12	0.1
	Air bubbling: last six days			SD (%)	10	4	2	10	6	3	37	41	31	4
				n	28	28	16	16	28	8	11	11	11	8

Source: Adapted with permission from Yigit et al. (2020).

10.4.2.1 Preventive Methods Relying on Inhibition of Oxidation of Sulfide Minerals

AMD generated by Cu-bearing chalcopyrite dissolution may create a pool of dissolved Cu in the environment. Moreover, microbial oxidation to chalcopyrite is used worldwide for Cu production, which may release large amounts of dissolved Cu ions and generate an acidic environment if the dissolution happens on the surface of the mining areas.

Hence, the amount of AMD generation can be reduced successfully through inhibition of biodissolution (Yang et al. 2020). Some coating substances, viz. clay, soil, plant cover, etc., can restrict the oxidation of sulfide minerals by preventing air and water. Yang et al. (2020) reported that biochar could be used effectively as a protective coating to passivate the dissolution of chalcopyrite. Biochar is well arranged with surface functional groups and adsorption sites where bacteria and a large number of Fe ions can get adsorbed (Li et al. 2017). Thus, the oxidation of chalcopyrite is retarded (Figure 10.10(ii)).

They investigated the effects of biochar on the biodissolution of chalcopyrite at different pyrolysis temperatures (biochar–300°C, biochar–400°C, and biochar–500°C) with *A. ferrooxidans* by adding $2\,g\,l^{-1}$ to the systems with biochar and reported that biochar can effectively inhibit the biodissolution of chalcopyrite (Figure 10.9(i)). The biodissolution rate of chalcopyrite was reduced from 57.1% to 39.4%, with the best inhibition effect at 300°C. The optimal concentration of biochar–300°C was recorded at $3\,g\,l^{-1}$ for inhibiting chalcopyrite biodissolution. The formation of a passivation layer (jarosite and Sn^{2-}/S^0) and adsorbing bacteria with the biochar was the prime cause of inhibition.

10.4.2.2 Bioremediation of Cu Relying on Sulfate Reduction

Many recent studies have reported the use of SRB in Cu removal from AMD. Qiu et al. (2009) extracted an eccentric SRB *Citrobacter* sp., a facultative anaerobe, from a mining area. The growth of the bacteria with the changes in Cu and sulfate concentrations was monitored. It was observed that the initial Cu concentration of 0.2 mM was decreased to traces in seven days and along with that, a decrease in sulfate concentration (via reduction to sulfide) of 15.0 mM was also recorded. 99% Bioremoval of Cu and reduction of 12.8 mM of sulfate was achieved in seven days with the initial Cu concentration of 0.4 mM. With further increase of Cu concentration to 0.60 and 0.80 mM, the activity of the organisms ceased, and Cu precipitation discontinued with no sulfate reduction in seven days.

Effective treatment of Cu-containing wastewater using SRB, further research came to light to improve the ability of SRB in hostile conditions with the addition of iron (Fe^0) (Bai et al. 2013). Continuous operation of the SRB system and the SRB + Fe^0 system were monitored. Gradual increase in Cu^{2+} from 0 to $125\,mg\,l^{-1}\,h^{-1}$ with $42\,mg\,l^{-1}\,h^{-1}$ sulfate loading bioreactor was run with at 25°C temperature, 4.0 pH and at 1.9 COD/SO_4^{2-} ratio. The result revealed that the SRB + Fe^0 system performed better in the Cu-containing wastewater treatment, and above 95%, removal efficiency of Cu^{2+} was achieved in SRB + Fe^0 system. The effluent from this system was good enough to be discharged directly.

Further study on Cu removal from wastewater with the application of SRB was done by Li et al. (2018a, b). A strain of *Desulfovibrio* sp. with different microalgae, viz. *Scenedesmus obliquus*, *Chlorella vulgaris*, *Selenastrum capricornutum*, and *Anabaena spiroides* as the carbon/energy source was used in the bioremediation of wastewater containing Cu(II) concentration as high as 60 and $600\,mg\,l^{-1}$ sulfates in batch experiments. Immobilized SRB beads, along with a nutrient source of low-cost microalgae, could remove 72.4–74.4% of sulfate,

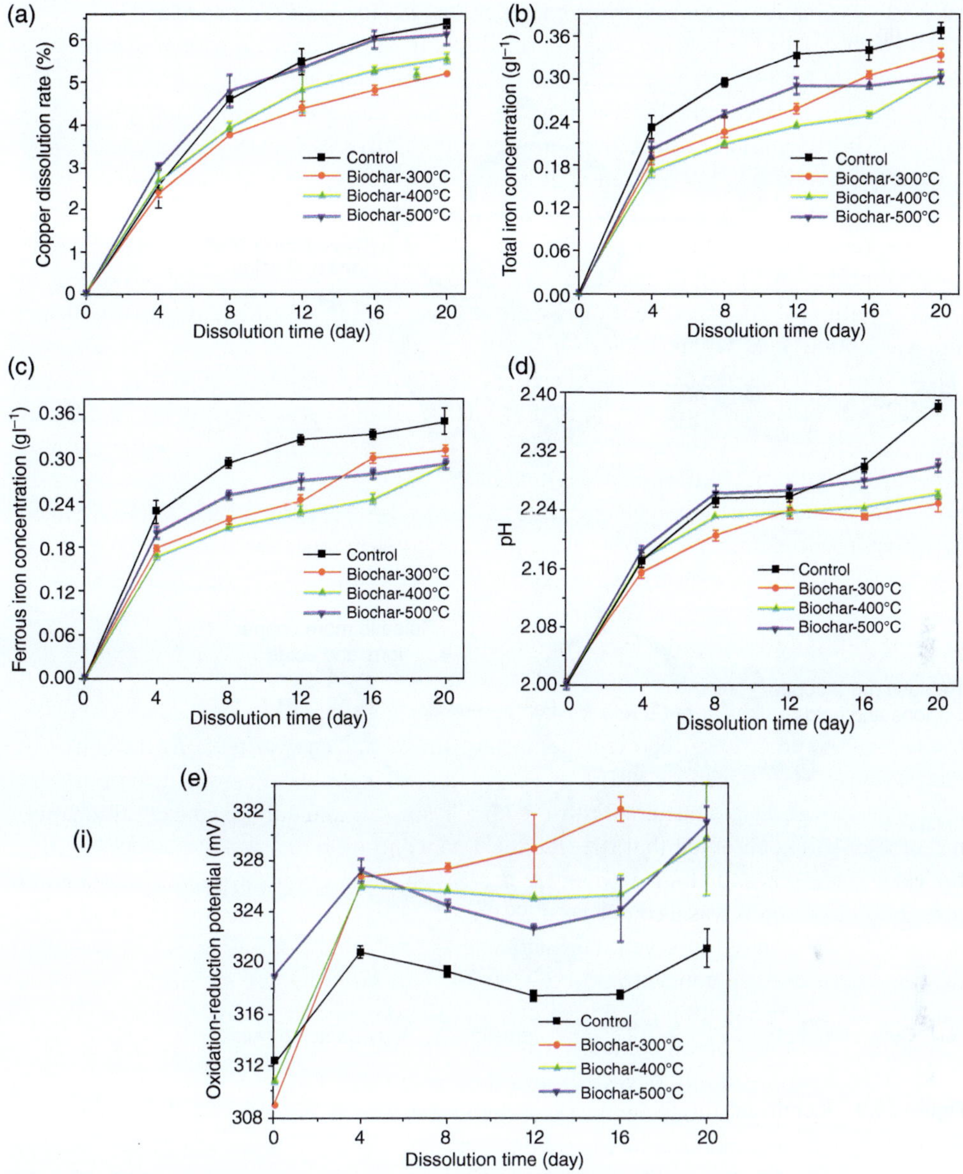

Figure 10.9 (i) Changes in the (a) Cu dissolution rate, (b) total Fe concentration, (c) ferrous Fe concentration, (d) Ph and (e) redox potential during chalcopyrite dissolution with different biochars. (ii) The mechanism of biochar inhibiting chalcopyrite biodissolution. *Source:* Adapted with permission from Yang et al. (2020)/Elsevier.

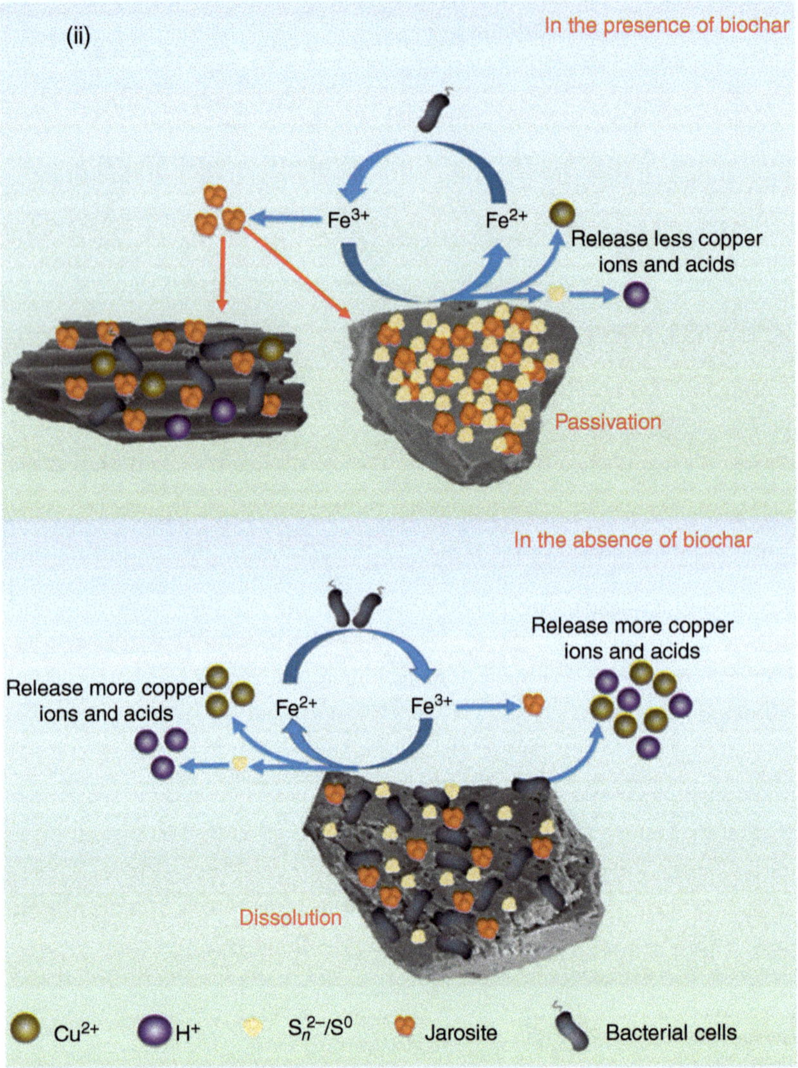

Figure 10.9 (Continued)

and over 91.7% of Cu(II) at initial pH 5.5, which indicates the use of SRB could be an efficient bioremediation alternative for AMD treatment. The action of the SRB was boosted in the presence of microalgae. Three sulfidogenic diffusive exchange system (SDES) reactors have been implemented by Pérez et al. (2017) to treat AMD with a soaring Cu concentration of 528 $mg\,l^{-1}$ using a screen tube and accomplished metal removal to the extent of 99%.

Accumulation of Cu at the bottom of the screen tubes reveals the prospect of recovering commercially valuable materials in passive reactors. In another study, Torregrosa et al. (2019) also used SDES permeable reactive barrier (PRB) bench-scale reactors to treat concentrated Cu in AMD (633.5 $mg\,l^{-1}$ Cu) and attained metal practically 100% removal efficiencies with high sulfate removal rates (0.44–0.47 $mol\,m^3\,day^{-1}$).

10.4.2.3 Bioremediation of Cu Relying on Selective Metal Recovery

AMD biological treatment not only facilitates the removal of Cu, but this is a promising and economical metal recovery (as CuS) option for the metallurgical process. Furthermore, acidophilic SRB is attaining importance due to the potentiality of valuable metal recovery (Silva et al. 2019). Synthesis of covellite crystals was proposed by using actual AMD (collected from a Cu mine in the Amazon region) in a sulfidogenic bioreactor. Covellite crystals being rare in natural conditions can be used in semiconductors (Yadav and Bajpai 2017), solar cells (Fang et al. 2018), adsorbents (Ayodhya and Veerabhadram 2018), electronic devices, use of various nanostructures, such as nanotubes, nanoplatelets, and nanowires (Tezuka et al. 2007).

The recovered Cu can be further used in industries and can be a value addition due to its high industrial demand. Selective Cu recovery is also possible from AMD with ion exchange adsorbents (mesoporous silica adsorbents) (Ryu et al. 2020). In addition, direct contact membrane distillation (DCMD) is an effective tool for reducing sewage generation by concentrating AMD and 80% high-quality water recovery can be achieved for reuse (Ryu et al. 2020; Silva et al. 2020). The use of polyethyleneimine (PEI), glutaraldehyde (GA), diatomaceous earth (DE) – PEI-GA-DE particles are also very effective in treating AMD solutions containing high Cu (80% recovery) (Larsson et al. 2018). Due to strong affinity and selectivity, GA-crosslinked PEI and DE particles were modified to form a Cu-extraction resin. The resin can be used for Cu removal of AMD, as well as for purifying Cu solutions.

10.4.2.4 Bioremediation of Cu by Using Microalgae and Fungi

Microalgae and Fungi can also be utilized in the remediation of heavy metal ions and are biosorbent (Romera et al. 2007; Bulgariu and Gavrilescu 2015; Leong and Chang 2020). The mechanism behind the immobilization of soluble metal ion species varies with the ionic charge and species. In Cu removal, various microalgae and fungi have been used widely by different researchers, such as *Chlorella minutissima* (Yang et al. 2015), *Nannochloropsis oculata* (Khairy 2009).

Martínez-Macias et al. (2019) used the microalgae *N. oculata* in Cu bioremediation. There was an increase in Cu^{2+} concentration in the algal tissue when it was propagated in a media containing 0.04, 0.1, 0.2, 0.3, and 0.4 μmol of Cu^{2+}. There was a reduction in growth rates, cell count, and chlorophyll-a content with the rise in exposure times and concentrations. Removal of the Cu up to $99.92 \pm 0.04\%$ in the culture medium was evidenced, of which $89.29 \pm 1.92\%$ was removed by metabolism reaction, and $10.70 \pm 1.92\%$ was removed by adsorption. The function of microalgae in metabolizing a large amount of C can be utilized widely in AMD to eliminate the high concentration of this metal ion. Fungi, having survival ability in the adverse environment, can be utilized for bioremediation as a cost-effective and eco-friendly alternative (Kapahi and Sachdeva 2017; Khan et al. 2019).

The use of plant pathogenic fungi is not appropriate for mycoremediation. Cu^{2+} is essential for metalloenzyme, directly participates in cellular metabolism, and also serves as a cofactor for some enzymes, including catalase and superoxide dismutase (Contreras-Cortés et al. 2020). Hence, many fungal species have been reported to tolerate high Cu concentrations (Ayele et al. 2021). Kumar and Dwivedi (2021) used *Trichoderma lixii CR700* (*T. lixii CR700*), considering its value as an organic pesticide and plant growth promoter (PGP) (Abhang and Kedar 2019; Katoch et al. 2019) and excellent Cu removal efficiency through cell wall adsorption and metabolic processes (Figure 10.10(i)) @of 84.6% at $10\,mg\,l^{-1}$ of Cu^{2+} concentration at

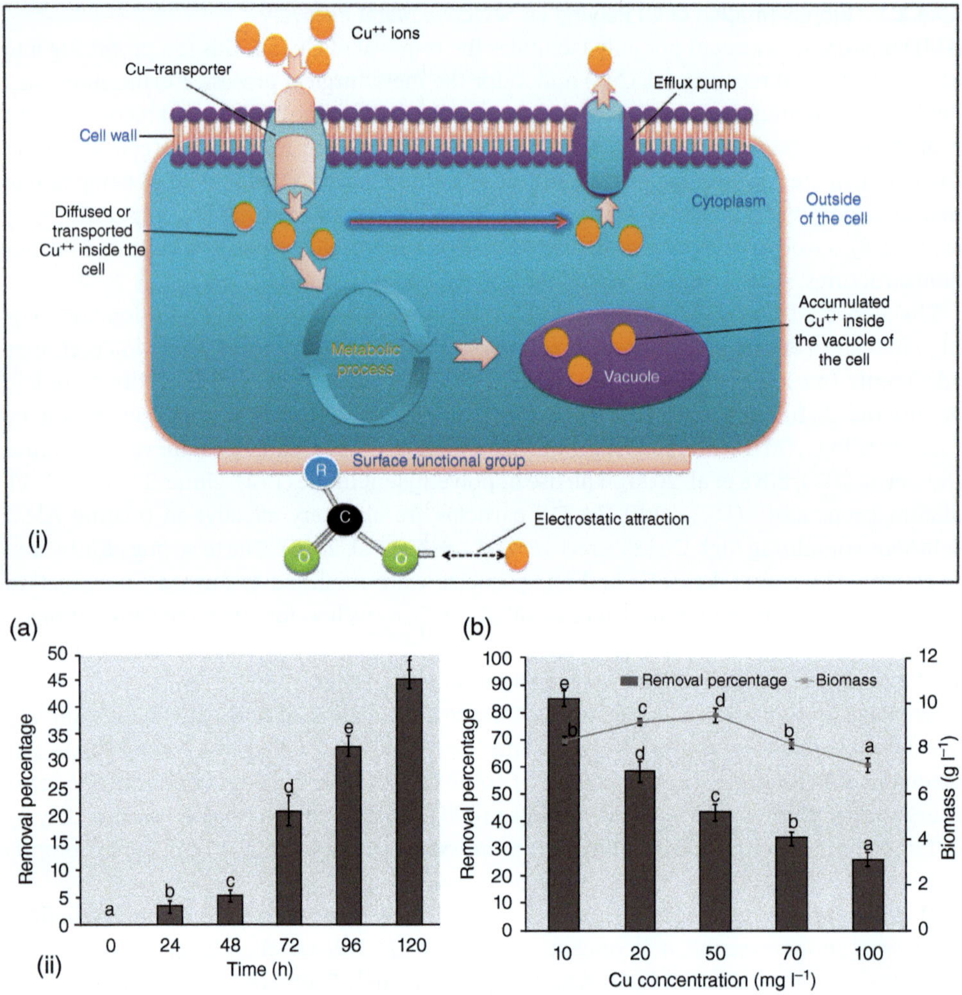

Figure 10.10 (i) Schematic representation of Cu removal mechanism by *T. lixii CR700*: Red spheres (smile) represent Cu^{2+} which is either transported by Cu transporter or directly diffused in the cytoplasm of the cell after interaction with the cell wall. Through the metabolic process, Cu^{2+} might get accumulated via compartmentation inside the vacuole and some amount may be transported from the cytoplasm to outside of the cell via an efflux mechanism to maintain cellular homeostasis. Another hand, after the interaction of Cu^{2+} with the cell wall, some amount of Cu^{2+} is also adsorbed on the surface of the cell wall through surface functional groups such as COO— via electrostatic attraction. Surface sorption of Cu^{2+} may also help in its accumulation inside the cell of the fungi as a first step mechanism; (ii) Effect of (a) incubation periods on Cu^{2+} removal by *T. lixii* CR700 at 50 mg l^{-1} of Cu^{2+} concentration, 28 ± 1 °C, pH 7.0 and 80 rpm of shaking condition and (b) effect of different concentration of Cu^{2+} on its removal by *T. lixii* CR700 and its biomass production at pH 7.0, 120 hours after incubation, 28 ± 1 °C and 80 rpm of shaking condition (the bar above the column and graph represent standard deviation and different letter denotes significant differences using DMRT ($p < 0.05$)). *Source:* Adapted with permission from Kumar and Dwivedi (2021)/Elsevier.

pH 7.0 (Figure 10.10(ii)). Even in the presence of various heavy metals, viz. Mn(II), Ni(II), and As(III), the fungi showed high Cu removal. Fungal species like *Aspergillus australensis*, *Nigrospora*, *Diaporthe*, and *Xylaria* have also been reported to show high tolerance to Cu (Wong et al. 2018).

10.4.3 Zinc, Cadmium, and Lead

Mining and industrial operations produce different heavy metals, among which Zn, Cd, and Pb are potentially toxic and cause damage to the ecological system to a great extent (Tchounwou et al. 2012; Ali et al. 2019; Yadav et al. 2020).

10.4.3.1 Bioremediation of Zinc with Sulfate-Reducing Bacterium

Sulfate-reducing bacterium, *Desulfotomaculum nigrificans*, can grow in Zn-containing environment with a Zn concentration of 12–210 $mg\,l^{-1}$ (Radhika et al. 2006). Through bio-precipitation with biosorption, this SRB can remove Zn from the system with 70% bioremoval efficiency at an initial Zn concentration of 210 $mg\,l^{-1}$ in 40 days. Lu et al. (2011) used a group of SRB with the genera *Clostridiaceae*, *Eubacterium*, *Pseudobutyrivibrio*, and *Clostridium* in a column experiment to treat the AMD using a sulfate-reducing reactor. Dissolved organic carbon was used as a nutrient source for the organisms. The result showed that the microorganisms could resist high acidic conditions in the media and the metal removal capacities were more than 98% and 85% for Cu and Zn, respectively, at influent pH of 2.0. It was also found that, under the highly acidic condition of AMD, SRB grows well in the presence of straw degraders.

10.4.3.2 Compost-Based Bioremediation of Zinc and Lead

Composts and organic matter have evolved as eco-friendly and economical means to remove the metal in AMD. Gibert et al. (2005) did a batch experiment for the removal of Zn. They found that metal removal could be attained by the collective result of precipitation and co-precipitation as metal (oxy)hydroxides and carbonates and biosorption onto the surface of the compost (Figure 10.11). River sediment was added as the source of SRB for which municipal compost was added as organic carbon or nutrient. The system was viable, and the removal of 60% of the influent Zn was attained by co-precipitation with Fe- and Al-(oxy)hydroxide and 40% by sorption on municipal compost. However, after a few days, the compost sorption site was saturated and co-precipitation with fresh Fe- and Al-(oxy)hydroxides was the dominant process established for Zn removal.

In another batch study, Muhammad et al. (2017) used multiple substrates like spent mushroom compost (SMC), limestone, activated sludge (AS), and woodchips (WC) along with limestone under anaerobic conditions suitable for bacterial sulfate reduction (BSR)] during bioremediation of heavy metal and sulfate-containing AMD. SMC played a crucial role in metal removal by supplying C-source to the organisms, and the substrate mixture of 40% CLS, 30% SMC, 20% AS, and 10% WC was the most potent. Microbial reduction helped in the removal of 89–100% Mn, Cu, Pb, and Zn, along with 65% Fe. Higher removal of Mn was evidenced over Fe, implying more sorption of Mn to SMC.

10.4.3.3 Cadmium Bioremediation by Bacterially Assembled Bio Polyester Nanobeads

Cd is one of the highly toxic (Bernard 2008) and persistent (Williams et al. 1986) heavy metals that are alarming for the livelihood on earth. This heavy metal is also found in high concentrations of AMD, adversely impacting the ecosystem (Kersten et al. 2017). Marques et al. (2020) used engineered *Escherichia coli* as Cd bioremediation agents by producing poly(3-hydroxybutyric acid) (PHB) beads coated with Cd-binding peptides. Two Cd binding peptides (cd1 and cd2) were constructed (Figure 10.11(i)) and applied in Cd removal.

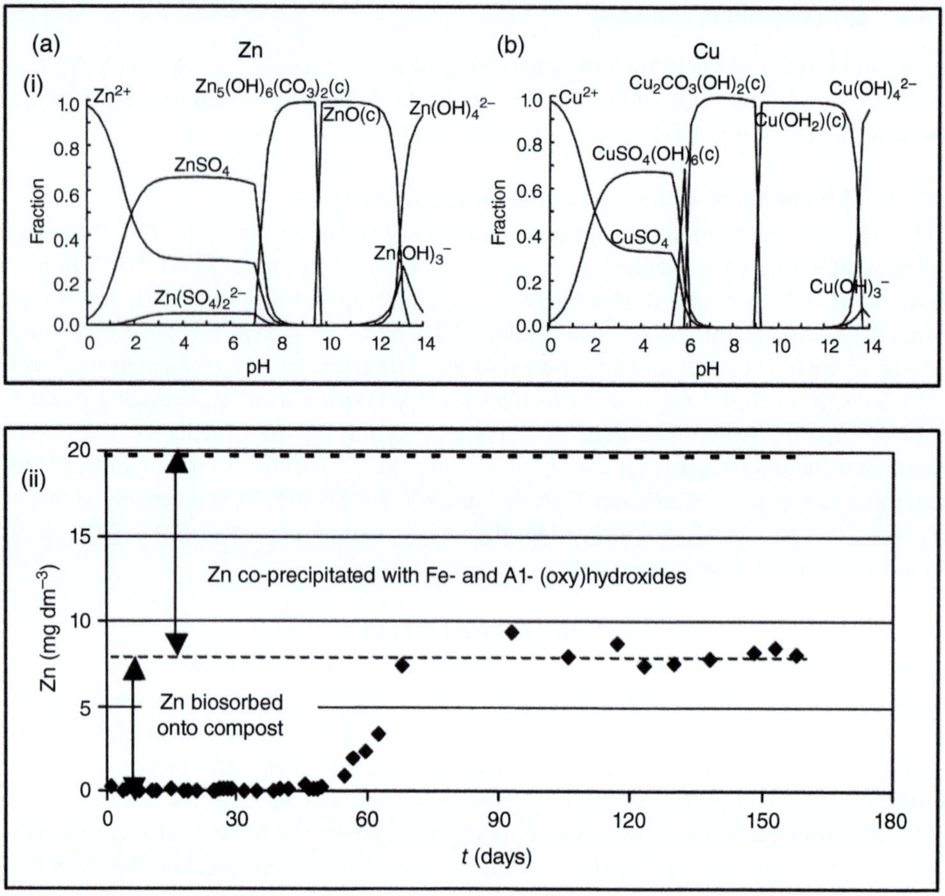

Figure 10.11 (i) Species distribution diagram as a function of pH for the system: (a) Zn^{2+}, CO_3^{2-}, SO_4^{2-}, H_2O, and (b) Cu^{2+}, CO_3^{2-}, SO_4^{2-}, H_2O, (ii) Contribution of co-precipitation with Fe- and Al-(oxy)hydroxides and sorption onto municipal compost in the Zn removal along the column experiment with calcite and municipal compost. *Source:* Adapted with permission from Gibert et al. (2005)/Elsevier.

10–41% of Cd was removed by cd1, and 47–89% of Cd was cleaned from the system by cd2 at pH 7 and 9, with the maximum removal at pH 9 (Figure 10.12(ii)). The main benefit of using nanobeads in the system is viable in a wide pH range (both acidic and alkaline reactions), which is a key prerequisite for the bioremediation of AMD effluents.

10.4.4 Bioremediation of Manganese and Iron

In Fe–Mn mines, low-grade ores are abandoned, resulting in large quantities of residues deposited in the abandoned mines (Isinkaye 2018). With rainfall and surface runoff, potentially toxic elements corresponding to the ore deposits may be released and contribute as a potential pollution source for surface water and groundwater (Kaksonen et al. 2006). In Mn ores, less than 60% is fully extracted, and many deposits of poor-grade Fe–Mn parent rock and rhodochrosite/pyrolusite are jilted (Li et al. 2019). In AMD, Fe and Mn co-occur and have an antagonistic effect on the removal/precipitation of Mn with Fe/Mn ratios >4 (Neculita and

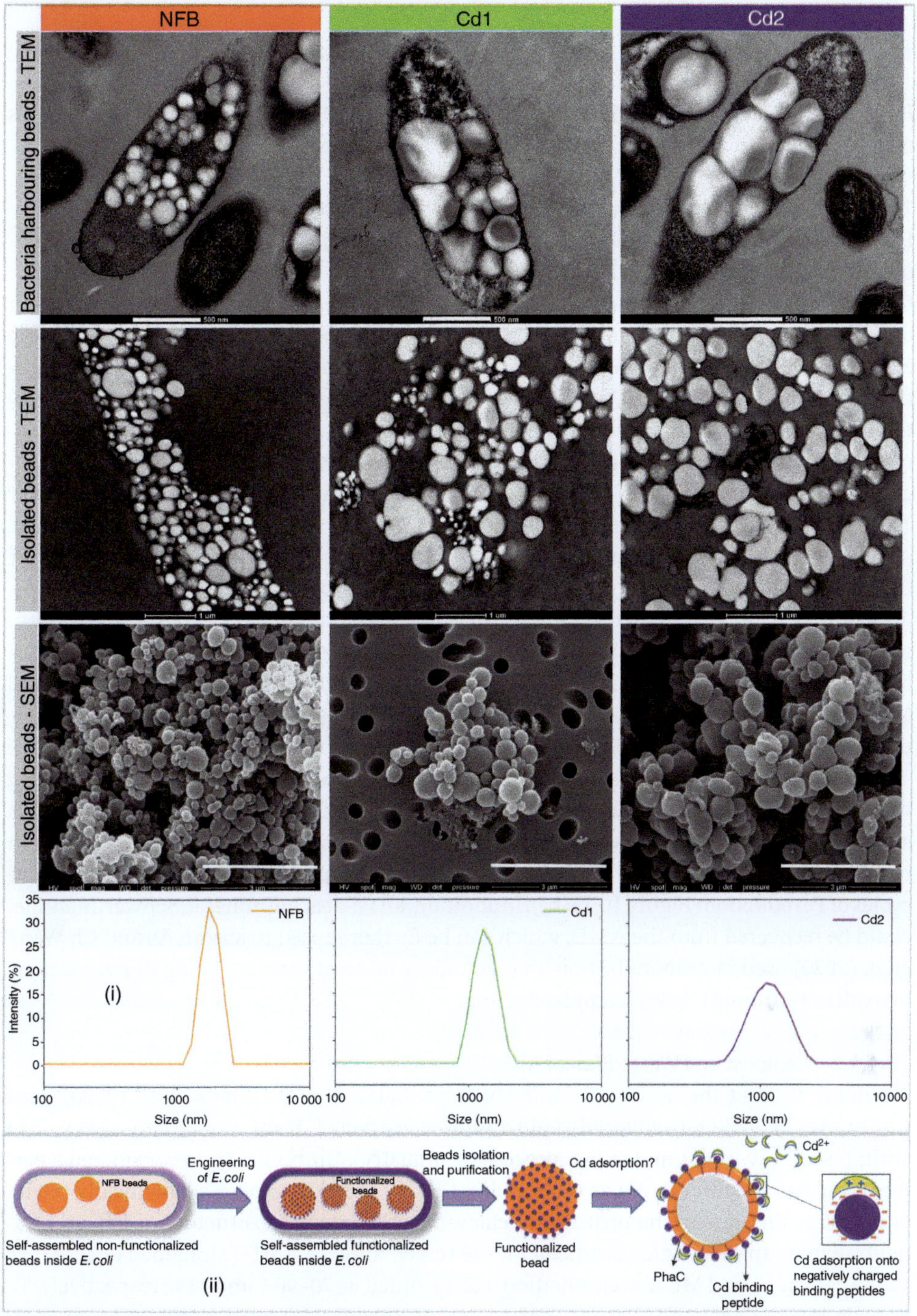

Figure 10.12 (i) Characterization of *Escherichia coli* inclusions of PHB beads and the isolated PHB beads. Top panel: TEM images of *Escherichia coli* BL21(DE3) containing self-assembled beads (NFB – non-functionalized beads; Cd1 – PhaC-linker-Cd1-displaying beads; Cd2 – Cd2-PhaC-displaying beads) (scale bar: 500 nm). Second from the top panel: TEM images of the respective isolated and purified beads (scale bar: 1 μm). Third from the top panel: SEM images of the purified beads (scale bar: 3 μm). Bottom panel: size distribution of the three beads as measured based on the dynamic light scattering. Note: cd1/Cd1 and cd2/Cd2 are used with nomenclatural purposes only and do not represent the name of deposited genes or proteins in available databases; (ii) Overview of synthesis and functionalization of PHB beads in engineered *Escherichia coli* for Cd adsorption. *Source:* Sibila Marques et al. 2020/Reproduced with permission from MDPI/CC BY 4.0.

Rosa 2019). Moreover, in higher Fe concentrations, the reductive dissolution of Mn oxides occurs, causing an increase in Mn concentration in the system (Petrunic et al. 2005).

10.4.4.1 Application of Iron- and Manganese-Oxidizing Bacteria (FMOB)

Fe- and Mn-oxidizing bacteria (FMOB) are the group of organisms which can grow and proliferate in the extreme condition of AMD medium. These bacteria can oxidize the metals from the environment and precipitate as Fe and Mn oxides, which can further bind other toxic heavy metals and pollutants from the aqueous solution (Figure 10.13(i)) (Wang et al. 2009; Larson et al. 2014). Hou et al. (2020) isolated Fe–Mn-oxidizing bacterial consortium from the AMD area in Qibaoshan Mine to put forward a stand in a method for AMD remediation. Bacterial community shifts were also studied during the bioremediation process by sequencing 16S rRNA gene amplicons. 95.4% and 72.3% Fe and Mn removal efficiencies were achieved, with a DO increase to 3.8 $mg\,l^{-1}$. At the later stage of treatment (after 12 days), maximum removal efficiencies of Fe (99.8%) and Mn (87.5%) were accomplished. An increase in pH during the treatment process benefited Fe–Mn removal rate and growth of the organisms like *Flavobacterium*, *Brevundimonas*, *Stenotrophomonans*, and *Thermomonas*.

Crafton et al. (2019) studied a 13-year (2002–2014)-old treatment plant in the Huff Run watershed of Eastern Ohio, United States, to analyze the long-term functionality of a biotreatment site. The metal removal was adequate, and the concentration of Fe, Al, and Mn was between 0.05 and 0.12 $mg\,l^{-1}$, 0.04 and 0.25 $mg\,l^{-1}$, and 0.05 and 0.34 $mg\,l^{-1}$, respectively. The maximum removal was evidenced for Fe and Al with partial removal of Mn. The system was found to be viable in the long-term, and metal removal was operative along with an increase of pH (Acidic [pH 3.66 in 2002 to 7.25 in 2014]). Jin et al. (2020) used an assorted culture with *A. multivorum JZ-6* and *A. ferrooxidans LX5* for treating AMD through biomineralization. Complete removal of Fe and 40% sulfate could be availed by only five cycles of Fe reduction (Figure 10.13(ii)). In addition, an Fe-bearing mineral, Schwertmannite, could be recovered from the AMD, which can be further reused to absorb As and Cr. Wang et al. (2020) used biomineralization and secondary mineral recovery using *A. ferrooxidans* to oxidize 160 $mmol\,l^{-1}$ Fe^{2+} completely.

10.4.4.2 Compost and Waste Biomaterials

Intending to treat the acidity, Fe and Mn from AMD, Núñez-Gómez et al. (2020) used Chitin, i.e. poly-β-(1,4)-N-acetyl-D-glucosamine extracted from crustacean shells, like shrimp shells, (waste from the fish-processing industry). Chitin has been considered a perfect remedial agent for AMD as this can bind heavy metal ions physicochemically (Datta et al. 2022), and due to the presence of calcium carbonate in its structure, it has an acid-neutralizing capacity, which is crucial for the remediation of AMD (Mohamed et al. 2015). In the case of Fe and Mn, the elimination was recorded as 70–90% and 41%, respectively. To remove toxic metals and sulfates, Kefeni and Mamba (2021) used a low-cost, naturally available, eco-friendly material, charcoal ash, in AMD treatment. Complete removal of metals was achieved within 25 min at $pH < 5$ along with 48–83% sulfate removal at a trend of $Ni^{2+} > Co^{2+} > Zn^{2+} > Mn^{2+}$ with the application of charcoal nanoparticles with AMD.

10.4.4.3 Application of SRB and Yeast in Bioremediation of AMD

Recent studies have revealed that SRB and yeasts can effectively remove Fe and Mn from the mine tailings. Dong et al. (2020) carried out a column experiment with coal gangue

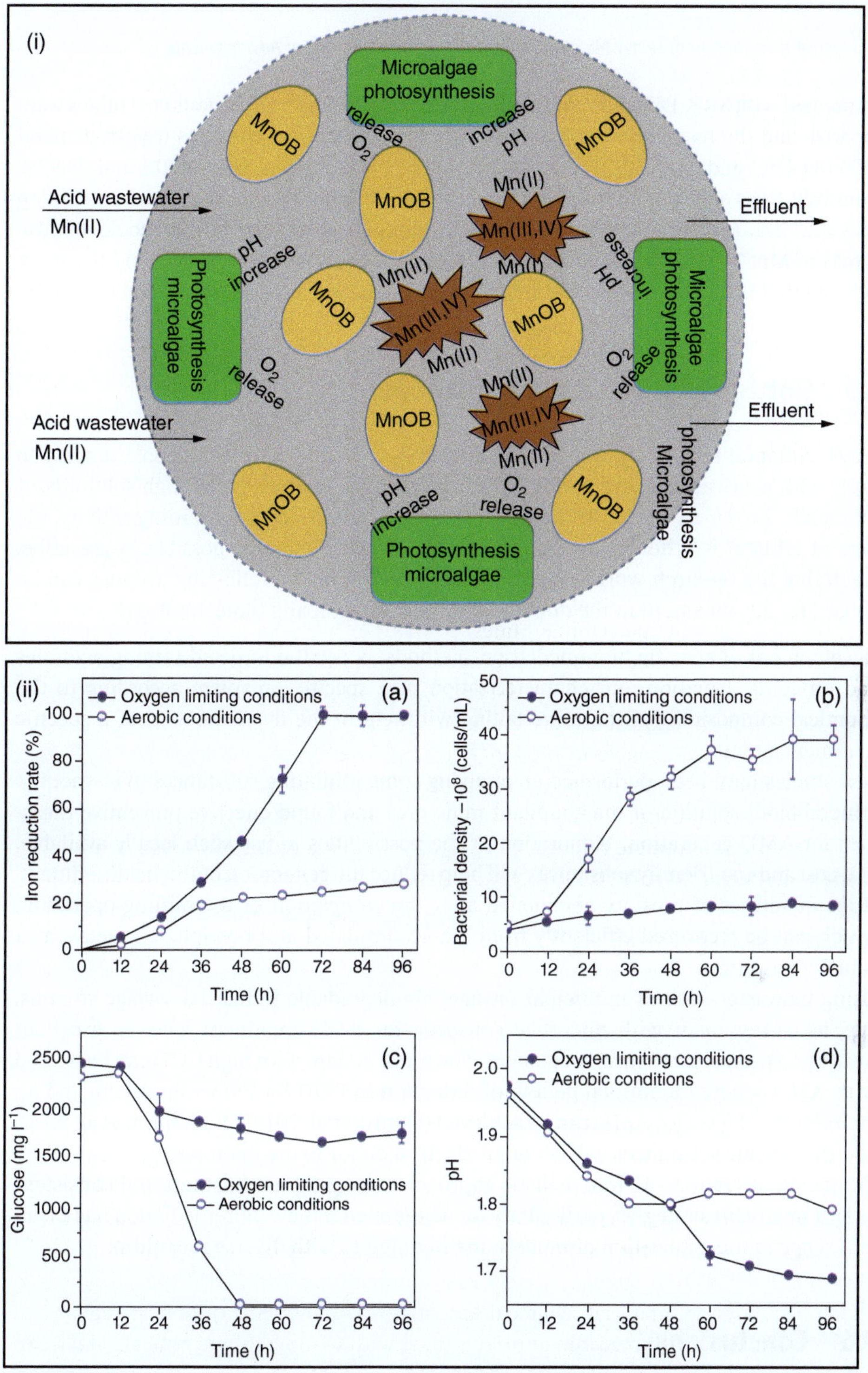

Figure 10.13 (i) Speculative action and mechanisms for an Mn mine drainage treatment with co-immobilized bacteria/microalgae beads. MnOB denotes Mn-oxidizing bacteria, (ii) Changes in Fe reduction rate (a), bacterial density (b), glucose concentration (c), and (d) the pH of the medium during the incubation of *Acidiphilium multivorum* JZ-6 under oxygen-limiting or aerobic conditions. Data are shown as the mean ± SD (n = 3). *Source:* Li et al. (2019)/MDPI/CC BY 4.0; (ii) Adapted with permission from Jin et al. (2020)/Elsevier.

incorporated with SRB. Different pollution loads in the columns from various coal mines were compared, and the result showed the maximum removal was for chemical oxygen demand (COD) (84.41%) and SO_4^{2-} (72.73%). For Fe and Mn, the removal was 98.70% and 79.97%, respectively. Ruas et al. (2020) isolated a yeast species *Rhodotorula mucilaginosa* from mining waters and evaluated the Mn-removal capacity through yeast. The species could remove 0.91 mM of Mn^{2+} (initial concentration was 140 mg l^{-1}, equivalent to 2.5 mM), and the study concluded that *R. mucilaginosa* could contribute immensely to the bioremediation of AMD.

10.5 Bottlenecks and Future Prospects

In environmental restoration and management, there is no “panacea” or sole remedy to comply with every aspect. In treating AMD, the primary concern is the high solubility of toxic metals, extremely acidic pH and sulfate concentration. In every mining activity, the nature of effluent is different, and no unique treatment method is possible to prescribe. Considering the research work done in the past few years, the following avenues can be proposed for advancement in the domain of bioremediation and biotechnology:

- Development of specific bioremediation methods as per the targeted mining activities and effluents. Isolation and characterization of a specific organism according to the chemical composition of the mine tailing will help in the determination of a specific treatment system.
- Few studies have been performed on applying some inhibiting substances to restrict the mineral biodissolution of the abundant mine ores and found effective preventive measures for AMD generation. Exploration of the possibilities to use such locally available, low-cost and eco-friendly substances will help reduce the generation of this acidic effluent.
- Huge quantities of ores are abandoned after the completion of the mining operation, which can be recovered efficiently from bio-accumulated and precipitated metals as a valuable resource.
- Using biowastes such as municipal sewage, bio-degradable industrial sewage streams, etc., as co-treatment with microbial consortia for AMD treatment have an excellent future perspective in waste management. The waste stream with high COD can be treated with AMD, where the mutual benefit of reduction in COD for the waste stream and an increase in pH level for AMD can be achieved (Cunha et al. 2018; Makhathini et al. 2020) and the organic substances will act as an electron donor to the microbes.
- Future studies require a more realistic approach to long-term robustness and consistent ex situ treatment strategies, particularly for abandoned mines. Bioremediation has excellent scope in the application of modern methodologies with diverse inoculum.

10.6 Conclusions

The significant problems in mining industries are the formation of AMD and the accumulation of tons of waste and tailings, leading to infertile land with low productivity and huge cost involvement in the remediation process. Researchers worldwide seek sustainable and economically viable strategies to ensure environmental protection and ecological balance. Research in the past few years on AMD indicates that biological remediation processes are very promising and widely used for their low initial and operational cost and effectiveness in treating the mine tailings. Both prokaryotes and eukaryotes (SRB-, Fe-, and Mn-oxidizing bacterial species, algae, fungi, and yeasts) have been used with single species or in mixed cultures to remove different metals and metalloids and have revealed new possibilities for biological treatment processes in AMD treatment. Bioengineered organisms, Fe–Mn-oxidizing consortium, biofilms, reactive barriers, and nano-based techniques have been reviewed in the remediation of Fe, Mn, Cu, Cd, Zn, Pd, and As from AMD. A reduction in biomass at higher concentrations of the heavy metal and a decline in growth and activity due to adverse acidic reactions of AMD can affect the treatment process. Despite these issues, bioremediation's success in treating AMD and recovering valuable raw materials and rare metal crystals in bio-based treatment has unfolded new avenues in environmental restoration.

Abbreviations

AMD	Acid mine drainage
AS	Limestone-activated sludge
BSR	Bacterial sulfate reduction
COD	Chemical oxygen demand
DCMD	Direct contact membrane distillation
DE	Diatomaceous earth
DO	Dissolved oxygen
FMOB	Iron (Fe)- and manganese (Mn)-oxidizing bacteria
GA	Glutaraldehyde
HRT	Hydraulic retention time
MBR	Membrane bioreactor
PEI	Polyethyleneimine
PHB	Poly(3-hydroxybutyric acid)
SDES	Sulfidogenic diffusive exchange system
SMC	Spent mushroom compost
SRB	Sulfate-reducing bacteria
WC	Woodchips

References

Abhang, P.B. and Kedar, S.S. (2019). Trichoderma: a bio-control agent for management soil born diseases. In: *Innovative Approaches in Agriculture and Environmental Development* (ed. P.B. Abhang and S.S. Kedar), 375. India: JPS Scientific Publications.

Abinandan, S., Subashchandrabose, S.R., Venkateswarlu, K., and Megharaj, M. (2018). Microalgae–bacteria biofilms: a sustainable synergistic approach in remediation of acid mine drainage. *Applied Microbiology and Biotechnology* 102 (3): 1131–1144.

Abinandan, S., Perera, I.A., Subashchandrabose, S.R. et al. (2020). Acid-adapted microalgae exhibit phenotypic changes for their survival in acid mine drainage samples. *FEMS Microbiology Ecology* 96 (11): fiaa113.

Abraham, M.R. and Susan, T.B. (2017). Water contamination with heavy metals and trace elements from Kilembe copper mine and tailing sites in Western Uganda; implications for domestic water quality. *Chemosphere* 169: 281–287.

Akcil, A. and Koldas, S. (2006). Acid mine drainage (AMD): causes, treatment and case studies. *Journal of Cleaner Production* 14 (12–13): 1139–1145.

Ali, H., Khan, E., and Ilahi, I. (2019). Environmental chemistry and ecotoxicology of hazardous heavy metals: environmental persistence, toxicity, and bioaccumulation. *Journal of Chemistry* 2019: 6730305.

Altun, M., Sahinkaya, E., Durukan, I. et al. (2014). Arsenic removal in a sulfidogenic fixed-bed column bioreactor. *Journal of Hazardous Materials* 269: 31–37.

Anawar, H.M. (2015). Sustainable rehabilitation of mining waste and acid mine drainage using geochemistry, mine type, mineralogy, texture, ore extraction and climate knowledge. *Journal of Environmental Management* 158: 111–121.

Ayele, A., Haile, S., Alemu, D. et al. (2021). Mycoremediation: fungal-based technology for biosorption of heavy metals–a review. In: *Strategies and Tools for Pollutant Mitigation*, 355–373. Cham: Springer.

Ayodhya, D. and Veerabhadram, G. (2018). A review on recent advances in photodegradation of dyes using doped and heterojunction based semiconductor metal sulfide nanostructures for environmental protection. *Materials Today Energy* 9: 83–113.

Azabou, S., Mechichi, T., Patel, B.K.C., and Sayadi, S. (2007). Isolation and characterization of a mesophilic bacteria heavy-metals-tolerant sulphate-reducing bacterium Desulfomicrobium sp. from an enrichment culture using phosphogypsum as a sulphate source. *Journal of Hazardous Materials* 140: 264–270.

Bachtlin, A. (2019). Acid mine drainage and the techniques to cleanup this issue. *DU Quark* 3 (2): 28–34.

Bai, H., Kang, Y., Quan, H. et al. (2013). Bioremediation of copper-containing wastewater by sulfate reducing bacteria coupled with iron. *Journal of Environmental Management* 129: 350–356.

Baker, B.J. and Banfield, J.F. (2003). Microbial communities in acid mine drainage. *FEMS Microbiology Ecology* 44 (2): 139–152.

Battaglia-Brunet, F., Itard, Y., Garrido, F. et al. (2006). A simple biogeochemical process removing arsenic from a mine drainage water. *Geomicrobiology Journal* 23 (3–4): 201–211.

Battaglia-Brunet, F., Crouzet, C., Burnol, A. et al. (2012). Precipitation of arsenic sulphide from acidic water in a fixed-film bioreactor. *Water Research* 46 (12): 3923–3933.

Bernard, A. (2008). Cadmium & its adverse effects on human health. *Indian Journal of Medical Research* 128 (4): 557.

Bolan, N., Kunhikrishnan, A., Thangarajan, R. et al. (2014). Remediation of heavy metal (loid) s contaminated soils–to mobilize or to immobilize? *Journal of Hazardous Materials* 266: 141–166.

Bostick, B.C., Fendorf, S., and Brown, G.E. (2005). In situ analysis of thioarsenite complexes in neutral to alkaline arsenic sulphide solutions. *Mineralogical Magazine* 69 (5): 781–795.

Boyd, E.S. and Druschel, G.K. (2013). Involvement of intermediate sulfur species in biological reduction of elemental sulfur under acidic, hydrothermal conditions. *Applied and Environmental Microbiology* 79 (6): 2061–2068.

Braga, J.K., de Melo Júnior, O.M., Rodriguez, R.P., and Sancinetti, G.P. (2020). Sulfate and metals removal from acid mine drainage in a horizontal anaerobic immobilized biomass (HAIB) reactor. *Journal of Environmental Science and Health, Part A* 55 (12): 1436–1449.

Bruneel, O., Mghazli, N., Sbabou, L. et al. (2019). Role of microorganisms in rehabilitation of mining sites, focus on sub Saharan African countries. *Journal of Geochemical Exploration* 205: 106327.

Bulgariu, L. and Gavrilescu, M. (2015). Bioremediation of heavy metals by microalgae. In: *Handbook of Marine Microalgae* (ed. S.-K. Kim), 457–469. Academic Press.

Buzzi, D.C., Viegas, L.S., Rodrigues, M.A.S. et al. (2013). Water recovery from acid mine drainage by electrodialysis. *Minerals Engineering* 40: 82–89.

Chandra, A.P. and Gerson, A.R. (2010). The mechanisms of pyrite oxidation and leaching: a fundamental perspective. *Surface Science Reports* 65 (9): 293–315.

Chen, Y.T., Li, J.T., Chen, L.X. et al. (2014). Biogeochemical processes governing natural pyrite oxidation and release of acid metalliferous drainage. *Environmental Science & Technology* 48 (10): 5537–5545.

Chen, G., Ye, Y., Yao, N. et al. (2021). A critical review of prevention, treatment, reuse, and resource recovery from acid mine drainage. *Journal of Cleaner Production* 329: 129666.

Colmer, A.R. and Hinkle, M.E. (1947). The role of microorganisms in acid mine drainage: a preliminary report. *Science* 106 (2751): 253–256.

Colmer, A.R., Temple, K.L., and Hinkle, M.E. (1950). An iron-oxidizing bacterium from the acid drainage of some bituminous coal mines. *Journal of Bacteriology* 59 (3): 317.

Contreras-Cortés, A.G., Almendariz-Tapia, F.J., Cortez-Rocha, M.O. et al. (2020). Biosorption of copper by immobilized biomass of *Aspergillus australensis*. Effect of metal on the viability, cellular components, polyhydroxyalkanoates production, and oxidative stress. *Environmental Science and Pollution Research* 27 (23): 28545–28560.

Coudert, L., Bondu, R., Rakotonimaro, T.V. et al. (2019). Treatment of As-rich mine effluents and produced residues stability: current knowledge and research priorities for gold mining. *Journal of Hazardous Materials* 386: 121920.

Crafton, E., Pritchard, C., Guo, L. et al. (2019). Dynamics of Mn removal in an acid mine drainage treatment system over 13 years after installation. *Environmental Earth Sciences* 78 (1): 1–9.

Cunha, M.P., Ferraz, R.M., Sancinetti, G.P., and Rodriguez, R.P. (2018). Long-term performance of a UASB reactor treating acid mine drainage: effects of sulfate loading rate, hydraulic retention time, and COD/SO4 2- ratio. *Biodegradation* 30: 47–58.

Da Costa, G., Hofmann, A., and Agangi, A. (2017). Provenance of detrital pyrite in Archean sedimentary rocks: examples from the Witwatersrand Basin. In: *Sediment Provenance* (ed. R. Mazumder), 509–531. Elsevier.

Daraz, U., Li, Y., Sun, Q. et al. (2021). Inoculation of Bacillus spp. modulate the soil bacterial communities and available nutrients in the rhizosphere of vetiver plant irrigated with acid mine drainage. *Chemosphere* 263: 128345.

Datta, S., Radhapyari, K., Saha, N., and Samanta, S.K. (2022). Recent trends in the application of biowaste for hazardous radioactive waste treatment. In: *Environmental Sustainability and Industries* (ed. P. Singh, J.P. Bassin, S. Rajkhowa, et al.), 159–192. Elsevier.

de Matos, L.P., Costa, P.F., Moreira, M. et al. (2018). Simultaneous removal of sulfate and arsenic using immobilized non-traditional SRB mixed culture and alternative low-cost carbon sources. *Chemical Engineering Journal* 334: 1630–1641.

Dold, B. and Fontboté, L. (2001). Element cycling and secondary mineralogy in porphyry copper tailings as a function of climate, primary mineralogy, and mineral processing. *Journal of Geochemical Exploration 74* (1-3): 3–55.

Dong, Y., Di, J., Yang, Z. et al. (2020). Study on the effectiveness of sulfate-reducing bacteria combined with coal gangue in repairing acid mine drainage containing Fe and Mn. *Energies* 13 (4): 995.

Edwards, K.J., Bond, P.L., Druschel, G.K. et al. (2000). Geochemical and biological aspects of sulfide mineral dissolution: lessons from Iron Mountain, California. *Chemical Geology* 169 (3–4): 383–397.

Fang, J., Zhang, P., Chang, H., and Wang, X. (2018). Hydrothermal synthesis of nanostructured CuS for broadband efficient optical absorption and high-performance photo-thermal conversion. *Solar Energy Materials and Solar Cells* 185: 456–463.

Fernandez-Rojo, L., Casiot, C., Laroche, E. et al. (2019). A field-pilot for passive bioremediation of As-rich acid mine drainage. *Journal of Environmental Management* 232: 910–918.

Ferris, F.G., Hallberg, R.O., Lyvén, B., and Pedersen, K. (2000). Retention of strontium, cesium, lead and uranium by bacterial iron oxides from a subterranean environment. *Applied Geochemistry* 15 (7): 1035–1042.

Florentino, A.P., Weijma, J., Stams, A.J., and Sánchez-Andrea, I. (2015). Sulfur reduction in acid rock drainage environments. *Environmental Science & Technology* 49 (19): 11746–11755.

Gibert, O., de Pablo, J., Cortina, J.L., and Ayora, C. (2005). Municipal compost-based mixture for acid mine drainage bioremediation: metal retention mechanisms. *Applied Geochemistry* 20 (9): 1648–1657.

Giordani, A., Hayashi, E.A., Rodriguez, R.P. et al. (2019). Potential of autochthonous sulfate-reducing microbial communities for treating acid mine drainage in a bench-scale sulfidogenic reactor. *Brazilian Journal of Chemical Engineering* 36: 733–751.

Gruzdev, E.V., Beletsky, A.V., Kadnikov, V.V. et al. (2020). Diversity of eukaryotic microorganisms in the drainage waters of a coal open-cast mine. *Microbiology* 89 (5): 641–646.

Gutiérrez, M., Mickus, K., and Camacho, L.M. (2016). Abandoned PbZn mining wastes and their mobility as proxy to toxicity: a review. *Science of the Total Environment* 565: 392–400.

Hallberg, K.B. (2010). New perspectives in acid mine drainage microbiology. *Hydrometallurgy* 104 (3–4): 448–453.

Han, Y.S., Youm, S.J., Oh, C. et al. (2017). Geochemical and eco-toxicological characteristics of stream water and its sediments affected by acid mine drainage. *Catena* 148: 52–59.

Hou, D., Zhang, P., Wei, D. et al. (2020). Simultaneous removal of iron and manganese from acid mine drainage by acclimated bacteria. *Journal of Hazardous Materials* 396: 122631.

Isinkaye, O.M. (2018). Distribution and multivariate pollution risks assessment of heavy metals and natural radionuclides around abandoned iron-ore mines in North Central Nigeria. *Earth Systems and Environment* 2 (2): 331–343.

Jang, J.H. and Dempsey, B.A. (2008). Coadsorption of arsenic (III) and arsenic (V) onto hydrous ferric oxide: effects on abiotic oxidation of arsenic (III), extraction efficiency, and model accuracy. *Environmental Science & Technology* 42 (8): 2893–2898.

Jin, D., Wang, X., Liu, L. et al. (2020). A novel approach for treating acid mine drainage through forming schwertmannite driven by a mixed culture of *Acidiphilium multivorum* and *Acidithiobacillus ferrooxidans* prior to lime neutralization. *Journal of Hazardous Materials* 400: 123108.

Johnson, D.B. and Hallberg, K.B. (2003). The microbiology of acidic mine waters. *Research in Microbiology* 154 (7): 466–473.

Johnson, D.B. and Hallberg, K.B. (2005). Acid mine drainage remediation options: a review. *Science of the Total Environment* 338 (1–2): 3–14.

Kaksonen, A.H., Plumb, J.J., Robertson, W.J. et al. (2006). The performance, kinetics and microbiology of sulfidogenic fluidized-bed treatment of acidic metal-and sulfate-containing wastewater. *Hydrometallurgy* 83 (1–4): 204–213.

Kapahi, M. and Sachdeva, S. (2017). Mycoremediation potential of Pleurotus species for heavy metals: a review. *Bioresources and Bioprocessing* 4 (1): 1–9.

Katoch, M., Singh, D., Kapoor, K.K., and Vishwakarma, R.A. (2019). *Trichoderma lixii* (IIIM-B4), an endophyte of *Bacopa monnieri* L. producing peptaibols. *BMC Microbiology* 19 (1): 1–10.

Kefeni, K.K. and Mamba, B.B. (2021). Charcoal ash leachate and its sparingly soluble residue for acid mine drainage treatment: waste for pollution remediation and dual resource recovery. *Journal of Cleaner Production* 320: 128717.

Kersten, G., Majestic, B., and Quigley, M. (2017). Phytoremediation of cadmium and lead-polluted watersheds. *Ecotoxicology and Environmental Safety* 137: 225–232.

Khairy, H.M. (2009). Toxicity and accumulation of copper in *Nannochloropsis oculata* (Eustigmatophyceae, Heterokonta). *World Applied Sciences Journal* 6 (3): 378–384.

Khan, I., Ali, M., Aftab, M. et al. (2019). Mycoremediation: a treatment for heavy metal-polluted soil using indigenous metallotolerant fungi. *Environmental Monitoring and Assessment* 191 (10): 1–15.

Kleinmann, R.L.P., Crerar, D.A., and Pacelli, R.R. (1981). Biogeochemistry of acid mine drainage and a method to control acid formation. *Mining Engineering* 33 (3): 300–305.

Konhauser, K.O., Lalonde, S.V., Planavsky, N.J. et al. (2011). Aerobic bacterial pyrite oxidation and acid rock drainage during the great oxidation event. *Nature* 478 (7369): 369–373.

Kumar, V. and Dwivedi, S.K. (2021). Bioremediation mechanism and potential of copper by actively growing fungus Trichoderma lixii CR700 isolated from electroplating wastewater. *Journal of Environmental Management* 277: 111370.

Kumar, N., Couture, R.M., Millot, R. et al. (2016). Microbial sulfate reduction enhances arsenic mobility downstream of zerovalent-iron-based permeable reactive barrier. *Environmental Science & Technology* 50 (14): 7610–7617.

Kumari, S., Udayabhanu, G., and Prasad, B. (2010). Studies on environmental impact of acid mine drainage generation and its treatment: an appraisal. *Indian Journal of Environmental Protection* 30 (11): 953–967.

Langmuir, D., Mahoney, J., and Rowson, J. (2006). Solubility products of amorphous ferric arsenate and crystalline scorodite (FeAsO4· 2H2O) and their application to arsenic behavior in buried mine tailings. *Geochimica et Cosmochimica Acta* 70 (12): 2942–2956.

Lapakko, K. (2002). Metal mine rock and waste characterization tools: an overview. *Mining, Minerals and Sustainable Development* 67: 1–30.

Larson, L.N., Sánchez-España, J., and Burgos, W. (2014). Rates of low-pH biological Fe (II) oxidation in the Appalachian bituminous coal basin and the Iberian pyrite belt. *Applied Geochemistry* 47: 85–98.

Larsson, M., Nosrati, A., Kaur, S. et al. (2018). Copper removal from acid mine drainage-polluted water using glutaraldehyde-polyethyleneimine modified diatomaceous earth particles. *Heliyon* 4 (2): e00520.

Le Pape, P., Battaglia-Brunet, F., Parmentier, M. et al. (2017). Complete removal of arsenic and zinc from a heavily contaminated acid mine drainage via an indigenous SRB consortium. *Journal of Hazardous Materials* 321: 764–772.

Lehner, S., Savage, K., Ciobanu, M., and Cliffel, D.E. (2007). The effect of As, Co, and Ni impurities on pyrite oxidation kinetics: an electrochemical study of synthetic pyrite. *Geochimica et Cosmochimica Acta* 71 (10): 2491–2509.

Leong, Y.K. and Chang, J.S. (2020). Bioremediation of heavy metals using microalgae: recent advances and mechanisms. *Bioresource Technology* 303: 122886.

Li, H., Dong, X., da Silva, E.B. et al. (2017). Mechanisms of metal sorption by biochars: biochar characteristics and modifications. *Chemosphere* 178: 466–478.

Li, X., Lan, S.M., Zhu, Z.P. et al. (2018a). The bioenergetics mechanisms and applications of sulfate-reducing bacteria in remediation of pollutants in drainage: a review. *Ecotoxicology and Environmental Safety* 158: 162–170.

Li, Y., Yang, X., and Geng, B. (2018b). Preparation of immobilized sulfate-reducing bacteria-microalgae beads for effective bioremediation of copper-containing wastewater. *Water, Air, & Soil Pollution* 229 (3): 1–13.

Li, Y., Xu, Z., Ma, H., and Hursthouse, A.S. (2019). Removal of manganese (II) from acid mine wastewater: a review of the challenges and opportunities with special emphasis on Mn-oxidizing bacteria and microalgae. *Water* 11 (12): 2493.

Liang, S., Zhang, L., and Jiang, F. (2016). Indirect sulfur reduction via polysulfide contributes to serious odor problem in a sewer receiving nitrate dosage. *Water Research* 100: 421–428.

Lottermoser, B.G. (2010). Introduction to mine wastes. In: *Mine Wastes*, 1–41. Berlin, Heidelberg: Springer.

Lounate, K., Coudert, L., Genty, T. et al. (2020). Performance of a semi-passive sulfate-reducing bioreactor for acid mine drainage treatment and prediction of environmental behavior of post-treatment residues. *Mine Water and the Environment* 39 (4): 769–784.

Lovley, D.R. (1997). Microbial Fe (III) reduction in subsurface environments. *FEMS Microbiology Reviews* 20 (3–4): 305–313.

Lu, J., Chen, T., Wu, J. et al. (2011). Acid tolerance of an acid mine drainage bioremediation system based on biological sulfate reduction. *Bioresource Technology* 102 (22): 10401–10406.

Lukhele, T., Selvarajan, R., Nyoni, H. et al. (2020). Acid mine drainage as habitats for distinct microbiomes: current knowledge in the era of molecular and omic technologies. *Current Microbiology* 77 (4): 657–674.

Ma, S.C., Zhang, H.B., Ma, S.T. et al. (2015). Effects of mine wastewater irrigation on activities of soil enzymes and physiological properties, heavy metal uptake and grain yield in winter wheat. *Ecotoxicology and Environmental Safety* 113: 483–490.

Maillot, F., Morin, G., Juillot, F. et al. (2013). Structure and reactivity of As(III)-and As(V)-rich schwertmannites and amorphous ferric arsenate sulfate from the Carnoules acid mine drainage, France: comparison with biotic and abiotic model compounds and implications for As remediation. *Geochimica et Cosmochimica Acta* 104: 310–329.

Majzlan, J., Drahota, P., and Filippi, M. (2014). Parageneses and crystal chemistry of arsenic minerals. *Reviews in Mineralogy and Geochemistry* 79 (1): 17–184.

Makhathini, T.P., Mulopo, J., and Bakare, B.F. (2020). Possibilities for acid mine drainage co-treatment with other waste streams: a review. *Mine Water and the Environment* 39 (1): 13–26.

Marques, C.R., Wibowo, D., Rubio-Reyes, P. et al. (2020). Bacterially assembled biopolyester nanobeads for removing cadmium from water. *Water Research* 186: 116357.

Martínez-Macias, M.D.R., Correa-Murrieta, M.A., Villegas-Peralta, Y. et al. (2019). Uptake of copper from acid mine drainage by the microalgae *Nannochloropsis oculata*. *Environmental Science and Pollution Research* 26 (7): 6311–6318.

Migaszewski, Z.M., Gałuszka, A., and Dołęgowska, S. (2019). Extreme enrichment of arsenic and rare earth elements in acid mine drainage: case study of Wiśniówka mining area (south-Central Poland). *Environmental Pollution* 244: 898–906.

Mohamed, S.H., El-Gendy, A.A., Abdel-Kader, A.H., and El-Ashkar, E.A. (2015). Removal of heavy metals from water by adsorption on chitin derivatives. *DerPharmaChemica* 7 (10): 275–283.

Morin, G., Juillot, F., Casiot, C. et al. (2003). Bacterial formation of tooeleite and mixed arsenic (III) or arsenic (V)– iron (III) gels in the Carnoulès acid mine drainage, France. A XANES, XRD, and SEM study. *Environmental Science & Technology* 37 (9): 1705–1712.

Muhammad, S.N., Kusin, F.M., Md Zahar, M.S. et al. (2017). Passive bioremediation technology incorporating lignocellulosic spent mushroom compost and limestone for metal-and sulfate-rich acid mine drainage. *Environmental Technology* 38 (16): 2003–2012.

Neculita, C.M. and Rosa, E. (2019). A review of the implications and challenges of manganese removal from mine drainage. *Chemosphere* 214: 491–510.

Nicol, M.J. (2020). A comparative study of the kinetics of the oxidation of iron (II) by oxygen in acidic media-mechanistic and practical implications. *Hydrometallurgy* 192: 105246.

Núñez-Gómez, D., Lapolli, F.R., Nagel-Hassemer, M.E., and Lobo-Recio, M.Á. (2020). Optimization of Fe and Mn removal from coal acid mine drainage (AMD) with waste biomaterials: statistical modeling and kinetic study. *Waste and Biomass Valorization* 11 (3): 1143–1157.

Paikaray, S. (2015). Arsenic geochemistry of acid mine drainage. *Mine Water and the Environment* 34 (2): 181–196.

Park, I., Tabelin, C.B., Jeon, S. et al. (2019). A review of recent strategies for acid mine drainage prevention and mine tailings recycling. *Chemosphere* 219: 588–606.

Pérez, N., Schwarz, A., Sanhueza, P., and Chaparro, G. (2017). Performance of three bench-scale diffusive exchange systems during treatment of acid mine drainage with high copper concentration. *Desalination and Water Treatment* 64: 21–30.

Petrunic, B.M., MacQuarrie, K.T.B., and Al, T.A. (2005). Reductive dissolution of Mn oxides in river-recharged aquifers: a laboratory column study. *Journal of Hydrology* 301 (1–4): 163–181.

Qiu, R., Zhao, B., Liu, J. et al. (2009). Sulfate reduction and copper precipitation by a Citrobacter sp. isolated from a mining area. *Journal of Hazardous Materials* 164 (2–3): 1310–1315.

Qiu, Y.Y., Guo, J.H., Zhang, L. et al. (2017). A high-rate sulfidogenic process based on elemental sulfur reduction: cost-effectiveness evaluation and microbial community analysis. *Biochemical Engineering Journal* 128: 26–32.

Radhika, V., Subramanian, S., and Natarajan, K.A. (2006). Bioremediation of zinc using *Desulfotomaculum nigrificans*: bioprecipitation and characterization studies. *Water Research* 40 (19): 3628–3636.

Raja Sathendra, E., Praveen Kumar, R., and Baskar, G. (2018). Microbial transformation of heavy metals. In: *Waste Bioremediation*, 249–263. Singapore: Springer.

Rambabu, K., Banat, F., Pham, Q.M. et al. (2020). Biological remediation of acid mine drainage: review of past trends and current outlook. *Environmental Science and Ecotechnology* 2: 100024.

Rittle, K.A., Drever, J.I., and Colberg, P.J. (1995). Precipitation of arsenic during bacterial sulfate reduction. *Geomicrobiology Journal* 13 (1): 1–11.

Rodriguez-Freire, L., Sierra-Alvarez, R., Root, R. et al. (2014). Biomineralization of arsenate to arsenic sulfides is greatly enhanced at mildly acidic conditions. *Water Research* 66: 242–253.

Rodriguez-Freire, L., Moore, S.E., Sierra-Alvarez, R. et al. (2016). Arsenic remediation by formation of arsenic sulfide minerals in a continuous anaerobic bioreactor. *Biotechnology and Bioengineering* 113 (3): 522–530.

Romera, E., González, F., Ballester, A. et al. (2007). Comparative study of biosorption of heavy metals using different types of algae. *Bioresource Technology* 98 (17): 3344–3353.

Rowe, O.F. and Johnson, D.B. (2008). Comparison of ferric iron generation by different species of acidophilic bacteria immobilized in packed-bed reactors. *Systematic and Applied Microbiology* 31 (1): 68–77.

RoyChowdhury, A., Sarkar, D., and Datta, R. (2015). Remediation of acid mine drainage-impacted water. *Current Pollution Reports* 1 (3): 131–141.

Ruas, F.A.D., Amorim, S.S., Leão, V.A., and Guerra-Sá, R. (2020). Rhodotorula mucilaginosa isolated from the manganese mine water in Minas Gerais, Brazil: potential employment for bioremediation of contaminated water. *Water, Air, & Soil Pollution* 231 (10): 1–14.

Ruehl, M.D. and Hiibel, S.R. (2020). Evaluation of organic carbon and microbial inoculum for bioremediation of acid mine drainage. *Minerals Engineering* 157: 106554.

Ryu, S., Naidu, G., Moon, H., and Vigneswaran, S. (2020). Selective copper recovery by membrane distillation and adsorption system from synthetic acid mine drainage. *Chemosphere* 260: 127528.

Sahinkaya, E., Isler, E., Yurtsever, A., and Coban, I. (2019). Sulfidogenic treatment of acid mine drainage using anaerobic membrane bioreactor. *Journal of Water Process Engineering* 31: 100816.

Sánchez-Andrea, I., Stams, A.J., Weijma, J. et al. (2016). A case in support of implementing innovative bio-processes in the metal mining industry. *FEMS Microbiology Letters* 363 (11): fnw106.

Serrano, J. and Leiva, E. (2017). Removal of arsenic using acid/metal-tolerant sulfate reducing bacteria: a new approach for bioremediation of high-arsenic acid mine waters. *Water* 9 (12): 994.

Shiyi, Y.U.A.N. and Qiang, W.A.N.G. (2018). New progress and prospect of oilfields development technologies in China. *Petroleum Exploration and Development* 45 (4): 698–711.

Siew, Y.W., Zedda, K.L., and Velizarov, S. (2020). Nanofiltration of simulated acid mine drainage: effect of pH and membrane charge. *Applied Sciences* 10 (1): 400.

Silva, P.M.P., Lucheta, A.R., Bitencourt, J.A.P. et al. (2019). Covellite (CuS) production from a real acid mine drainage treated with biogenic H2S. *Metals* 9 (2): 206.

Silva, M.R., Reis, B.G., Grossi, L.B., and Amaral, M.C.S. (2020). Improving the energetic efficiency of direct-contact membrane distillation in mining effluent by using the waste-heat-and-water process as the cooling fluid. *Journal of Cleaner Production* 260: 121035.

Singer, P.C. and Stumm, W. (1970). Acidic mine drainage: the rate-determining step. *Science* 167 (3921): 1121–1123.

Stafilov, T., Aliu, M., and Sajn, R. (2010). Arsenic in surface soils affected by mining and metallurgical processing in K. Mitrovica Region, Kosovo. *International Journal of Environmental Research and Public Health* 7 (11): 4050–4061.

Straif, K., Benbrahim-Tallaa, L., Baan, R. et al. (2009). A review of human carcinogens—part C: metals, arsenic, dusts, and fibres. *The Lancet Oncology* 10 (5): 453–454.

Stumm, W. and Morgan, J.J. (2012). *Aquatic Chemistry: Chemical Equilibria and Rates in Natural Waters*, vol. Vol. 126. Wiley.

Sun, J., Li, L., Zhou, G. et al. (2018). Biological sulfur reduction to generate H2S as a reducing agent to achieve simultaneous catalytic removal of SO2 and NO and sulfur recovery from flue gas. *Environmental Science & Technology* 52 (8): 4754–4762.

Sun, J., Hong, Y., Guo, J. et al. (2019a). Arsenite removal without thioarsenite formation in a sulfidogenic system driven by sulfur reducing bacteria under acidic conditions. *Water Research* 151: 362–370.

Sun, W., Xiao, E., Krumins, V. et al. (2019b). Comparative analyses of the microbial communities inhabiting coal mining waste dump and an adjacent acid mine drainage creek. *Microbial Ecology* 78 (3): 651–664.

Tabelin, C.B., Veerawattananun, S., Ito, M. et al. (2017). Pyrite oxidation in the presence of hematite and alumina: I. Batch leaching experiments and kinetic modeling calculations. *Science of the Total Environment* 580: 687–698.

Tabelin, C.B., Igarashi, T., Villacorte-Tabelin, M. et al. (2018). Arsenic, selenium, boron, lead, cadmium, copper, and zinc in naturally contaminated rocks: a review of their sources, modes of enrichment, mechanisms of release, and mitigation strategies. *Science of the Total Environment* 645: 1522–1553.

Tabelin, C.B., Corpuz, R.D., Igarashi, T. et al. (2020). Acid mine drainage formation and arsenic mobility under strongly acidic conditions: importance of soluble phases, iron oxyhydroxides/oxides and nature of oxidation layer on pyrite. *Journal of Hazardous Materials* 399: 122844.

Tchounwou, P.B., Yedjou, C.G., Patlolla, A.K., and Sutton, D.J. (2012). Heavy metal toxicity and the environment. In: *Molecular, Clinical and Environmental Toxicology*, 133–164. Basel: Springer.

Tezuka, K., Sheets, W.C., Kurihara, R. et al. (2007). Synthesis of covellite (CuS) from the elements. *Solid State Sciences* 9 (1): 95–99.

Tong, L., Fan, R., Yang, S., and Li, C. (2021). Development and status of the treatment technology for acid mine drainage. *Mining, Metallurgy & Exploration* 38 (1): 315–327.

Torregrosa, M., Schwarz, A., Nancucheo, I., and Balladares, E. (2019). Evaluation of the bio-protection mechanism in diffusive exchange permeable reactive barriers for the treatment of acid mine drainage. *Science of the Total Environment* 655: 374–383.

Vieira, B.F., Rodriguez, R.P., de Paula, E.C. et al. (2020). Sustainable treatment of real-mine drainage using crude glycerol and brewery waste as electron donors in a micro-aerobic system. *Journal of Water Process Engineering* 36: 101297.

Wang, W., Shao, Z., Liu, Y., and Wang, G. (2009). Removal of multi-heavy metals using biogenic manganese oxides generated by a deep-sea sedimentary bacterium–Brachybacterium sp. strain Mn32. *Microbiology* 155 (6): 1989–1996.

Wang, X., Li, Q., Liao, Q. et al. (2020). Arsenic (III) biotransformation to tooeleite associated with the oxidation of Fe (II) via Acidithiobacillus ferrooxidans. *Chemosphere 248*: 126080.

Williams, K.A., Green, D.W., Pascoe, D., and Gower, D.E. (1986). The acute toxicity of cadmium to different larval stages of *Chironomus riparius* (Diptera: Chironomidae) and its ecological significance for pollution regulation. *Oecologia* 70 (3): 362–366.

Wong, C., Tan, L.T., Mujahid, A. et al. (2018). Biosorption of copper by endophytic fungi isolated from Nepenthes ampullaria. *Letters in Applied Microbiology* 67: 384–391. https://doi.org/10.1111/lam.13049.

Xu, R., Li, B., Xiao, E. et al. (2020). Uncovering microbial responses to sharp geochemical gradients in a terrace contaminated by acid mine drainage. *Environmental Pollution* 261: 114226.

Xuwen, H. and Fuqin, L. (2010). New technology and development tendency of mine water treatment. *Coal Science and Technology* 38 (11): 17–22.

Yadav, S. and Bajpai, P.K. (2017). Synthesis of copper sulfide nanoparticles: pH dependent phase stabilization. *Nano-Structures & Nano-Objects* 10: 151–158.

Yadav, H., Kumar, R., and Sankhla, M.S. (2020). Residues of pesticides and heavy metals in crops resulting in toxic effects on living organism. *Journal of Seybold Report* 1533: 9211.

Yang, J., Cao, J., Xing, G., and Yuan, H. (2015). Lipid production combined with biosorption and bioaccumulation of cadmium, copper, manganese and zinc by oleaginous microalgae *Chlorella minutissima* UTEX2341. *Bioresource Technology* 175: 537–544.

Yang, W., Tang, Q., Wei, J. et al. (2016). Enhanced removal of Cd(II) and Pb(II) by composites of mesoporous carbon stabilized alumina. *Applied Surface Science* 369: 215–223.

Yang, B., Luo, W., Wang, X. et al. (2020). The use of biochar for controlling acid mine drainage through the inhibition of chalcopyrite biodissolution. *Science of the Total Environment* 737: 139485.

Yigit, E., Yurtsever, A., Basaran, S.T., and Sahinkaya, E. (2020). Optimization of arsenic removal from an acid mine drainage in an anaerobic membrane bioreactor. *Environmental Technology and Innovation* 18: 100712.

Yuan, Z., Zhang, G., Ma, X. et al. (2019). Rapid abiotic As removal from As-rich acid mine drainage: effect of pH, Fe/As molar ratio, oxygen, temperature, initial As concentration and neutralization reagent. *Chemical Engineering Journal* 378: 122156.

Yuan, Z., Zhang, G., Ma, X. et al. (2021). A combined abiotic oxidation-precipitation process for rapid As removal from high-As (III)-Mn (II) acid mine drainage and low As-leaching solid products. *Journal of Hazardous Materials* 401: 123360.

Zhang, G., Ouyang, X., Li, H. et al. (2016). Bioremoval of antimony from contaminated waters by a mixed batch culture of sulfate-reducing bacteria. *International Biodeterioration & Biodegradation* 115: 148–1575.

Zhang, L., Lin, X., Zhang, Z. et al. (2018a). Elemental sulfur as an electron acceptor for organic matter removal in a new high-rate anaerobic biological wastewater treatment process. *Chemical Engineering Journal* 331: 16–22.

Zhang, Y., Zhang, L., Li, L. et al. (2018b). A novel elemental sulfur reduction and sulfide oxidation integrated process for wastewater treatment and sulfur recycling. *Chemical Engineering Journal* 342: 438–445.

Zhang, X., Tang, S., Wang, M. et al. (2019). Acid mine drainage affects the diversity and metal resistance gene profile of sediment bacterial community along a river. *Chemosphere* 217: 790–799.

Zhong, T., Xue, D., Zhao, L., and Zhang, X. (2018). Concentration of heavy metals in vegetables and potential health risk assessment in China. *Environmental Geochemistry and Health* 40 (1): 313–322.

11

Phytoremediation and Microbe-Assisted Removal of Heavy Metals

Sathya Albert Manoharan and Priya Dharshini Veeraragavan

Department of Bioinformatics, School of Chemical and Biotechnology, SASTRA University, Thanjavur, Tamil Nadu, India

11.1 Introduction

Heavy metals have penetrated the realms of land, water, and life of man over the yonder years of past as a consequence of rapid industrialization and speedy urbanization, alarming the common man and countries around the globe (Suman et al. 2018; Ashraf et al. 2019; Hao et al. 2014; Malik 2004).

Lead (Pb), Cadmium (Cd), Nickel (Ni), Zinc (Zn), Chromium (Cr), Mercury (Hg), Arsenic (As), and Copper (Cu) are well-known persistent heavy metals which are acclaimed to have high atomic number and density (Tchounwou et al. 2012; Bizily et al. 2000; Choudhury and Srivastava, 2001).

According to Cempel and Nikel (2006), there exist two groups of heavy metals which are classified as "essential heavy metals" and "non-essential heavy metals". Cu, Zn, Mn, Ni, and Fe fall into "essential heavy metals" class as many metabolic processes of plant life are regulated by them, but unfortunately, these heavy metals turn out to be toxic if their availability exceeds the necessity. Fasani et al. (2018) classified Pb, Cd, As, and Hg into "non-essential class" as these metals have not been favorably claimed for a significant role in plant life, but their excess accumulation results in infertility of the soil ending up with reduction in crop productivity (Clemens 2006; Barkay et al. 2003; Bruins et al. 2000).

The major notorious carcinogenic effects are due to the mutagenicity and teratogenicity properties of the non-biodegradable nature of both of the above two classes of heavy metals, which are persistent in the soil for many years and also pose threat to human health (Ahemad and Malik 2011; Ahemad and Kibret 2013; Ali et al. 2013).

Many approaches have been adopted as remediation techniques for reclaiming the soils and sites contaminated with such non-biodegradable heavy metals. Some of the approaches include removal of heavy metals by thermal, electrical, physical (washing and removal), and chemical (solidification) methods (Sheoran et al. 2011; Wuana and Okieimen 2011a, b; DalCorso et al. 2019).

Remediation of Heavy Metals: Sustainable Technologies and Recent Advances, First Edition. Edited by Rangabhashiyam Selvasembian, Binota Thokchom, Pardeep Singh, Ali H. Jawad, and Willis Gwenzi.

Although such physicochemical technologies are adopted to decontaminate heavy metal-laden soils, the costs of implementation of such technologies are escalating, which is a setback to such efforts of decontamination (Hashim et al. 2011). Apart from high costs, while reflecting on the ecological threats posed by toxic wastes of heavy metals, Hansda et al. (2022) deliberates about the pros and cons of physicochemical remediation methods of heavy metals by citing the drawbacks of inefficiency of such methods demanding high energy requirements, release of toxic wastes, and the problems of their disposal as observed by Mahar et al. (2016).

Moreover, the texture and structural components of organics in the soils are also adversely affected during the use of physicochemical methods of decontamination (Rajkumar et al. 2010). These alterations in the structure of soil affect the fertility conditions of the soil layers. Hashim et al. (2011) emphasize that considering these negative impacts and issues on sustainability and ethics of environment, the shift to employ biological processes for heavy metal decontamination (Shahpiri and Mohammadzadeh 2018) seems to be more befitting. The biological processes of decontamination of heavy metals are promising and ecologically safe and cheap alternatives in contrast to chemical technologies with adverse effects on soil layer structures.

Normal physiological as well as metabolic functions of plants are also hindered by the heavy metal ions. More specifically, the metabolisms of enzymes and pigments are disrupted as the functional groups of such phytochemicals are obstructed by the heavy metals and also they distort the plasma membrane architecture of the plant cells (Ali et al. 2013; Emamverdian et al. 2015; Gupta et al. 2015). Emamverdian et al. (2015) and Krumova et al. (2016) have cited about dislodging of amino acids and damaging of functional protein structures in plants due to bonding with heavy metal ions and sulfhydryl groups. As a consequence, many crucial metabolic functions are hampered, affecting the normal functioning of the plant.

The biological intervention in cleaning up the heavy metal-contaminated soils could be effected even at a very low concentration of the contaminants as the main biological components are endowed with special adaptation of absorbing the ionic compounds even at a very minimum level itself.

Among the various bio-based interventions in decontaminating heavy metal-polluted environments, the use of plants as "phytoremediation" is a promising methodology which is popular owing to its cost-effective mechanism. Simultaneously, there is a drawback associated with this technique since it operates at a slow rate and thereby a time-consuming process. Also, Ali et al. (2013) skeptically claim that efficacy of plants in decontaminating heavily polluted soils of toxic and persistent heavy metals is low. In spite of it, there are reports by Jacob et al. (2018) and DalCorso et al. (2019) that the extensive rhizosphere of the plants penetrates various soil layers and uptakes the heavy metals, trapping them in their roots and rootlets, thereby relieving the soil free of the heavy metal contaminants and facilitating for microbial proliferation enriching the soil and augmenting the bioavailability of plant nutrients.

In addition to floral use in phytoremediation, the conjoining of microbial assistance with plants in phytoremediation synergistically expedites the amelioration of heavy metal toxicity (Mastretta et al. 2009) through phytostabilization (reducing the heavy metal level of contamination through immobilization (Gadd 2000; Gillespie and Philip 2013) and phytoextraction (reducing heavy metal level of contamination through mobilization) of heavy metal ion species (Dushenkov et al. 1995; Robinson et al. 2003). The vegetative and reproductive growth and vigor of the phytoremediating plants are accelerated, assisted by synergistic microbial decontamination of heavy metals (Khan et al. 2009).

11.2 Popular Floral Profiles in Phytoremediation

Arbaoui et al. (2014) and Osman et al. (2017) have highlighted that edible plants could not be employed for such phytoremediation efforts for the concern that potentially toxic heavy metals could possibly be introduced into the food pyramid. Obviously, then the non-food crops, which may include "fiber crops," are ideal for phytoremediation purposes.

Osman et al. (2017) also emphasize that the entire phenomenon of phytoremediation is dependent on the plant's anatomical features involved in the uptake mechanism of ions physiologically. His views project that if a crop calendar of sowing, growing, and harvesting of hyperaccumulating plants of heavy metals is carried out in a sequential manner, then the benchmark in reducing the toxicity of heavy metals in the soil can be mitigated systematically.

Kenaf (*Hibiscus cannabinus* L.) is one such fiber crop of Malvaceae family which can be cultivated on appropriate crop calendar basis as the best candidate for phytoremediation. Moreover, its fibers are of good value in the paper and textile industry as a means for manufacturing biomaterials. Thereby, it can be assured that the threat of cumulative accumulation of heavy metals across the levels of food pyramids could be avoided by employing non-food, industrially beneficial plants (Cartoga et al. 2005; Arbaoui et al. 2016).

Table 11.1 surmises the list of plants acclaimed for phytoremediation.

11.2.1 Heavy Metal Defense Mechanisms in Plants

Thakur et al. (2016) suggest that detoxification is one of the mandatory mechanisms of plants as a means of phytoremediation. Plants adapt defense strategies to safeguard themselves from intrusion of problematic heavy metals. The defense approaches involve avoidance as well as tolerance. These mechanisms support the plants to manage and regulate the level of heavy metals well below the threshold of toxicity (Hall 2016; Becerra-Castro et al. 2009). There are certain limitations in these mechanisms too though not very prominent. In the process of avoidance of heavy metals, some of the major and minor mineral nutrients might be missed to be uptaken by plants. This may have a negative impact on the regulation of metabolic functions of the plants.

11.2.1.1 Avoidance

Plants reduce the uptake of heavy metals into their tissues by virtue of ability of avoidance (Dalvi and Bhalerao 2013). They also have cited that the mechanisms of metal exclusions, metallic ion precipitations, and root sorption are the main defense systems in this method. In their effort to minimize the penetration of heavy metals into plant cells, the plants attempt to reduce the accumulation of heavy metals and thereby prevent entry of heavy metals into symplast (Malachowska-Jutsz and Gnid 2015; Beveridge 1989; Rabêlo et al. 2017). In attempting to prevent heavy metal entry, absorption of certain essential metals may also be prevented, which may be an undesirable co-effect.

Moreover, according to Memon and Schröder (2009), plants embed heavy metals in their cell walls preventing further penetration into the cells, which is also a system of avoidance of heavy metals. In such instances, the cell wall components play a crucial role in trapping the heavy metals and immobilizing them. Ernst et al. (1992) makes a note that the pectinates

Table 11.1 Prospective plants in heavy metal remediation.

Plants	Common name	Heavy metal	References
Allium schoenoprasum L.	Chive	Ni, Co, Cd	Goland-Goldhirsh (2006)
Alyssum bertolonii	Sweet alison	Ni	Mengoni et al. (2012)
Alyssum murale	Yellow tuft	Ni	Broadhurst and Chaney (2016)
Arabidopsis halleri	—	Zn, Cd	Zhang et al. (2017)
Berkheya coddii	—	Ni	Slatter (2013)
Brassica juncea L.	Mustard	Pb, Cd, Zn, Cu	Belimov et al. (2005), Singh and Fulekar (2012), Turan and Esringu (2007), and Takeda et al. (2006)
Brassica napus L.	Canola	Cd, Zn, Cu, Pb	Dell'Amico et al. (2008), Turan and Esringu (2007), and Sheng and Xia (2006)
Cajanus cajan L.	Pigeon pea	Cd, As,	Garg et al. (2014)
Cicer arietinum L.	Chickpea	Cd, Cr, Cu, Pb	Dasgupta et al. (2011), Wani et al. (2007), and Kambhampati and Vu (2013)
Chrysopogon zizanioides	Vetiver	Cd, Cu, Fe, Pb	Sebastián et al. (2021)
Helianthus annuus	Sunflower	Pb, Zn, Cd	Angelova et al. (2016) and Fulekar (2016)
Jatropha curcas L.	Physic nut	Fe, Hg, Al, Zn, Cu, Mn, As, Cr	Jamil et al. (2009), Marrugo-Negrete et al. (2015), and Yadav et al. (2009)
Lantana camara L.	Lantana	Pb	Alaribe and Agamuthu (2015)
Lens culinaris	Lentil	Pb	Wani and Khan (2012)
Lepidium sativum L.	Cress	As, Pb, Fe, Cd, Hg	Gunduz et al. (2012) and Smolinska and Szczodrowska (2016)
Lactuca sativa L.	Lettuce	Cu, As, Mn, Pb, Zn, Cd, Co, Fe, Ni	Gunduz et al. (2012), Achakzai et al. (2011), Rashid et al. (2014), and Quainoo et al. (2015)
Medicago sativa L.	Alfalfa	Cd	Ghnaya et al. (2015)
Minuartia verna		Zn, Cd, Pb	Bothe (2011)
Oryza sativa L.	Rice	Cd, Cu	Li et al. (2008)
Pisum sativum L.	Pea	Cu, Fe, Pb, Cd, Ni, Zn, As, Cr	Malecka et al. (2008), Wani et al. (2008), Hegedusova et al. (2009), and Sharma et al. (2010)
Pteris vittata	Chinese Brake Fern	As, Hg	Rathinasabapathi (2011), Xie et al. (2009), Datta et al. (2017), and Su et al. (2008)
Rapanus sativus L.	Radish	Fe, Cu, As, Cd, Pb	Hatano et al. (2016)

Table 11.1 (Continued)

Plants	Common name	Heavy metal	References
Spartina argentinensis		Cr	Nalla et al. (2012)
Spinacia oleracea L.	Spinach	Pb, Cd, Fe, Ni, Zn, Cu, Cr	Patel and Subramanian (2006), Salaskar et al. (2011), Pathak et al. (2013), Jahanbakhshi et al. (2014), and Abhilash et al. (2016)
Solanum nigrum L.	Black nightshade	Cd	Wei et al. (2006), Wei et al. (2010), and Ji et al. (2011)
Sorghum bicolor L.	Sorghum	Zn, Cd, Fe, Cu	Pinto et al. (2004)
Viola boashanensis		Cd, Pb, Zn	Zhuang et al. (2005)
Zea mays L.	Corn	Cd, Pb, Zn, Cu	Mojiri (2011)

of cell walls have a carboxyl group containing polygalacturonic acids, which bind heavy metals owing to their negative change. Thus, the cell wall components effectively restrict the to and fro passage of heavy metal ions as a cation exchanger.

11.2.1.2 Tolerance

Manara (2012) emphasizes that plants need to detoxify the accumulated heavy metal ions when it exceeds the plant threshold levels, thereby preventing toxicity injuries. So, after the entry and passage of heavy metal ions into the cell, there are a series of second lines of defense to manage with the heavy metal ion presence, which includes compartmentalization which restricts movement of heavy metal ions from crossing into neighboring cells. Inactivation of heavy metal ions and chelation with heavy metal ions are other modes of trapping the heavy metal ions against further advance spreading in plant systems (Dalvi and Bhalerao 2013). As viewed earlier in avoidance mechanism, here too while tolerating the heavy metals by compartmentalization, the essential metals and minerals may also get trapped without being available to the respective plant parts by transport.

11.2.2 Major Phytoremediation Pipelines by Plants

Tangahu et al. (2011) opine that the methodological mechanisms of heavy metal remediation are inclusive of plant-centered processes such as phytoextraction, phytoaccumulation, phytostabilization, phytovolatilization, and phytofiltration. As mentioned earlier, phytoaccumulation and phytoextraction are synonymous in the sense that the heavy metals are systematically taken up and transported (Silver and Phung 1996) from roots to shoots which can also be referred to as phytoabsorption and phytosequestration (Van Dillewijn et al. 2007). Ali et al. (2013) proactively suggest uprooting of such roots and shoots which have phytoextracted heavy metal ions from the soil by the harvesting operation enabling complete decontamination of the phytoremediated sites. The following table summarizes the pipelines of phytoremedial modes (Table 11.2).

Table 11.2 Pipeline of phytoremediation mechanisms.

Phytoremediation types	Definition	Mode of function	References
Phytoimmobilization	Uptake of metal contaminants from the substrates (soil/water) through root absorption and their translocation and sequestration in aerial tissues of plants	Restriction of movement and spread of heavy metal contaminants in the soil layers	Wuana and Okieimen (2011a, b)
Phytostabilization	In-place inactivation	The bioavailability of heavy metals is reduced to optimum level by the plants	USEPA (2000) and Marques et al. (2009)
Phytoextraction	Uptake and translocation from soil to aerial parts of the plant	Plants removing the heavy metal ions from the soil layers by uptake	Ernst (2005)
Phytovolatilization	Conversion into less toxic volatiles	Volatilizing the absorbed heavy metals from the soil and releasing them into the atmosphere	Awa and Hadibarata (2020)
Phytofiltration	Restricting movement of heavy metals in water bodies	Absorption and adsorption of heavy metal ions from water bodies by the plants	Jovancicevic et al. (1986) and Mesjasz-Przybyłowicz et al. (2004)

11.2.3 Sequential Process of Phytoimmobilization

The sequential series of steps in phytoimmobilization can be summarized as follows (Figure 11.1):

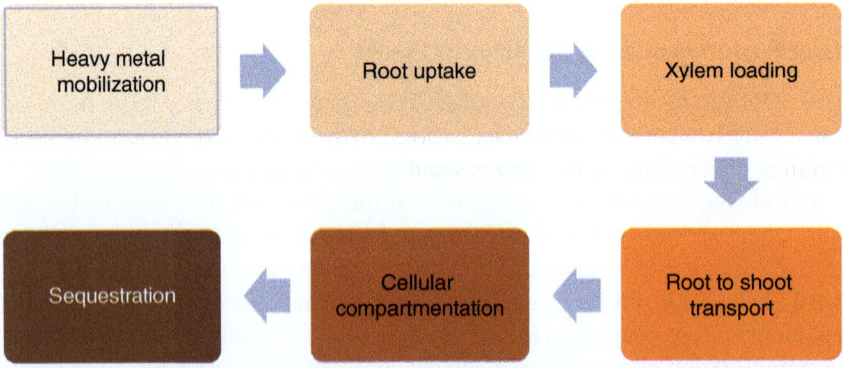

Figure 11.1 Phytoimmobilization sequential steps.

Peer et al. (2005) delineate that the heavy metal ions are available in insoluble forms between the layers of soil and so these are not in a readily bioavailable form to the plants. Initially, the biosorption of heavy metal ions (Romera et al. 2006) occurs on the surface of roots and gradually it passes across the cell membrane into the cells of the root (Ahamed 2012). There are two uptake pathways, namely apoplastic and symplastic. Apoplastic pathway is based on "passive diffusion," whereas symplastic pathway is based on "active diffusion." The passive diffusion is along the electrochemical gradient, whereas the active diffusion is against the electrochemical gradient across the cellular membrane (Sloof et al. 1995). Carriers of metal ions or complex biomolecules mediate the uptaking of heavy metals at the expense of energy in the symplastic pathway. Flavio et al. (2017) have reported about the symplastic and apoplastic pathways of uptake of one of the heavy metals – Cd influenced by high and low concentrations of sulfur. According to their observation, presence of sulfur increases the uptake of Cd by Massai grass which could possibly be immobilized.

11.2.4 Phytostabilization

Wong (2003) and Marques et al. (2009) state that metal-tolerant species of plants immobilize the accumulated heavy metals from the soil and reduce the bioavailability of the same by hindering their seepage into the ecosystem and to the ultimate food chain and such a means of balance of immobilization in the layers of soil and prevention of passage into the ecosystem is termed as "phytostabilization." Ginn et al. (2008), Gerhardt et al. (2017), and Kumpiene et al. (2012) observe that root cells and root tissues play a crucial role in the bisequential events of absorption and sequestration in effecting the phytostabilization of heavy metal ions and minimizing their availability in the rhizosphere.

Phytostabilization could be rendered using appropriate plant species with adaptive features for heavy metal remediation. Marques et al. (2009) and also Berti and Cunningham (2000) share their specifications on the adaptive features of phytostabilizers as follows:

1) The phytostabilizing plant species need to have heavy metal tolerance.
2) The root architecture of such plants needs to be dense for prolific growth in immobilizing the heavy metals in their roots, rootlets, and root hairs.
3) The root system needs to be thicker to hold the soil structure against erosional effects.
4) The shoots of the phytostabilizers need to have voluminous growth with rapid growth rate for broader quick ground coverage.
5) Above all, the phytostabilizers need to be of "easy to grow & maintain" class.

Burges et al. (2018) have identified many such phytostabilizing plant species which have been identified with such characteristics for effectively stabilizing heavy metal-contaminated sites.

11.2.5 Phytoextraction

Plants capable of translocating and accumulating the contaminants from the soil and air media can be the best choice for the mechanism of phytoextraction. The translocated and accumulated contaminants are trapped in the plant tissues above the ground (Jacob et al. 2018). It has been emphasized by Ali et al. (2013) and Sarwar et al. (2017) that one of the most significant phytoremediation methods for reclaiming heavy metal-contaminated patches is "phytoextraction."

In general, the rate of accumulation is a crucial factor during the operational steps of extraction, uptake, and sequestration of heavy metals by plants. Based on this, the plants have been categorized into "hyperaccumulators and hypoaccumulators." It is understood that hyperaccumulators are plants that are able to rapidly accumulate heavy metal components from the soil and compartmentalize them in their upper aerial parts without any symptoms or signs of toxicity. In nature, the rate of accumulation of hyperaccumulators is 100-fold higher than hypoaccumulators and non-accumulators under similar environmental conditions (Rascio and Navari-Izzo 2011).

In a note of high precision, it is appropriate to classify any plant as "hyperaccumulator" if the following specifications are satisfied (McGrath and Zhao 2003; Marques et al. 2009).

1) As a sign of efficient heavy metal transportation ability from root to shoots, the heavy metal load accumulation ratio between the root and shoot is to be more than 1.
2) The uptake ability is assessed by calculating shoot to soil ratio of concentration of heavy metal accumulation. If this ratio exceeds 1, then it indicates rapid and efficient uptake efficiency of heavy metals by the plants.
3) More particularly, Baker and Brooks (1989) ascertain that the hyperaccumulators show higher levels of concentration of respective heavy metals accumulated in their aerial parts, which is as follows: >10 ppm for Hg, >100 ppm for Se and Cd, 1000 ppm for Co, Cr, Cu, Pb, and Ni and 10,000 ppm for Zn and Mn

11.2.6 Phytovolatilization

Plants are also able to convert toxic heavy metals and organic pollutants into lesser toxic volatiles via their foliage. Especially heavy metals such as Selenium (Se), Hg, and As are converted into such less toxic volatiles as a measure of phytoremediation. Awa and Hadibarata (2020) share about few noteworthy examples of phytovolatilizers, namely *Astragalus racemosus* which converts Se into dimethyl diselenide. They also highlight that *Arabdiopsis thaliana* could convert Hg^{2+} to HgO. Sakakibara et al. (2010) also record about the metabolization by *Pteris vittata* in converting As into volatile forms of As. Yan et al. (2020) also affirm that phytovolatilizers disperse the accumulated heavy metals in the form of gaseous compounds to spread and to be dissipated in the atmosphere.

11.2.7 Rhizo/Phytofiltration

Another interesting phenomenon of phytoremediation included is "phytofiltration." The heavy metal pollutants from water bodies could be filtered out by using three modes. That is, "rhizo filtration" (roots), "caulofiltration" (shoots) and "biofiltration" (seedlings). According to Mesjasz-Przybyłowicz et al. (2004), roots trap heavy metals from soil either by absorption or adsorption from water bodies (Lin and Lin 2005). The microzone pH conditions around the rhizosphere regions are influenced by the exudations from roots, which is one of the prime factors which helps to precipitate the heavy metals on plant roots which in turn prevents dispersion into the underground water tables (Javed et al. 2019; Awa and Hadibarata 2020).

11.3 Assistance of Microorganisms in Phytoremediation

Egamberdieva et al. (2008) report that the diverse and dense population of microbes ranges to about 30,000 genera and their communal density is 1011 microbes g^{-1} of the roots. The same notion is affirmed by Mendes et al. (2011). With such a wide spectrum of microbial presence in the rhizosphere (Egamberdieva 2012), the "Rhizospheric Microbial Community" are influenced by the chemical attributes and interactions of the rhizospheric exudates (Sorensen 1997; Jaeger et al. 1999). The phytoexudates comprising citrates, malates, fumarates, isocitrates, succinates, oxalates, lactates, acetates, etc., provide attractive media for fungal and bacterial colonies to flourish and sprawl to occupy the rhizosphere through cell-to-cell commune technique called "quorum sensing" (Ozyigit and Dogan 2015; Bhattacharyya and Jha 2012) which enables for biofilm formation which begins in transformation of the heavy metals around the root zones of plants through symbiotic relationship (Mathew et al. 2015).

The combination of exudates of roots along with microbial population which constitute the rhizosphere ecology, influences the bioavailability levels of metals as well as nutrients in the soil (Janoušková et al. 2006). As a perfect symbiotic role, the root exudates provide copious supply of energy and nutrients to the proliferating microbes and the microbes in turn stimulate the plant roots for more secretion of root exudates. As a process of co-evolution, microbes and plants either mutually support each other or compete with each other in the effort for survival and establishment (Spain and Alm 2003; Viti and Giovannetti 2003). On a positive note, root exudates accelerate the mobility of metals as well as nutrients by the following processes:

1) Hydrogen ion acidification of metals and nutrients or formation of complexes of metal/ amino acid/organic complexes
2) Binding/chelation with intracellular compounds (amino acids, organic acids, and phytochelatins) (Jones 1998)
3) Electron transfer mechanisms by phytoenzymes through mechanisms like redox reactions
4) Indirect stimulation of growth and proliferation of rhizospheric microbial communities (Ström et al. 2002; Pérez-Esteban et al. 2013; Sessitsch et al. 2013; Bach et al. 2016; Rouch et al. 1995)

Although some plants possess inherent nature to detoxify the heavy metals, the transforming skills of microbes are imperative as an augmentation to plants (Joutey et al. 2015; Zhou et al. 2012). The softening process of organic toxins in fuels which contain hydrocarbons and heavy metals is mediated by microbes (Adesodun et al. 2010; Wang and Chen 2006). Metal sequestration by synergistic microbial colonies increases the tolerance ability of the plants to heavy metals (laneva 2009; Velásquez and Dussan 2009).

Arbuscular mycorrhizal fungi (AMF) secrete a type of conjugated protein – "glomalin." This type of protein complexes with metals as a mechanism of chelation. Also, the microbial communities around the rhizoplane also mediate the phytoremediation by protecting the plants from harmful effects of heavy metals and this assistance by microbes is reciprocated by the plants by the offer of nutrients for microbial processes. The organic exudates serve as functional foods to the microbes which contribute to the upscaling of the establishment and proliferation of microbial assemblages in the soil. Lebeau (2011) cites that colonies of a single or group of microbial species with chemical activities transform the heavy metal-contaminated sites and build the biodegradation, eventually resulting in "bioaugmentation." Rui et al. (2022) highlight those assemblages of microbial strains and species are advantageous economically

(White et al. 2016) as it pools up the integration of individual transformational roles of each species with the augmented result of detoxification.

Based on the studies as reported by De Souza et al. (1999) and Stout et al. (2010), the results have indicated high concentrations of heavy metal cumulation by the plants whenever the rhizosphere of plants under investigation were grown with axenic microorganisms compared to control or with antibiotics which depletes such microbes.

11.4 Microbial and Plant Symbiosis in Phytoremediation

Plants accumulate and store the pollutants such as heavy metals that can be uprooted or harvested later. But in case of microbial-assisted heavy metal phtyo extraction, conversion of heavy metals to less toxic and stable form is plausible (Naik et al. 2012; Pratush et al. 2018). These non-toxic stable forms of heavy metals are easily uptaken by plant roots. Bell et al. (2015) also quote about this form of reworking of contaminants limiting the volatilization by the microorganisms. The plant root exudates of sugars, enzymes, aromatics, and aliphatics nourish for the thriving and proliferation mechanisms of microbial assemblages and the microbes in turn augment the plant's ability to decontaminate heavy metals (Knee et al. 2001; Ma et al. 2013). Thereby Lin et al. (2008) rightly point out that the symbiotic benefit from microbes propels the plant roots to penetrate soil sublayers, increasing aeration and water circulation and optimizing oxidation–reduction potential, pH, O_2 and CO_2 levels, providing a comfortable zone for their microbial colonies for such mutual symbiotic relational rewards.

As we perceive well that the augmented bioavailability of non-degradable heavy metals by microbial populations increases efficiency of biosorption of heavy metals by the plants (Idris et al. 2004), there are also other benefits mutually shared by the plants and microbes indirectly via assisting phytoremediation of heavy metals. This includes release of oxygen into surrounding rhizosphere in favor of microbial sustenance, and in turn, these species of microbes suppress the pathogenic microbes which could harm such phytoremediating plants (Babalola 2010; Guittonny-Philippe et al. 2014).

Following is the table consolidating the phytomicrobial symbiotic remediation of heavy metals (Table 11.3):

Table 11.3 Microbes-assisted heavy metal decontamination.

Plant	Microorganisms	Heavy metal uptaken	Mechanism	Reference
Alfalfa	*Cellulosimicrobium* sp.	Cr, Cu, Zn	↑ Plant growth ↑ Metal uptake (phytoextraction)	Tirry et al. (2018)
Brassica juncea	*Paenibacillus macerans NBRFT5, Bacillus endophyticus NBRFT4*	Cu, Ni, Zn	Increases metal uptake, phytoextraction	Tiwari et al. (2012)
	Rhizobium leguminosarum	Zn	Increases plant growth and biomass	Adediran et al. (2015)

Table 11.3 (Continued)

Plant	Microorganisms	Heavy metal uptaken	Mechanism	Reference
	Streptomyces pactum (Act 12)	Zn, Pb, Cd, Cu	Increases metal uptake, phytoextraction	Ali et al. (2017)
	Enterobacter sp. *NBRI K28, mutant NBRI K28 SD1*	Ni, Zn, Cr	↑ Plant biomass, protein and chlorophyll content, and uptake of HM (phytoextraction)	Kumar et al. (2008)
	Pseudomonas aeruginosa KUCd1	Cd	↑ Plant length, fresh and dry weight, and chlorophyll	Sinha and Mukherjee (2008)
	Pseudomonas sp. *PsA, Bacillus* sp. *Ba32*	Cr	↑ Plant growth (phytostabilization)	Rajkumar et al. (2006)
	Pseudomonas aeruginosa, Alcaligenes faecalis, and Bacillus subtilis	Cd, Cr, Ni	↑ Plant growth ↑ Metal accumulation	Ndeddy Aka and Babalola (2016)
Brassica oxyrrhina	*Pseudomonas libanensis*	Cu, Zn	↑ Plant growth ↑ Metal uptake (phytoextraction)	Ma et al. (2016)
Brassica napus L.	*Burkholderia cepacian J62*	Cu	↑ Plant biomass ↑ Cu uptake (phytoextraction)	Ren et al. (2019)
	Pseudomonas fluorescens G10, Microbacterium	Pb	↑ Root length	Sheng et al. (2008)
	Pseudomonas sp. *RJ10, Bacillus* sp. *RJ16*	Cd	↑ Plant dry weight ↑ Cd uptake (phytoextraction)	Sheng and Xia (2006)
	Ralstonia metalidurans, Pseudomonas aeruginosa	Cr, Pb	↑ Accumulation of Cr in shoots	Braud et al. (2009)
Broussonetia papyrifera	*Bacillus cereus HM5 and Bacillus thuringiensis HM7*	Mn	Increases metal uptake, phytoextraction	Huang et al. (2020)
Catharanthus roseus	*Burkholderia cepacian CS8*	Cd	↑ Plant growth ↑ Cd uptake (phytoextraction)	Khan et al. (2018)
	Pseudomonas fluorescens RB4 and Bacillus subtilis 189	Cu, Pb	↑ Plant growth ↑Metal uptake (phytoextraction)	Khan et al. (2017)
Calendula officinalis	*Glomus mossea and G. intraradices*	Pb, Cd	Phytoextraction	Tabrizi et al. (2015)

(*Continued*)

Table 11.3 (Continued)

Plant	Microorganisms	Heavy metal uptaken	Mechanism	Reference
Eruca sativa	*Pseudomonas putida*	Cr	↑ Plant growth ↑ Cr uptake (phytoextraction)	Kamran et al. (2017)
Helianthus annuus	*Ralstonia eutropha, and Chrysiobacterium humi*	Cd and Zn	Bioaugmentation, and phytostabilization	Marques et al. (2013)
	Pseudomonas libanensis TR1, Claroideoglomus claroideum BEG210	Ni	Phytoextraction	Ma et al. (2019)
	Micrococcus sp. *MU1, Klebsiella* sp. *BAM1*	Cd	↑ Root length ↑ Bioaugmentation	Prapagdee et al. (2012)
Orychophragmus violaceus	*B. subtilis, B. cereus, Flavobacterium* sp. *and P. aeruginosa*	Zn	Bioaugmentation, and phytoextraction	He et al. (2010)
Polygonum pubescens	*Enterobacter* sp. *JYX7, Klebsiella* sp. *JYX10*	Cd, Pb, and Zn	Increases metal uptake, phytoextraction	Jing et al. (2014)
Wedelia trilobata	*Phosphate solubilizing bacteria (PSB)*	Cu	Bioaugmentation and phytoextraction	Lin et al. (2018)
Solanum nigrum	*Glomerella truncate and Phomopsis fukushii*	Cd	Myco remediation and phytoextraction	Khan et al. (2017)
Zea mays, Lycopersicon esculentum	*Burkholderia* sp. *J62*	Cd, Pb	↑ Root length ↑ Rhizo remediation of Cd	Guo et al. (2015) Ganesan (2008)
Z. mays	*Microbacterium* sp. *Strain SUCR140*	Cr(VI)	↑ Plant growth	Soni et al. (2014)
O. violaceus	*B. subtilis, B. cereus, Flavobacterium* sp., *P. aeruginosa*	Zn	↑ Shoot biomass ↑ Zn accumulation	He et al. (2010)
Vigna radiata var. K851	*Bradyrhizobium* sp. *RM8*	Ni, Zn	↑ Plant growth ↓ Uptake of heavy	Wani et al. (2007)
L. esculentum	*Pseudomonas* sp. *RJ10, Bacillus* sp. *RJ16*	Cd, Pb	↑ Cd and Pb uptake (phytoextraction)	He et al. (2009)
Z. mays	*Bacillus* sp. *CIK-517, CIK-519, Klebsiella* sp. *CIK-518, Leifsonia* sp., *CIK-521, and Enterobacter* sp. *CIK-521R*	Cd	↑ Plant growth ↑ Cd uptake (phytoextraction)	Ahmad et al. (2016)
Sesbania sesban	*Bacillus xiamenensis PM14*	Cr	↑ Plant growth ↑ Chlorophyll content	Din et al. (2020)

Table 11.3 (Continued)

Plant	Microorganisms	Heavy metal uptaken	Mechanism	Reference
Triticum aestivum L.	*Enterobacter ludwigii CDP-14*	Zn	↑ Plant growth ↑ Zn uptake (phytoextraction)	Islam et al. 2014 Singh et al. (2018)
Sedum alfredii	*P. fluorescens Sasm05 and IAA*	Cd	↑ Plant growth ↑ Chlorophyll content ↑ Cd uptake (phytoextraction) ↑ Plant growth ↑ Cd uptake (phytoextraction)	Chen et al. (2017)
Vetiveria zizanioides	*Aeromonas* sp.	Cd	↑ Plant growth ↑ Cd uptake (phytoextraction)	Itusha et al. (2019)
Sesbania bispinosa	*Providencia* spp, *Moganella* sp, *Stenotrophomonas* sp. and *Bacillus* spp.	Cd	↑ Plant growth ↑ Cd uptake (phytoextraction)	Kartik et al. (2016)
Spinacia oleracea L. *var. Matador*	*Pseudomonas* sp., *Bacillus* sp.	Cd, Pb, Zn	↑ Plant growth ↓ Metal accumulation (phytostabilization)	Shilev et al. (2020)
Napier grass	*Micrococcus* sp.	Cd	Cd accumulation in the roots and shoots	Wiangkham and Prapagdee (2018)

11.5 Phyto-Microbe Contributory Roles

As we concise the contributory roles of plant roots and microbial components, mutualism between them can be appreciated while taking sort of their symbiotic support. In particular, the root exudates provide C & N chemoattractants for bacterial colonization for microbial remediation (Geva et al. 2016) of heavy metals (Bais et al. 2006; Compant et al. 2010). The root exudates build adaptive capacity of plants to stress induced by heavy metals by preventing heavy metal ions passage. In addition, the root exudates stimulate allelopathic effects of secondary phytochemicals on adverse microbes in the root zone. Chiang et al. (2011) and Luo et al. (2014) reported that the root exudates enhance the binding of metal ions for chelation. The root exudates also assist in detoxification of heavy metals through adsorption → chelation → transformation → inactivation of metals (Gupta et al. 2016). Nedjimi (2021) reiterates that the solubility of metals or their mobility could be modified by root exudates of certain specific plant species.

With respect to contributory roles of microbes to plants, Reinhold-Hurek and Hurek (2011) mention about the shaping of diversity of microbial colonization in lieu of

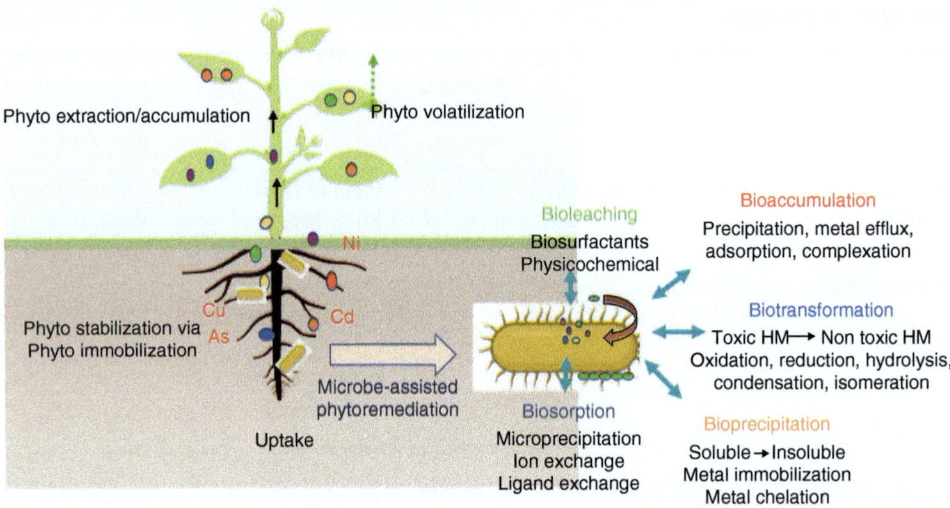

Figure 11.2 Phytoremedial mechanisms and microbial assisted phytoremediation.

host plant architecture. Also, the microbes get allured by phytochemicals, move, anchor, spread, and disperse for spatial coverage for decontamination of heavy metals. Microbes promote infestation of inner tissues of host plants as endophytes for assisted phytoremediation (Ma et al. 2015; Wang et al. 2016). Kamnev (2008) highlights the enhanced adherence of microbes to root surfaces through bacterial cell wall components such as proteins, glycoproteins and glycophospholipids and thereby augmenting conversion of heavy metal ions and compounds into bioavailable form through softening by microbial metabolic processes. Sarah and Thaura (2021) make a special mention of the biochemical and molecular mechanisms of *Klebsiella* and *Enterobacter* genera of bacteria which enable them to be top scorers of tolerance to Cd and Pb implying their contributory roles to the plants (Figure 11.2).

11.6 Conclusion

Meticulous study across various reports and studies clearly portrays the roles adopted by plant species in phytoremediation of heavy metals. Phytoremediation is rendered efficiently both by the root as well as shoot components of plants. The aerial parts phytoremediate by phytoaccumulation/phytoextraction, phytovolatilization, phytofiltration, and phytostabilization mechanisms. Whereas, it can be captured that the subaerial and underground root system of phytoremediating plants effect decontamination and detoxification of heavy metal-polluted sites by means of phytoimmobilization and phytostabilization through biotransformation with the aid of microbes. In addition, the efficacy of phytoremediation is enhanced by the assistance of microbes via biosorption, bioleaching, bioprecipitation, biotransformation, and bioaccumulation. The unique symbiotic associational relationships in phytomicrobes augment the adaptive capacity of plants to heavy metal stress, which naturally supplies nutrient-rich root exudates to feed microbes for more assisted biotransformations of tough heavy metals into bioavailable forms for the host plants to flourish as the best mechanism of phytoremediation. This chapter draws a clear-cut framework on roles played by plants and their associated mechanisms in phytoremediation with a callous understanding of the role of microbes in assisting for uptake of heavy metals. Exhaustive research reports are available with respect to specific biochemical and molecular mechanisms of phytoremediation in association with microbes which can be compiled to capture a holistic picture of these mechanisms as a book or book chapter exclusively as a future perspective.

References

Abhilash, M.R., Srikantaswamy, S., Shiva Kumar, D. et al. (2016). Phytoremediation of heavy metal industrial contaminated soil by *Spinacia oleracea* L. and *Zea mays* L. *International Journal of Applied Sciences* 4 (1): 192–199.

Achakzai, A.K.K., Bazai, Z.A., and Kayani, S.A. (2011). Accumulation of heavy metals by lettuce (*Lactuca sativa* L.) irrigated with different levels of wastewater of Quetta City. *Pakistan Journal of Botany* 43 (6): 2953–2960.

Adediran, G.A., Ngwenya, B.T., Mosselmans, J.F.W. et al. (2015). Mechanisms behind bacteria induced plant growth promotion and Zn accumulation in *Brassica juncea*. *Journal of Hazardous Materials* 283: 490–499.

Adesodun, J.K., Atayese, M.O., Agbaje, T.A. et al. (2010). Phytoremediation potentials of sunflowers (*Tithonia diversifolia* and *Helianthus annuus*) for metals in soils contaminated with zinc and lead nitrates. *Water, Air, and Soil Pollution* 207: 195–201.

Ahemad, M. (2012). Implications of bacterial resistance against heavy metals in bioremediation: a review. *Journal of Institute of Integrative Omics and Applied Biotechnology* 3: 39–46.

Ahemad, M. and Khan, M.S. (2012). Biotoxic impact of fungicides on plant growth promoting activities of phosphate-solubilizing Klebsiella sp. isolated from mustard (*Brassica compestris*) rhizosphere. *Journal of Pest Science* 85: 29–36.

Ahemad, M. and Kibret, M. (2013). Mechanisms and applications of plant growth promoting rhizobacteria: current perspective. *Journal of King Saud University – Science* 26 (1): 1–20. ISSN: 1018-3647.

Ahemad, M. and Malik, A. (2011). Bioaccumulation of heavy metals by zinc resistant bacteria isolated from agricultural soils irrigated with wastewater. *Bacteriology Journal* 2: 12–21.

Ahmad, I., Akhtar, M.J., Asghar, H.N. et al. (2016). Differential effects of plant growth-promoting rhizobacteria on maize growth and cadmium uptake. *Journal of Plant Growth Regulation* 35 (2): 303–315.

Alaribe, F.O. and Agamuthu, P. (2015). Assessment of phytoremediation potentials of *Lantana camara* in Pb impacted soil with organic wasted additives. *Ecological Engineering* 83: 513–520.

Ali, H., Khan, E., and Sajad, M.A. (2013). Phytoremediation of heavy metals-concepts and applications. *Chemosphere* 91: 869–881.

Ali, S., Charles, T.C., and Glick, B.R. (2017). Endophytic phytohormones and their role in plant growth promotion. In: *Functional Importance of the Plant Microbiome: Implications for Agriculture, Forestry and Bioenergy*, 89–105. Cham: Springer International Publishing.

Angelova, V.R., Perifanova-Nemska, M., Uzunova, G. et al. (2016). Potential of sunflower (*Helianthus annuus* L.) for phytoremediation of soils contaminated with heavy metals. *World Journal of Engineering Research and Technology* 10: 1–11.

Arbaoui, S., Campanella, B., Rezgui, S. et al. (2014). Bioaccumulation and photosynthetic activity response of kenaf (*Hibicus cannabinus* L.) to cadmium and zinc. *Greener Journal of Agricultural Sciences* 4: 91–100.

Arbaoui, S., Soufi, S., Rogsumaner, P., and Bettaieb, T. (2016). Phytoremediation of trace metal polluted soil with fiber crop: Kenaf (*Hibiscus cannabinus* L). *International Journal of Advances in Agricultural and Environmental Engineering* 3: 367–370.

Ashraf, S., Ali, Q., Zahir, Z.A. et al. (2019). Phytoremediation: environmentally sustainable way for reclamation of heavy metal polluted soils. *Ecotoxicology and Environmental Safety* 174: 714–727. https://doi.org/10.1016/j.ecoenv.2019.02.068.

Awa, S.H. and Hadibarata, T. (2020). Removal of heavy metals in contaminated soil by phytoremediation mechanism: a review. *Water, Air, & Soil Pollution* 231: 47.

Babalola, O.O. (2010). Beneficial bacteria of agricultural importance. *Biotechnology Letters* 32: 1559–1570.

Bach, E., dos Santos Seger, G.D., de Carvalho Fernandes, G. et al. (2016). Evaluation of biological control and rhizosphere competence of plant growth promoting bacteria. *Applied Soil Ecology* 99: 141–149.

Bais, H.P., Weir, T.L., Perry, L.G. et al. (2006). The role of root exudates in rhizosphere interactions with plants and other organisms. *Annual Review of Plant Biology* 57: 233–266. https://doi.org/10.1146/annurev.arplant.57.032905.105159.

Baker, A. and Brooks, R. (1989). Terrestrial higher plants which hyperaccumulate metallic elements. A review of their distribution, ecology and phytochemistry. *Biorecovery* 1: 81–126. https://doi.org/10.1080/01904168109362867.

Barkay, T., Miller, S.M., and Summers, A.O. (2003). Bacterial mercury resistance from atoms to ecosystems. *FEMS Microbiology Reviews* 27 (2–3): 355–384.

Becerra-Castro, C., Monterroso, C., Garcia-Leston, M. et al. (2009). Rhizosphere microbial densities and trace metal tolerance of the nickel hyperaccumulator *Alyssum serpyllifolium* subsp. *lusitanicum*. *International Journal of Phytoremediation* 11 (6): 525–541.

Belimov, A.A., Hontzeas, N., Safranova, V.I. et al. (2005). Cadmium-tolerant plant growth[1]promoting bacteria associated with the roots of Indian mustard (*Brassica juncea* L. Czern.). *Soil Biology and Biochemistry* 37: 241–250.

Bell, T.H., Joly, S., Pitre, F.E., and Yergeau, E. (2015, 2014). Increasing phytoremediation efficiency and reliability using novel omics approaches. *Trends in Biotechnology* 32: 271–280.

Berti, W.R. and Cunningham, S.D. (2000). Phytostabilization of metals. In: *Phytoremediation of Toxic Metals: Using Plants to Clean-up the Environment* (ed. I. Raskin and B.D. Ensley), 71–88. New York, NY: Wiley.

Beveridge, T.J. (1989). Role of cellular design in bacterial metal accumulation and mineralization. *Annual Review of Microbiology* 43: 147–171.

Bhattacharyya, P.N. and Jha, D.K. (2012). Plant growth-promoting rhizobacteria (PGPR): emergence in agriculture. *World Journal of Microbiology and Biotechnology* 28 (4): 1327–1350.

Bizily, S.P., Rugh, C.L., and Meagher, R.B. (2000). Phytodetoxification of hazardous organomercurials by genetically engineered plants. *Nature Biotechnology* 18: 213–217.

Bothe, H. (2011). Plants in heavy metal soils. In: *Detoxification of Heavy Metals*, vol. 30 (ed. I. Sherameti and A. Varma), 35–57. Berlin/Heidelberg: Springer-Verlag.

Braud, A., Jézéquel, K., Bazot, S., and Lebeau, T. (2009). Enhanced phytoextraction of an agricultural Cr- and Pb-contaminated soil by bioaugmentation with siderophore-producing bacteria. *Chemosphere* 74 (2): 280–286. https://doi.org/10.1016/j.chemosphere.2008.09.013.

Broadhurst, C.L. and Chaney, R.L. (2016). Growth and metal accumulation of an *Alyssum murale* nickel hyperaccumulator ecotype co-cropped with *Alyssum montanum* and perennial ryegrass in serpentine soil. *Frontiers in Plant Science* 2016 (7): 451.

Bruins, M.R., Kapil, S., and Oehme, F.W. (2000). Microbial resistance to metals in the environment. *Ecotoxicology and Environmental Safety* 45 (3): 198–207. https://doi.org/10.1006/mesa.1999.1860.

Burges, A., Alkorta, I., Epelde, L., and Garbisu, C. (2018). From phytoremediation of soil contaminants to photo management of ecosystem services in metal-contaminated sites. *International Journal of Phytoremediation* 20: 384–397. https://doi.org/10.1080/15226514.2017.1365340.

Cartoga, A., Fernando, A., and Oliveira, J.S. (2005). Effects on growth, productivity and biomass quality of kenaf of soils contaminated with heavy metals. *Sci Ref Cosmos* 4: 1–4.

Cempel, M. and Nikel, G. (2006). Nickel: a review of its sources and environmental toxicology. *Polish Journal of Environmental Studies* 15: 375–382.

Chen, B., Luo, S., Wu, Y. et al. (2017). The effects of the endophytic bacterium *Pseudomonas fluorescens* Sasm05 and IAA on the plant growth and cadmium uptake of *Sedum alfredii* hance. *Frontiers in Microbiology* 8: 2538.

Chiang, P.N., Chiu, C.Y., Wang, M.K., and Chen, B.T. (2011). Low-molecularweight organic acids exuded by millet (*Setaria italica* (L.) Beauv.) roots and their effect on the remediation of cadmium-contaminated soil. *Soil Science* 176: 33–38.

Choudhury, R. and Srivastava, S. (2001). Zinc resistance mechanisms in bacteria. *Current Science* 81 (7): 768–775.

Clemens, S. (2006). Toxic metal accumulation, responses to exposure and mechanisms of tolerance in plants. *Biochimie* 88: 1707–1719. https://doi.org/10.1016/j.biochi.2006.07.003.

Compant, S., Clément, C., and Sessitsch, A. (2010). Plant growth-promoting bacteria in the rhizo-and endosphere of plants: their role, colonization, mechanisms involved and prospects for utilization. *Soil Biology and Biochemistry* 42 (5): 669–678.

DalCorso, G., Fasani, E., Manara, A. et al. (2019). Heavy metal pollutions: state of the art and innovation in phytoremediation. *International Journal of Molecular Sciences* 20: 3412.

Dalvi, A.A. and Bhalerao, S.A. (2013). Response of plants towards heavy metal toxicity: an overview of avoidance, tolerance and uptake mechanism. *Annals of Plant Sciences* 2: 362–368.

Dasgupta, S., Satvat, P.S., and Mahinrakar, A.B. (2011). Ability of *Cicer arientinum* (L.) for bioremoval of lead and chromium from soil. *International Journal of Technology and Engineering System* 2 (3): 338–341.

Datta, R., Das, P., Tappero, R. et al. (2017). Evidence for exocellular arsenic in fronds of *Pteris vittata*. *Scientific Reports* 7: 2839.

de Souza, M.P., Chu, D., Zhao, M. et al. (1999). Rhizosphere bacteria enhance selenium accumulation and volatilization by Indian mustard. *Plant Physiology* 119 (2): 565–574.

Dell'Amico, E., Cavalva, L., and Andreoni, V. (2008). Improvement of *Brassica napus* growth under cadmium stress by cadmium-resistant rhizobacteria. *Soil Biology and Biochemistry* 40: 74–84.

Din, B.U., Amna, Rafique, M. et al. (2020). Assisted phytoremediation of chromium spiked soils by Sesbania Sesban in association with *Bacillus xiamenensis* PM14: a biochemical analysis. *Plant Physiology and Biochemistry* 146: 249–258.

Dushenkov, V., Kumar, P.B.A.N., Motto, H., and Raskin, I. (1995). Rhizofiltration: the use of plants to remove heavy metals from aqueous streams. *Environmental Science & Technology* 29 (5): 1239–1245.

Egamberdieva, D. (2012). *Pseudomonas chlororaphis*: a salt-tolerant bacterial inoculant for plant growth stimulation under saline soil conditions. *Acta Physiologiae Plantarum* 34 (2): 751–756.

Emamverdian, A., Ding, Y., Mokhberdoran, F., and Xie, Y. (2015). Heavy metal stress and some mechanisms of plant defense response. *Scientific World Journal* 2015: 756120.

Egamberdieva, D., Kamilova, F., Validov, S. et al. (2008). High incidence of plant growth-stimulating bacteria associated with the rhizosphere of wheat grown on salinated soil in Uzbekistan. *Environmental Microbiology* 10 (1): 1–9.

Ernst, W.H. (2005). Phytoextraction of mine wastes – options and impossibilities. *Chemie der Erde-Geochemistry* 65: 29–42.

Ernst, W.H., Verkleij, J., and Schat, H. (1992). Metal tolerance in plants. *Acta Botanica Neerlandica* 41: 229–248. https://doi.org/10.1111/j.1438-8677.1992.tb01332.x.

Fasani, E., Manara, A., Martini, F. et al. (2018). The potential of genetic engineering of plants for the remediation of soils contaminated with heavy metals. *Plant, Cell & Environment* 41: 1201–1232. https://doi.org/10.1111/pce.12963.

Flávio S.R, Luiz Tadeu J and Jose L.J. (2017). A glimpse into the symplastic and apoplastic Cd uptake by Massai grass modulated by sulfur nutrition: Plants well-nourished with S as a strategy for phytoextraction. *Plant Physiology and Biochemistry.* 121.10.1016/j.plaphy.2017.10.018.

Fulekar, M. (2016). Phytoremediation of heavy metals by Helianthus annuus in aquatic and soil environment. *International Journal of Current Microbiology and Applied Sciences* 5: 392–404.

Gadd, G.M. (2000). Bioremedial potential of microbial mechanisms of metal mobilization and immobilization. *Current Opinion in Biotechnology* 11: 271–279.

Ganesan, V. (2008). Rhizoremediation of cadmium soil using a cadmium-resistant plant growth-promoting rhizopseudomonad. *Current Microbiology* 56 (4): 403–407.

Garg, N., Singla, P., and Bhandari, P. (2014). Metal uptake, oxidative metabolism, and mycorrhization in pegeon pea and pea under arsenic and cadmium stress. *Turkish Journal of Agriculture and Forestry* 39: 234–250.

Gerhardt, K.E., Gerwing, P.D., and Greenberg, B.M. (2017). Opinion: taking phytoremediation from proven technology to accepted practice. *Plant Science* 256: 170–185.

Geva, P., Kahta, R., Nakonechny, F. et al. (2016). Increased copper bioremediation ability of new transgenic and adapted *Saccharomyces cerevisiae* strains. *Environmental Science and Pollution Research* 23: 19613–19625.

Ghnaya, T., Mnassri, M., Ghabriche, R. et al. (2015). Nodulation by *Sinorhizobium meliloti* originated from a mining soil alleviates Cd toxicity and increases Cd-phytoextraction in *Medicago sativa* L. *Frontiers in Plant Science* 6: 1–10.

Gillespie, I.M.M. and Philp, J.C. (2013). Bioremediation, an environmental remediation technology for the bioeconomy. *Trends in Biotechnology* 31: 329–332.

Ginn, B.R., Szymanowski, J.S., and Fein, J.B. (2008). Metal and proton binding onto the roots of *Fescue rubra*. *Chemical Geology* 253: 130–135. https://doi.org/10.1016/j.chemgeo.2008.05.001.

Giri, B.E., Okoduwa, S.I.R., Idoko, G.O. et al. (2018). Toxicity and bioremediation of heavy metals contaminated ecosystem from tannery wastewater: a review. *Journal of Toxicology* 2018: 2568038.

Goland-Goldhirsh, A. (2006). Plant tolerance to heavy metals, a risk for food toxicity or a means for food fortification with essential metals: the *Allium schoenoprasum* model. In: *Soil and Water Pollution Monitoring, Protection and Remediation* (ed. I. Twardowska, H.E. Allen, and M.M. Haggblom), 479–486. Amsterdam: Springer.

Guittonny-Philippe, A., Masotti, V., Höhener, P. et al. (2014). Constructed wetlands to reduce metal pollution from industrial catchments in aquatic Mediterranean ecosystems: a review to overcome obstacles and suggest potential solutions. *Environment International* 64: 1–16.

Gunduz, S., Uygur, F.N., and Kahramanoglu, I. (2012). Heavy metal phytoremediation potentials of *Lepidium sativum* L., *Lactuca sativa* L., *Spinacia oleracea* L. and *Raphanus sativus* L. *Herald Journal of Agriculture and Food Science Research* 1 (1): 1–5.

Guo, J.K., Ding, Y.Z., Feng, R.W. et al. (2015). *Burkholderia metalliresistens* sp. nov., multiple metal-resistant and phosphate-solubilising species isolated from heavy metal-polluted soil in Southeast China. *Antonie Van Leeuwenhoek* 107 (6): 1591–1598.

Gupta, V.K., Nayak, A., and Agarwal, S. (2015). Bio adsorbents for remediation of heavy metals: current status and their future prospects. *Environmental Engineering Research* 2015 (20): 1–18.

Gupta, A., Joia, J., Sood, A. et al. (2016). Microbes as potential tool for remediation of heavy metals: a review. *Journal of Microbial and Biochemical Technology* 8: 364–372.

Hall, J. (2016). Cellular mechanisms for heavy metal detoxification and tolerance. *Journal of Experimental Botany* 53: 1–11. https://doi.org/10.1093/jexbot/53.366.1.

Hansda, A. et al. (2022). Plant-microbe association to improve phytoremediation of heavy metal. In: *Advances in Microbe-assisted Phytoremediation of Polluted Sites*, 113–146. Elsevier (Chapter 6).

Hao, X., Mohamad, O.A., Xie, P. et al. (2014). Removal of zinc from aqueous solution by metal resistant symbiotic bacterium *Mesorhizobium amorphae*. *Separation Science and Technology* 49 (3): 376–387.

Hashim, M.A., Mukhopadhyay, S., Sahu, J.N., and Sengupta, B. (2011). Remediation technologies for heavy metal contaminated groundwater. *Journal of Environmental Management* 92: 2355–2388.

Hatano, K., Kanazawa, K., Tomura, K. et al. (2016). Molases melanoidin promotes copper uptake for radish sprouts: the potential for an accelerator of phytoextraction. *Environmental Science and Pollution Research* 23: 17656–17663.

He, L.Y., Chen, Z.J., Ren, G.D. et al. (2009). Increased cadmium and lead uptake of a cadmium hyperaccumulator tomato by cadmium-resistant bacteria. *Ecotoxicology and Environmental Safety* 72 (5): 1343–1348.

He, L.Y., Zhang, Y.F., Ma, H.Y. et al. (2010). Characterization of copper-resistant bacteria and assessment of bacterial communities in rhizosphere soils of copper-tolerant plants. *Applied Soil Ecology* 44 (1): 49–55.

Hegedusova, A., Jakabova, S., Vargova, A. et al. (2009). Use of phytoremediation techniques for elimination of lead from polluted soils. *Nova Biotechnologica et Chimica* 9 (2): 125–132.

Huang, H., Zhao, Y., Fan, L. et al. (2020). Improvement of manganese phytoremediation by *Broussonetia papyrifera* with two plant growth promoting (PGP) Bacillus species. *Chemosphere* 260: 127614.

Ianeva, O.D. (2009). Mechanisms of bacteria resistance to heavy metals. *Mikrobiolohichnyĭ Zhurnal* 71 (6): 54.

Idris, R., Trifonova, R., Puschenreiter, M. et al. (2004). Bacterial communities associated with flowering plants of the Ni hyperaccumulator *Thlaspi goesingense*. *Applied and Environmental Microbiology* 70 (5): 2667–2677.

Islam, F., Yasmeen, T., Ali, Q. et al. (2014). Influence of *Pseudomonas aeruginosa* as PGPR on oxidative stress tolerance in wheat under Zn stress. *Ecotoxicology and Environmental Safety* 104: 285–293.

Itusha, A., Osborne, W.J., and Vaithilingam, M. (2019). Enhanced uptake of Cd by biofilm forming Cd resistant plant growth promoting bacteria bioaugmented to the rhizosphere of *Vetiveria zizanioides*. *International Journal of Phytoremediation* 21 (5): 487–495.

Jacob, J.M., Karthik, C., Saratale, R.G. et al. (2018). Biological approaches to tackle heavy metal pollution: a survey of literature. *Journal of Environmental Management* 217: 56–70.

Jaeger, C.H., Lindow, S.E., Miller, W. et al. (1999). Mapping of sugar and amino acid availability in soil around roots with bacterial sensors of sucrose and tryptophan. *Applied and Environmental Microbiology* 65 (6): 2685–2690.

Jahanbakhshi, S., Rezaei, M.R., and Sayyari-Zahan, M.H. (2014). Optimization of phytoremediation in cd-contaminated soil by using Taguchi method in *Spinacia oleracea*. In: *Proceedings of the International Academy of Ecology and Environmental Sciences*, vol. vol. 4 (ed. W. Zhang), 185–193. Hong Kong: International Academy of Ecology and Evironmental Sciences.

Jamil, S., Abhilash, P.C., Singh, N., and Sharma, P.N. (2009). *Jatropha curcas*: a potential crop for coal fly ash. *Journal of Hazardous Materials* 172: 269–275.

Janoušková, M., Pavlíková, D., and Vosátka, M. (2006). Potential contribution of arbuscular mycorrhiza to cadmium immobilisation in soil. *Chemosphere* 65 (11): 1959–1965.

Javed, M.T., Tanwir, K., Akram, M.S. et al. (2019). Phytoremediation of cadmium-polluted water/sediment by aquatic macrophytes: role of plant-induced pH changes. In: *Cadmium Toxicity and Tolerance in Plants* (ed. M. Hasanuzzaman, M.N.V. Prasad, and M. Fujita), 495–529. London: Academic Press 10.1016/B978-0-12-814864-8.00020-6 (Chapter 20).

Ji, P., Sun, T., Song, Y. et al. (2011). Strategies for enhanching the phytoremediation of cadmium[1]contaminated agricultural soils by *Solanum nigrum* L. *Environmental Pollution* 159: 762–768.

Jing, Y.X., Yan, J.L., He, H.D. et al. (2014). Characterization of bacteria in the rhizosphere soils of *Polygonum pubescens* and their potential in promoting growth and Cd, Pb, Zn uptake by *Brassica napus*. *International Journal of Phytoremediation* 16 (4): 321–333.

Jones, D.L. (1998). Organic acids in the rhizosphere—a critical review. *Plant and Soil* 205 (1): 25–44.

Joutey, N.T., Sayel, H., Bahafid, W., and El Ghachtouli, N. (2015). Mechanisms of hexavalent chromium resistance and removal by microorganisms. *Reviews of Environmental Contamination and Toxicology* 233: 45–69.

Jovancicevic, V., Bockris, J., Carbajal, J.L. et al. (1986). Adsorption and absorption of chloride ions on passive iron systems. *Journal of the Electrochemical Society* 133: 2219.

Kambhampati, M.S. and Vu, V.T. (2013). EDTA enhanced phytoremediation of copper contaminated soils using chickpea (*Cicer aeritinum* L.). *Bulletin of Environmental Contamination and Toxicology* 91: 310–313.

Kamnev, A.A. (2008). FTIR spectroscopic studies of bacterial cellular responses to environmental factors, plant-bacterial interactions and signaling. *Spectroscopy* 22: 83–95. https://doi.org/10.3233/SPE-2008-0329.

Kamran, M.A., Bibi, S., Xu, R.k. et al. (2017). Phyto-extraction of chromium and influence of plant growth promoting bacteria to enhance plant growth. *Journal of Geochemical Exploration* 182: 269–274.

Kartik, V.P., Jinal, H.N., and Amaresan, N. (2016). Characterization of cadmium-resistant bacteria for its potential in promoting plant growth and cadmium accumulation in *Sesbania bispinosa* root. *International Journal of Phytoremediation* 18 (11): 1061–1066.

Khan, M.S., Zaidi, A., Wani, P.A., and Oves, M. (2009). Role of plant growth promoting rhizobacteria in the remediation of metal contaminated soils. *Environmental Chemistry Letters* 7: 1.

Khan, W.U., Ahmad, S.R., Yasin, N.A. et al. (2017). Effect of *Pseudomonas fluorescens* RB4 and *Bacillus subtilis* 189 on the phytoremediation potential of *Catharanthus roseus* (L.) in Cu and Pb-contaminated soils. *International Journal of Phytoremediation* 19 (6): 514–521.

Khan, W.U., Yasin, N.A., Ahmad, S.R. et al. (2018). Role of *Burkholderia cepacia* CS8 in Cd-stress alleviation and phytoremediation by *Catharanthus roseus*. *International Journal of Phytoremediation* 20 (6): 581–592.

Knee, E.M., Gong, F.C., Gao, M. et al. (2001). Root mucilage from pea and its utilization by rhizosphere bacteria as a sole carbon source. *Molecular Plant-Microbe Interactions* 14 (6): 775–784.

Krumova, E., Kostadinova, N., Miteva-Staleva, J. et al. (2016). Cellular response to Cu-and Zn-induced oxidative stress in aspergillus fumigatus isolated from polluted soils in Bulgaria. *Clean: Soil, Air, Water* 2016 (44): 657–666.

Kumar, K.V., Singh, N., Behl, H.M., and Srivastava, S. (2008). Influence of plant growth promoting bacteria and its mutant on heavy metal toxicity in *Brassica juncea* grown in fly ash amended soil. *Chemosphere* 72 (4): 678–683.

Kumpiene, J., Fitts, J.P., and Mench, M. (2012). Arsenic fractionation in mine spoils 10 years after aided phytostabilization. *Environmental Pollution* 166: 82–88.

Lebeau, T. (2011). Bioaugmentation for in situ soil remediation: how to ensure the success of such a process. In: *Bioaugmentation, Biostimulation and Biocontrol* (ed. A. Singh et al.), 129–186. Berlin: Springer.

Li, P., Wang, X., Zhang, T. et al. (2008). Effect of several amendments on rice growth and uptake of copper and cadmium from a contaminated soil. *Journal of Environmental Sciences* 20: 449–455.

Lin, C.-C. and Lin, H.-L. (2005). Remediation of soil contaminated with the heavy metal (Cd2+). *Journal of Hazardous Materials* 122: 7–15.

Lin, X., Li, X., Li, P. et al. (2008). Evaluation of plant microorganism synergy for the remediation of diesel fuel contaminated soil. *Bulletin of Environmental Contamination and Toxicology* 81: 19–24.

Lin, M., Jin, M., Xu, K. et al. (2018). Phosphate-solubilizing bacteria improve the phytoremediation efficiency of *Wedelia trilobata* for Cu-contaminated soil. *International Journal of Phytoremediation* 20 (8): 813–822.

Luo, Q., Sun, L., Hu, X., and Zhou, R. (2014). The variation of root exudates from the hyperaccumulator *Sedum alfredii* under cadmium stress: metabonomics analysis. *PLoS One* 9: e115581.

Ma, Y., Rajkumar, M., Luo, Y., and Freitas, H. (2013). Phytoextraction of heavy metal polluted soils using *Sedum plumbizincicola* inoculated with metal mobilizing *Phyllobacterium myrsinacearum* RC6b. *Chemosphere* 93: 1386–1392.

Ma, Y., Oliveira, R.S., Nai, F. et al. (2015). The hyper accumulator *Sedum plumbizincicola* harbors metal-resistant endophytic bacteria that improve its phytoextraction capacity in multi-metal contaminated soil. *Journal of Environmental Management* 156: 62–69.

Ma, Y., Oliveira, R.S., Freitas, H., and Zhang, C. (2016). Biochemical and molecular mechanisms of plant-microbe- metal interactions: relevance for phytoremediation. *Frontiers in Plant Science* 7: 918.

Ma, Y., Rajkumar, M., Oliveira, R.S. et al. (2019). Potential of plant beneficial bacteria and arbuscular mycorrhizal fungi in phytoremediation of metal-contaminated saline soils. *Journal of Hazardous Materials* 379: 120813. https://doi.org/10.1016/j.jhazmat.2019.120813.

Mahar, A., Wang, P., Ali, A. et al. (2016). Challenges and opportunities in the phytoremediation of heavy metals contaminated soils: a review. *Ecotoxicology and Environmental Safety* 126: 111–121.

Malachowska-Jutsz, A. and Gnid, A. (2015). Mechanisms of stress avoidance and tolerance by plants used in phytoremediation of heavy metals. *Archives of Environmental Protection* 41 (4): 104–114.

Malecka, A., Piechalak, A., and Morkunas, I. (2008). Accumulation of lead in root cells of *Pisum sativum*. *Acta Physiologiae Plantarum* 30: 629–637.

Malik, A. (2004). Metal bioremediation through growing cells. *Environment International* 30: 261–278. https://doi.org/10.1016/j.envint.2003.08.001.

Manara, A. (2012). Plant responses to heavy metal toxicity. In: *Plants and Heavy Metals* (ed. A. Furini), 27–53. Dordrecht: Springer.

Marques, A.P., Rangel, A.O., and Castro, P.M. (2009). Remediation of heavy metal contaminated soils: phytoremediation as a potentially promising clean-up technology. *Critical Reviews in Environment Science and Technology* 39: 622–654. https://doi.org/10.1080/10643380701798272.

Marques, A.P., Moreira, H., Franco, A.R. et al. (2013). Inoculating *Helianthus annuus* (sunflower) grown in zinc and cadmium contaminated soils with plant growth promoting bacteria – effects on phytoremediation strategies. *Chemosphere* 92: 74–83.

Marrugo-Negrete, J., Durango-Hernandez, J., Pinedo-Hernandez, J. et al. (2015). Phytoremediation of mercury-contaminated soils by *Jatropha curcas*. *Chemosphere* 127: 58–63.

Mastretta, C., Taghavi, S., Van Der Lelie, D. et al. (2009). Endophytic bacteria from seeds of *Nicotiana tabacum* can reduce cadmium phytotoxicity. *International Journal of Phytoremediation* 11 (3): 251–267.

Mathew, D.C., Ho, Y.N., Gicana, R.G. et al. (2015). A rhizosphere-associated symbiont, Photobacterium spp. strain MELD1, and its targeted synergistic activity for phytoprotection against mercury. *PLoS One* 10 (3): e0121178.

McGrath, S.P. and Zhao, F.-J. (2003). Phytoextraction of metals and metalloids from contaminated soils. *Current Opinion in Biotechnology* 14: 277–228.

Memon, A.R. and Schröder, P. (2009). Implications of metal accumulation mechanisms to phytoremediation. *Environemental Science and Pollution Research* 16: 162–175.

Mendes, R., Kruijt, M., De Bruijn, I. et al. (2011). Deciphering the rhizosphere microbiome for disease-suppressive bacteria. *Science* 332 (6033): 1097–1100.

Mengoni, A., Cecchi, L., and Gonnelli, C. (2012). Nickel hyperaccumulating plants and *Alyssum bertolonii*: model systems for studying biogeochemical interactions in serpentine soils. In: *Bio-Geo Interactions in Metal-Contaminated Soils*, vol. Vol. 31 (ed. E. Kothe and A. Varma), 279–296. Berlin/Heidelberg: Springer.

Mesjasz-Przybyłowicz, J., Nakonieczny, M., Migula, P. et al. (2004). Uptake of cadmium, lead nickel and zinc from soil and water solutions by the nickel hyperaccumulator *Berkheya coddii*. *Acta Biologica Cracoviensia Series Botanica* 46: 75–85.

Mojiri, A. (2011). The potencial of corn (*Zea mays*) for phytoremediation of soil contaminated with cadmium and lead. *Journal of Biological and Environmental Sciences* 5 (13): 17–22.

Naik, M.M., Pandey, A., and Dubey, S.K. (2012). Pseudomonas aeruginosa strain WI-1 from Mandovi estuary possesses metallothionein to alleviate lead toxicity and promotes plant growth. *Ecotoxicology and Environmental Safety* 79: 129–133.

Nalla, S., Hardaway, C.J., and Sneddon, J. (2012). Phytoextraction of selected metals by the first and second growth seasons of *Spartina alterniflora*. *Instrumentation Science and Technology* 40: 17–28.

Ndeddy Aka, R.J. and Babalola, O.O. (2016). Effect of bacterial inoculation of strains of *Pseudomonas aeruginosa*, *Alcaligenes feacalis* and *Bacillus subtilis* on germination, growth and heavy metal (Cd, Cr, and Ni) uptake of *Brassica juncea*. *International Journal of Phytoremediation* 18 (2): 200–209. https://doi.org/10.1080/1522651 4.2015.1073671.

Nedjimi, B. (2021). Phytoremediation: a sustainable environmental technology for heavy metals decontamination. *SN Applied Sciences* 3: 286. https://doi.org/10.1007/s42452-021-04301-4.

Osman, M.E.H., El-Feky, S.S., Elshahawy, M.I., and Shaker, E.M. (2017). Efficiency of flax (*Linum usitatissimum* L.) as a phytoremediator plant for the contaminated soils with heavy metals. *International Journal of Agriculture and Environmental Research* 3: 3577–3600.

Ozyigit, I.I. and Dogan, I. (2015). Plant–microbe interactions in phytoremediation. In: *Soil Remediation and Plants*, 255–285. Elsevier Inc. (Chapter 9).

Patel, M. and Subramanian, R.B. (2006). Effect of a chelating agent on lead uptake by *Spinacia olearea*. *Pollution Research* 25 (1): 77–79.

Pathak, C., Chopra, A.K., and Srivastava, S. (2013). Accumulation of heavy metals in *Spinacia oleracea* irrigated with paper mill effluent and sewage. *Environmental Monitoring and Assessment* 185: 7343–7352.

Peer, W.A., Baxter, I.R., Richards, E.L. et al. (2005). Phytoremediation and hyperaccumulator plants. In: *Molecular Biology of Metal Homeostasis and Detoxification* (ed. M.J. Tamas and E. Martinoia), 299–340. Berlin: Springer 10.1007/4735_100.

Pérez-Esteban, J., Escolástico, C., Moliner, A., and Masaguer, A. (2013). Chemical speciation and mobilization of copper and zinc in naturally contaminated mine soils with citric and tartaric acids. *Chemosphere* 90: 276–283. https://doi.org/10.1016/j.chemosphere.2012.06.065.

Pinto, A.P., Mota, A.M., de Varennes, A., and Pinto, F.C. (2004). Influence of organic matter on uptake of cadmium, zinc, copper and iron by shorghum plant. *Science of the Total Environment* 326: 239–247.

Prapagdee, B., Chumphonwong, N., Khonsue, N., and Mongkolsuk, S. (2012). Influence of cadmium resistant bacteria on promoting plant root elongation and increasing cadmium mobilization in contaminated soil. *Fresenius Environmental Bulletin* 21 (5): 1186–1191.

Pratush, A., Kumar, A., and Hu, Z. (2018). Adverse effect of heavy metals (As, Pb, Hg, and Cr) on health and their bioremediation strategies: a review. *International Microbiology* 21: 97–106.

Quainoo, A.K., Konadu, A., and Kumi, M. (2015). The potential of shea nut shells in phytoremediation of heavy metals in contaminated soil using lettuce (*Lactuca sativa*) as a test crop. *Journal of Bioremediation & Biodegradation* 6 (1): 1–7.

Rabêlo, F.H.S., Jordão, L.T., and Lavres, J. (2017). A glimpse into the symplastic and apoplastic Cd uptake by Massai grass modulated by sulfur nutrition: plants well-nourished with S as a strategy for phytoextraction. *Plant Physiology and Biochemistry* 121: 48–57.

Rajkumar, M., Nagendran, R., Lee, K.J. et al. (2006). Influence of plant growth promoting bacteria and Cr6+ on the growth of Indian mustard. *Chemosphere* 62 (5): 741–748. https://doi.org/10.1016/j.chemosphere.2005.04.117.

Rajkumar, M., Ae, N., Prasad, M.N.V., and Freitas, H. (2010). Potential of siderophore-producing bacteria for improving heavy metal phytoextraction. *Trends in Biotechnology* 28: 142–149.

Rascio, N. and Navari-Izzo, F. (2011). Heavy metal hyperaccumulating plants: how and why do they do it? And what makes them so interesting? *Plant Science* 180: 169–181. https://doi.org/10.1016/j.plantsci.2010.08.016.

Rashid, A., Mahmood, T., Mehmood, F. et al. (2014). Phytoaccumulation, competitive adsorption and evaluation of chelators-metal interaction in lettuce plant. *Environmental Engineering and Management Journal* 13 (10): 2683–2692.

Rathinasabapathi, B. (2011). Arsenic hyperaccumulator fern *Pteris vittata*: Utilities for arsenic phytoremediation and plant biotechnology. In: *Working with Ferns* (ed. H. Fernández, A. Kumar, and M.A. Revilla), 261–269. New York, NY: Springer.

Reinhold-Hurek, B. and Hurek, T. (2011). Living inside plants: bacterial endophytes. *Current Opinion in Plant Biology* 14 (4): 435–443.

Ren, X.M., Guo, S.J., Tian, W. et al. (2019). Effects of plant growth-promoting bacteria (PGPB) inoculation on the growth, antioxidant activity, Cu uptake, and bacterial community structure of rape (Brassica napus L.) grown in cu-contaminated agricultural soil. *Frontiers in Microbiology* 10: 1455. https://doi.org/10.3389/fmicb.2019.01455.

Robinson, B., Fernández, J.E., Madejón, P. et al. (2003). Phytoextraction: an assessment of biogeochemical and economic viability. *Plant and Soil* 249: 117–125.

Romera, E., Gonzalez, F., Ballester, A. et al. (2006). Biosorption with algae: a statistical review. *Critical Reviews in Biotechnology* 26: 223–235.

Rouch, D.A., Lee, B.T.O., and Morby, A.P. (1995). Understanding cellular responses to toxic agents: a model for mechanism-choice in bacterial metal resistance. *Journal of Industrial Microbiology* 14 (2): 132–141.

Rui, L., Wang, B., Niu, A. et al. (2022). Application of biochar immobilized microorganisms for pollutants removal from wastewater: a review. *Science of the Total Environment* **837**.

Sakakibara, M., Watanabe, A., Inoue, M. et al. (2010). Phytoextraction and phytovolatilization of arsenic from As-contaminated soils by *Pteris vittata*. *Proceedings of the Annual International Conference on Soils, Sediments, Water and Energy* 12 (1): 26.

Salaskar, D., Shrivastava, M., and Kale, S.P. (2011). Bioremediation potential od spinach (*Spinacia oleracea* L.) for decontamination of cadmium in soil. *Current Science* 101 (10): 1359–1363.

Sarah, G.H. and Thaura, G.-H. (2021). Heavy metals in soils and the remediation potential of bacteria associated with the plant microbiome. *Frontiers in Environmental Science* 9: 604216. https://doi.org/10.3389/fenvs.2021.604216.

Sarwar, N., Imran, M., Shaheen, M.R. et al. (2017). Phytoremediation strategies for soils contaminated with heavy metals: modifications and future perspectives. *Chemosphere* 171: 710–721. https://doi.org/10.1016/j.chemosphere.2016.12.116.

Sebastián, A., Kuffner, M., Kidd, P. et al. (2021). The role of plant-associated bacteria in the mobilization and phytoextraction of trace elements in contaminated soils. *Soil Biology and Biochemistry* 60: 182–194, ISSN 0038-0717, https://doi.org/10.1016/j.soilbio.2013.01.012.

Sessitsch, A., Kuffner, M., Kidd, P. et al. (2013). The role of plant-associated bacteria in the mobilization and phytoextraction of trace elements in contaminated soils. *Soil Biology and Biochemistry* 60: 182–194.

Shahpiri, A. and Mohammadzadeh, A. (2018). Mercury removal by engineered *Escherichia coli* cells expressing different rice metallothionein isoforms. *Annales de Microbiologie* 68: 145–152.

Sharma, S., Sharma, P., and Mehrotra, P. (2010). Bioaccumulation of heavy metals in *Pisum sativum* L. gowing in fly ash amended soil. *Journal of American Science* 6 (6): 43–50.

Sheng, X.F. and Xia, J.J. (2006). Improvement of rape (*Brassica napus*) plant growth and cadmium uptake by cadmium-resistant bacteria. *Chemosphere* 64: 1036–1042.

Sheng, X.F., Xia, J.J., Jiang, C.Y. et al. (2008). Characterization of heavy metal-resistant endophytic bacteria from rape (*Brassica napus*) roots and their potential in promoting the growth and lead accumulation of rape. *Environmental Pollution* 156 (3): 1164–1170. https://doi.org/10.1016/j.envpol.2008.04.007.

Sheoran, V., Sheoran, A., and Poonia, P. (2011). Role of hyperaccumulators in phytoextraction of metals from contaminated mining sites: a review. *Critical Reviews in Environment Science and Technology* 41: 168–214. https://doi.org/10.1080/10643380902718418.

Shilev, S., Babrikova, I., and Babrikov, T. (2020). Consortium of plant growth-promoting bacteria improves spinach (*Spinacea oleracea* L.) growth under heavy metal stress conditions. *Journal of Chemical Technology and Biotechnology* 95 (4): 932–939. https://doi.org/10.1002/jctb.6077.

Silver, S. and Phung, L.T. (1996). Bacterial heavy metal resistance: new surprises. *Annual Review of Microbiology* 50 (1): 753–789. https://doi.org/10.1146/annurev.micro.50.1.753.

Singh, A. and Fulekar, M.H. (2012). Phytoremediation of heavy metals by *Brassica juncea* in aquatic and terrestrial environment. In: *The Plant Family Brassicaceae: Contribution Towards Phytoremediation* (ed. N.A. Anjum, I. Ahmad, M. Eduarda Pereira, et al.), 153–169. Amsterdam: Springer Science+Business Media.

Singh, R.P., Mishra, S., Jha, P. et al. (2018). Effect of inoculation of zinc-resistant bacterium *Enterobacter ludwigii* CDP-14 on growth, biochemical parameters and zinc uptake in wheat (*Triticum aestivum* L.) plant. *Ecological Engineering* 116: 163–173. https://doi.org/10.1016/j.ecoleng.2017.12.033.

Sinha, S. and Mukherjee, S.K. (2008). Cadmium-induced siderophore production by a high Cd-resistant bacterial strain relieved Cd toxicity in plants through root colonization. *Current Microbiology* 56 (1): 55–60. https://doi.org/10.1007/s00284-007-9038-z.

Slatter, K.A. (2013). *Nickel Accumulation and Tolerance in Berkheya codii and Its Application in Phytoremediation*. Master's Thesis. Natal: University of Kwazulu.

Sloof, W., van Beelen, P., Annema, J.A., and Janus, J.A. (ed.) (1995). *Integrated Criteria Document Mercury, RIVM Report 601014008*. Bilthoven: National Institute of Public Health and Environmental Protection.

Smolinska, B. and Szczodrowska, A. (2016). Antioxidative response of *Lepidium sativum* L. during assisted phytoremediation of Hg contaminated soil. *New Biotechnology* 38: 74–83.

Soni, S.K., Singh, R., Awasthi, A., and Kalra, A. (2014). A Cr(VI)-reducing Microbacterium sp. strain SUCR140 enhances growth and yield of *Zea mays* in Cr(VI) amended soil through reduced chromium toxicity and improves colonization of arbuscular mycorrhizal fungi. *Environmental Science and Pollution Research* 21 (3): 1971–1979.

Sorensen, J. (1997). The rhizosphere as a habitat for soil microorganisms. In: *Modern Soil Microbiology* (ed. J.D. Van Elsas, J.T. Trevors, and E.M.H. Wellington), 21–45. New York, NY: Marcel Dekker, Inc.

Spain, A. and Alm, E. (2003). Implications of microbial heavy metal tolerance in the environment. *Reviews in Undergraduate Research* 2: 1–6.

Stout, L.M., Dodova, E.N., Tyson, J.F., and Nusslein, K. (2010). Phytoprotective influence of bacteria on growth and cadmium accumulation in the aquatic plant lemna minor. *Water Research* 44 (17): 4970–4979.

Ström, L., Owen, A.G., Godbold, D.L., and Jones, D. (2002). Organic acid mediated P mobilization in the rhizosphere and upt Tchounwou ake by maize roots. *Soil Biology and Biochemistry* 34: 703–710.

Su, Y., Han, F.X., Chen, J. et al. (2008). Phytoextraction and accumulation of mercury in three plant species: Indian mustard (*Brassica juncea*), beard grass (*Polypogon monospeliensis*), and Chinese brake fern (*Pteris vittata*). *International Journal of Phytoremediation* 10: 547–560.

Suman, J., Uhlik, O., Viktorova, J., and Macek, T. (2018). Phytoextraction of heavy metals: a promising tool for clean-up of polluted environment? *Frontiers in Plant Science* 9: 1476. https://doi.org/10.3389/fpls.2018.01476.

Tabrizi, L., Mohammadi, S., Delshad, M., and Moteshare Zadeh, B. (2015). Effect of arbuscular mycorrhizal fungi on yield and phytoremediation performance of pot marigold (*Calendula officinalis* L.) under heavy metals stress. *International Journal of Phytoremediation* 17: 1244–1252.

Takeda, R., Sato, Y., Yoshimura, R. et al. (2006). Accumulation of heavy metals by cucumber and *Brassica juncea* under different cultivation conditions. In: *Proc. Ann. Int. Conf. on Soil Sediments Water Energy (Massachusetts)*, vol. Vol. 11, 293–299. California: The Barkeley Electronic Press.

Tangahu, B.V., Sheikh Abdullah, S.R., Basri, H. et al. (2011). A review on heavy metals (As, Pb, and Hg) uptake by plants through phytoremediation. *International Journal of Chemical Engineering* 2011.

Tchounwou, P.B., Yedjou, C.G., Patlolla, A.K., and Sutton, D.J. (2012). Heavy metal toxicity and the environment. *Experientia. Supplementum* 101: 133–164.

Thakur, S., Singh, L., Wahid, Z.A. et al. (2016). Plant-driven removal of heavy metals from soil: uptake, translocation, tolerance mechanism, challenges, and future perspectives. *Environmental Monitoring and Assessment* 188: 206.

Tirry, N., Tahri Joutey, N., Sayel, H. et al. (2018). Screening of plant growth promoting traits in heavy metals resistant bacteria: Prospects in phytoremediation. *Journal, Genetic Engineering & Biotechnology* 16: 613–619.

Tiwari, S., Singh, S.N., and Garg, S.K. (2012). Stimulated phytoextraction of metals from fly ash by microbial interventions. *Environmental Technology (United Kingdom)* 33 (21): 2405–2413.

Turan, M. and Esringu, A. (2007). Phytoremediation based on canola (*Brassica napus* L.) and Indian mustard (*Brassica juncea* L.) planted on spiked soil by aliquot amount of Cd, Cu, Pb, and Zn. *Plant, Soil and Environment* 53 (1): 7–15.

USEPA (2000). *Introduction to Phytoremediation*" Tech. Rep. EPA 600/R-99/107. Cincinnati, OH: United States Environmental Protection Agency, Office of Research and Development.

van Dillewijn, P., Caballero, A., Paz, J.A. et al. (2007). Bioremediation of 2,4,6-trinitrotoluene under field conditions. *Environmental Science & Technology* 41 (4): 1378–1383.

Velásquez, L. and Dussan, J. (2009). Biosorption and bioaccumulation of heavy metals on dead and living biomass of *Bacillus sphaericus*. *Journal of Hazardous Materials* 167: 713–716.

Viti, C., Pace, A., and Giovannetti, L. (2003). Characterization of Cr(VI)-resistant bacteria isolated from chromium-contaminated soil by tannery activity. *Current Microbiology* 46 (1): 1–5.

Wang, J. and Chen, C. (2006). Biosorption of heavy metals by *Saccharomyces cerevisiae*: a review. *Biotechnology Advances* 24: 427–451.

Wang, W., Zhai, Y., Cao, L. et al. (2016). Endophytic bacterial and fungal microbiota in sprouts, roots and stems of rice (*Oryza sativa* L.). *Microbiological Research* 188: 1–8.

Wani, P.A. and Khan, M.S. (2012). Bioremediation of lead by a plant growth promoting Rhizobium species RL9. *Bacteriology Journal* 2 (4): 66–78.

Wani, P.A., Khan, M.S., and Zaidi, A. (2007). Effect of metal tolerant plant growth promoting Bradyrhizobium sp. (vigna) on growth, symbiosis, seed yield and metal uptake by greengram plants. *Chemosphere* 70 (1): 36–45.

Wani, P.A., Khan, M.S., and Zaidi, A. (2007). Impact of heavy metal toxicity on plant growth, symbiosis, seed yield and nitrogen and metal uptake in chickpea. *Australian Journal of Experimental Agriculture* 47: 712–720.

Wani, P.A., Khan, M.S., and Zaidi, A. (2008). Effects of heavy metal toxicity on growth, symbiosis, seed yield and metal uptake in pea grown in metal amended soil. *Bulletin of Environmental Contamination and Toxicology* 81: 152–158.

Wei, S., Zhou, Q., and Koval, P.V. (2006). Flowering stage characteristic of cadmium hyperaccumulator *Solanum nigrum* L. and their significance to phytoremediation. *Science of the Total Environment* 369: 441–446.

Wei, S., Li, Y., Zhou, Q. et al. (2010). Effect of fertilizer amendments on phytoremediation of Cd-contaminated soil by newly discovered hyperaccumulator *Solanum nigrum* L. *Journal of Hazardous Materials* 176: 269–273.

White, G.F., Edwards, M.J., Gomez-Perez, L. et al. (2016). Mechanisms of bacterial extracellular electron exchange. *Advances in Microbial Physiology* 68: 87–138. https://doi.org/10.1016/bs.ampbs.2016.02.002.

Wiangkham, N. and Prapagdee, B. (2018). Potential of Napier grass with cadmium-resistant bacterial inoculation on cadmium phytoremediation and its possibility to use as biomass fuel. *Chemosphere* 201: 511–518.

Wong, M.H. (2003). Ecological restoration of mine degraded soils, with emphasis on metal contaminated soils. *Chemosphere* 50: 775–778.

Wuana, R. and Okieimen, F.E. (2011a). Heavy metals in contaminated soils: a review of sources, chemistry, risks and best available strategies for remediation. *ISRN Ecology* 2011: 2356–7872.

Wuana, R.A. and Okieimen, F.E. (2011b). Heavy metals in contaminated soils: a review of sources, chemistry, risks and best available strategies for remediation. *ISRN Ecology* 2011: 402647.

Xie, Q.-E., Yan, X.-L., Liao, X.-Y., and Li, X. (2009). The arsenic hyperaccumulator fern *Pteris vittata* L. *Environmental Science & Technology* 43: 8488–8495.

Yadav, S.K., Juwarkar, A.S., Kumar, P. et al. (2009). Bioaccumulation and phyto-translocation of arsenic, chromium, and zinc by *Jatropha curcas* L.: impact of dairy sludge and biofertilizer. *Bioresource Technology* 100: 4616–4622.

Yan, A., Wang, Y., Tan, S.N. et al. (2020). Phytoremediation: a promising approach for revegetation of heavy metal-polluted land. *Frontiers in Plant Science* 11: 359.

Zhang, Z., Wen, X., Huang, Y. et al. (2017). Higher accumulation capacity of cadmium than zinc by *Arabidopsis halleri* ssp. Germmifera in the field using different sowing strategies. *Plant and Soil* 2017 (418): 1–12.

Zhou, W., Zhang, Y., Ding, X. et al. (2012). Magnetotactic bacteria: promising biosorbents for heavy metals. *Applied Microbiology and Biotechnology* 95: 1097–1104.

Zhuang, P., Ye, Z., Lan, C. et al. (2005). Chemically assisted phytoextraction of heavy metal contaminated soils using three plant species. *Plant and Soil* 276: 153–162.

12

Recycling and Disposal of Spent Metal(loid)-Laden Adsorbents: Current and Emerging Technologies, and Future Directions

Willis Gwenzi[1,2], Jerikias Marumure[3,4], Zakio Makuvara[3,4], and Tinoziva T. Simbanegavi[5]

[1] *Faculty of Organic Agricultural Sciences, Universität Kassel, Grassland Science and Renewable Plant Resources, Witzenhausen, Germany*
[2] *Department of Technology Assessment, Leibniz Institute for Agricultural Engineering and Bioeconomy (ATB), Potsdam, Germany*
[3] *Department of Physics, Geography and Environmental Sciences, School of Natural Sciences, Great Zimbabwe University, Masvingo, Zimbabwe*
[4] *Department of Life and Consumer Sciences, School of Agriculture and Life Sciences, College of Agriculture and Environmental Sciences, University of South Africa, Pretoria, South Africa*
[5] *Department of Soil Science and Environment, Faculty of Agriculture Environment and Food Systems, University of Zimbabwe, Harare, Zimbabwe*

12.1 Introduction

Organic and inorganic contaminants of geogenic and anthropogenic origins occur widely in soils and aquatic systems. Organic contaminants include: (1) legacy or currently regulated ones, such as pesticides and natural organic matter, and (2) emerging ones including personal care products, pharmaceuticals, and illicit drugs (Styszko et al. 2021). Inorganic contaminants include metalloids (e.g. arsenic), toxic metals, halides (e.g. F), radionuclides (e.g. U), and rare earth metals used in consumer and high-technology industrial products (Zhang et al. 2019; Gwenzi et al. 2022). The environmental occurrence and persistence of organic and inorganic pollutants present serious ecological, human health, and environmental risks (Gwenzi and Chaukura 2018).

Adsorption using natural or engineered biomaterials or geomaterials is a prominent, cheap, and effective method for the remediation of organic and inorganic contaminants (Chaukura et al. 2016; Gwenzi et al. 2017, 2021). Adsorbents include carbonaceous or biomaterials, (bi)metals (e.g. metallic iron), metal oxides/hydroxides, and metal–organic frameworks (MOFs). To date, adsorption has been applied to remediate several organic and inorganic pollutants in soil and aqueous media (Mohan et al. 2014; Chaukura et al. 2016).

In principle, adsorption binds pollutants from the mobile phase to the solid or immobile phase (Rathi and Kumar 2021). Adsorption also generates spent adsorbents laden with toxic contaminants such as metals and metalloids. Without proper disposal, contaminants

Remediation of Heavy Metals: Sustainable Technologies and Recent Advances, First Edition. Edited by Rangabhashiyam Selvasembian, Binota Thokchom, Pardeep Singh, Ali H. Jawad, and Willis Gwenzi.

from spent adsorbents may undergo desorption and subsequent release into the environment. The released contaminants may contribute to ecological, human health, and environmental risks, thereby defeating the initial purpose of remediation. Thus, safe and proper disposal technologies are needed for spent adsorbents. To date, the bulk of the studies on adsorption has focused on the following: (1) development or synthesis of adsorbents including their activation, (2) the remediation applications for removal of organic, inorganic, and microbial contaminants, (3) various contaminant removal mechanisms such as size exclusion, precipitation, occlusion, and ion exchange, (4) adsorption kinetics, thermodynamics, and isotherm modeling, and (5) regeneration of adsorbents using physical, thermal, and chemical methods (Chaukura et al. 2016; Rathi and Kumar 2021). Several review papers exist on various aspects of adsorption using many different adsorbents (Chaukura et al. 2016; Muisa et al. 2020; Rathi and Kumar 2021).

Compared to applications in environmental remediation, the safe disposal of spent adsorbents has only received cursory research attention in the literature on adsorption. To bridge this gap, the present chapter discusses the technologies for the disposal and recycling of spent adsorbents used for the remediation of metals, metalloids, and radionuclides. Therefore, the objectives of this chapter are: (1) to highlight the properties, nature, and potential health risks of spent adsorbents, (2) to present and critique existing disposal technologies, including landfilling, soil application, and cement stabilization/solidification, (3) to discuss emerging or novel applications, including catalysis, construction materials, novel raw materials, solid fuels for carbonaceous adsorbents, and re-purposed or re-engineered adsorbents, and (4) to highlight future perspectives including opportunities, challenges, and research needs. Although the present chapter focuses on metal(loid)-laden spent adsorbents, the results are generic and are applicable to other contaminant-laden spent contaminants.

To address the objectives, literature on spent adsorbents, current and emerging disposal practices, and recycling technologies was retrieved and reviewed. The present chapter first outlines the nature and characteristics of the various adsorbents. Then, existing or conventional methods of disposal and recycling of spent adsorbents and their associated human health and environmental risks are discussed. This is then followed by a discussion of emerging or novel methods for the valorization and recycling of spent adsorbents. Lastly, the opportunities, constraints, and future research directions on the recycling of spent adsorbents are highlighted.

12.2 Nature and Health Concerns/Risks of Spent/Used Adsorbents

12.2.1 Nature

Several natural and engineered materials have been applied as adsorbents for the removal of metalloids and metals in aqueous environments (Chaukura et al. 2016; Selvasembian et al. 2021; Ghosh et al. 2022). Briefly, these include those based on: (1) carbonaceous materials or biosorbents, (2) metal/metal oxide adsorbents, (3) metal/metal oxide-organic (nano)composites, and (4) metal–organic frameworks (MOFs), among others. A detailed

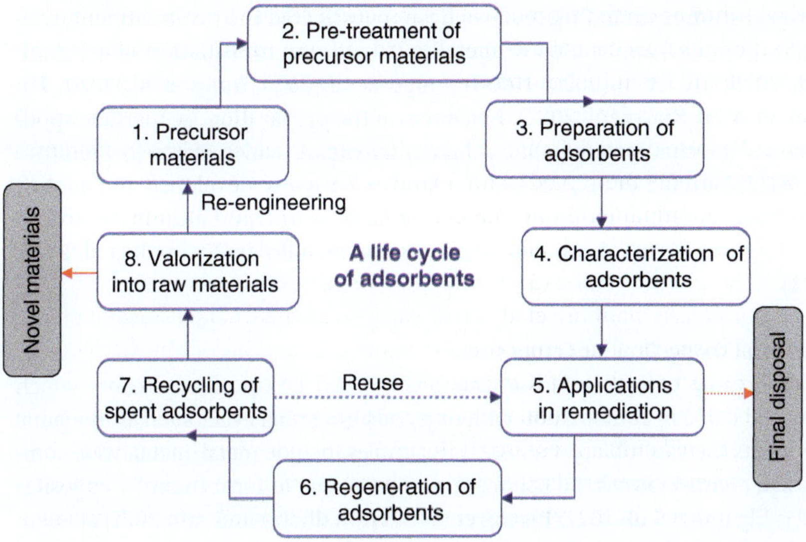

Figure 12.1 A conceptual depiction of a life cycle of an adsorbent from precursor material to final disposal or re-purposing into novel materials.

discussion of the adsorbents is presented in other chapters of this book and in several earlier reviews (Chaukura et al. 2016; Gwenzi et al. 2017, 2021). Therefore, comprehensive discussions of the various adsorbents are beyond what is presented in this chapter. Figure 12.1 indicates a summarized conceptual depiction of the life cycle of adsorbents from identification of precursor materials to final disposal or repurposing into novel materials.

12.2.1.1 Carbonaceous Adsorbents

Carbonaceous adsorbents or biosorbents are prepared from biomass as the precursor material or feedstock (Gorbounov et al. 2022). These include dead or live biomass and activated biosorbents such as activated carbon, biochars, and their engineered derivatives (Chaukura et al. 2016; Adegoke et al. 2022). These are materials rich in carbon and carbon compounds, characterized by high surface area and microporosity for adsorption of contaminants, including metals and metalloids. A detailed discussion of carbonaceous adsorbents, including their nature, properties, adsorption mechanisms, and regeneration, is presented in earlier reviews (Gorbounov et al. 2022; Mahesh et al. 2022).

12.2.1.2 Metal/Metal Oxide and Their Composite Adsorbents

Several metal-/metal oxide-based adsorbents have also been developed and applied specifically to remove metals, metalloids, and other contaminants in aqueous systems (Dehmani et al. 2022; Ghosh et al. 2022; Konadu-Amoah et al. 2022; Kumar et al. 2022). These metal/metal oxide adsorbents entail the use of trace or transition metals/metal oxides and rare earth elements and their oxides or compounds (Dehmani et al. 2022; Ghosh et al. 2022; Konadu-Amoah et al., 2022; Kumar et al. 2022). Metallic or zerovalent iron is one of the most prominent metal adsorbents reported in the literature (Konadu-Amoah et al. 2022).

For instance, several studies including reviews have investigated the pre-treatment, activation, characterization, and application of metallic iron for the remediation of contaminants (Hu et al. 2019; Hu et al. 2020; Ndé-Tchoupé et al. 2020; Yang et al. 2020; Hu et al. 2021; Tao et al. 2022; Xiao et al. 2022). Evidence on the preparation, properties, applications, and removal mechanisms of contaminants by metal oxides exists in literature (Dehmani et al. 2022; Fatimah et al. 2022). Other studies have also developed and applied metal oxide adsorbents based on natural geomaterials such as iron and aluminium oxides or hydroxides in clays or soils as adsorbents for metals and metalloids (Fatimah et al. 2022; Salehi et al. 2022).

12.2.1.3 Metal/Metal Oxide-Organic Composites

Metal/metal oxide-organic composite adsorbents are a broad group of adsorbents which include a metal/metal oxide in combination with other adsorbent materials such as inorganic or organic components (Selvasembian et al. 2021). Examples include metal/metal oxide combined with chitosan, alginate, synthetic polymers, and graphene to form (nano) composites (Badiei et al. 2021; Elgarahy et al. 2021; Huang et al. 2021; Malhotra and Jain 2021; Naseem and Waseem 2021; Selvasembian et al. 2021). For example, using polyurethane foam (PUF) rubber as a synthetic polymer, several adsorbents based on PUF have been manufactured and applied in contaminants' remediation (Selvasembian et al. 2021). These PUF-based composite adsorbents include PUF-alginate, PUF-chitosan, and PUF-graphene, among others (Selvasembian et al. 2021). Similarly, adsorption of contaminants, including metals and metalloids, occurs on both the metal/metal oxide as well as the organic or polymeric material. In this regard, metal/metal oxide-organic composite adsorbents act as dual or hybrid adsorbents. Dual or hybrid adsorbents are often more efficient than conventional adsorbents relying on adsorption sites on one matrix for the adsorption of contaminants. This is because contaminant adsorption occurs on both matrices – the metal/metal oxides and the organic component.

12.2.1.4 Metal–Organic Frameworks

The development and application of MOFs is a relatively emerging area of research in adsorption and separation science (Zhang et al. 2022). MOFs are a class of engineered adsorbents comprising metal atoms or ions or their clusters forming a one-, two-, or three-dimensional complex structure via coordination with organic components in the form of ligands (Zhang et al. 2022). They are often characterized by high porosity and surface area, which account for their application in adsorption of contaminants. Several studies have developed MOFs and assessed their efficacy in adsorbing metals from aqueous systems (Abdollahi et al. 2022; Adil et al. 2022; Zhang et al. 2022). Several reviews also exist on the preparation, characterization, application, and contaminant removal mechanisms of MOFs (Khan et al. 2013; Ren et al. 2017; Vaitsis et al. 2019; Du et al. 2021; Qian et al. 2021; Zhang et al. 2022).

12.2.2 Potential Environmental Health Risks

A major criticism of adsorption is that it transfers contaminants such as metals and metalloids from the dissolved to solid phase without transforming them to benign or non-toxic forms. In addition, the disposal of the spent adsorbents laden with metals or metalloids poses potential environmental and human health risks. Post-disposal changes in physicochemical and biogeochemical conditions may induce the desorption and subsequent release of contaminants from spent adsorbents. For example, changes in media pH, redox potential, and ligand type and concentration may cause previously adsorbed contaminants to undergo desorption and release into the dissolved or soluble phase, and ultimately into the environment. Once in the environment, dissolved or soluble contaminants may become bioavailable and bioaccessible to terrestrial organisms and humans. In turn, this may pose human exposure and health, and ecological risks. Yet surprisingly, these aspects are rarely addressed by the research community working on adsorption. Hence, the health risks posed by the disposal of spent adsorbents are a key research gap in adsorption.

12.3 Current Recycling and Disposal Technologies

Several methods exist for the disposal and recycling of spent adsorbents. Broadly, these are classified into two groups: (1) current or conventional methods, and (2) novel or emerging methods (Figure 12.2). Figure 12.2 summarizes a depiction of the current and emerging methods for the disposal and recycling of spent adsorbents. Briefly, current or conventional methods include: (1) regeneration and recycling, (2) land/soil application, (3) landfilling, and (4) cement stabilization or solidification. Emerging or novel methods include the use of spent adsorbents for the development of the following industrial products: (1) catalysts, (2) construction materials, (3) ceramics, and (4) solid fuels.

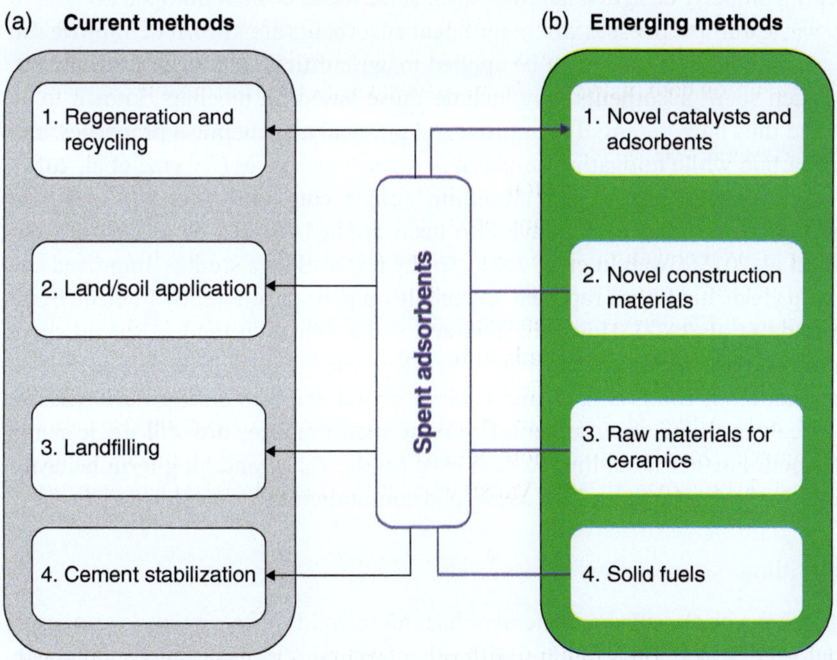

Figure 12.2 A summary depiction of the current and emerging methods for the disposal and recycling of spent adsorbents.

12.3.1 Regeneration and Recycling as Adsorbents

Before final disposal, contaminant-laden spent adsorbents are often regenerated to restore their adsorption capacity (Chaukura et al. 2016; Gwenzi et al. 2017, 2021). Adsorbent regeneration is based on several methods, and broadly, they include thermal, chemical, and physical techniques. Specific regeneration processes under the various groups are outlined and examined elsewhere in the literature (Chaukura et al. 2016; Selvasembian et al. 2021). Chemical methods using acids, ligands, and alkaline solutions are among the most prominent regeneration processes (Chaukura et al. 2016; Gwenzi et al. 2017). Thermal methods include heating the contaminant-laden spent adsorbents to facilitate the release of contaminants from the adsorption sites. Physical methods include shaking using ultrasonic techniques, and squeezing (Selvasembian et al. 2021). Studies show that the adsorption capacity and the mechanical properties of adsorbents tend to decline with the increasing number of regeneration cycles (Chaukura et al. 2016; Vakili et al. 2019; Selvasembian et al. 2021). Hence, after a number of regeneration cycles, the spent adsorbents have to be finally disposed of. Several options currently exist for the final disposal of spent adsorbents, and these are discussed below.

12.3.2 Land/Soil Application

Similar to other solid waste materials such as biosolids, spent adsorbents may be considered as general or hazardous solid waste. Hence, spent adsorbents can be disposed of and mixed with other solid waste from household and industrial processes. In most low-income countries lacking properly designed landfills, such solid waste is often dumped on land in non-sanitary waste dumps. In cases where the spent adsorbents are known to improve soil quality and crop productivity, they can be applied to agricultural soils as an amendment. Examples of such spent adsorbents may include those based on biochars, known to be carbon-rich, and thus have potential to improve soil physical and chemical properties, and sequester soil carbon while mitigating emission of greenhouse gases (Gwenzi et al. 2015). Other spent adsorbents, such as alum or aluminum sulfate, commonly used as a coagulant in water and wastewater treatment, have also been applied to soils as an amendment (Dassanayake et al. 2015; Gwandu et al. 2021). In the few available studies, improved soil quality and crop yields have been reported, especially on poor or inherently infertile tropical soils (Gwandu et al. 2021). However, land and soil application tend to spread spent adsorbents and their associated contaminants over large spatial areas. This creates the risk for the dissemination of contaminants and environmental and human exposure risks. In addition, the ecological and environmental risks of such practices are still unclear and poorly understood. Furthermore, limited data exist on the short- and long-term behavior and fate of the spent adsorbents and the associated contaminants.

12.3.3 Landfilling

Spent adsorbents can be classified as general or hazardous solid waste, similar to municipal and industrial solid waste. Therefore, as with other general and hazardous solid waste, landfilling is a common practice for the final disposal of spent adsorbents (Verbinnen et al. 2015; Kasiuliene et al. 2019). Landfilling is particularly possible in developed countries with properly designed sanitary landfills with hydraulic liners and leachate collection and treatment systems. However, the practice is not feasible in low-income countries lacking properly designed and operated landfills. In such cases, disposal in non-sanitary waste dumps and land and soil application is most common. The disposal of spent adsorbents in landfills has been criticized due to high cost and the depletion of landfill capacity (Verbinnen et al. 2015). In addition, a potential risk also exists for contaminants to leach into groundwater systems when the hydraulic liner and leachate collection and treatment systems fail (Anand and Palani 2022). In addition, contaminants and spent adsorbents may undergo dissemination mediated by wind and surface run-off into other environmental compartments such as surface water and air. In turn, this may pose environmental and human health risks to surrounding communities.

12.3.4 Cement Stabilization/Solidification

Cement stabilization or solidification is a common method for the remediation of hazardous solid wastes, including those containing toxic metals and metalloids (Batchelor 2006; Malviya and Chaudhary 2006; Silva et al. 2007; Kundu and Gupta 2008). The technology has been widely applied and literature (in the form of reviews) exists on this particular subject (Batchelor 2006; Malviya and Chaudhary 2006; Silva et al. 2007). For example, several studies elucidated and validated the factors controlling the cement stabilization/solidification process (Malviya and Chaudhary 2006). Silva et al. (2007) investigated the cement stabilization of metal-enriched wastewater treatment sludge and its subsequent disposal. Similarly, spent adsorbents laden with toxic metals and metalloids can be remediated through cement stabilization and solidification. Several studies have reported the cement stabilization or solidification of spent adsorbents, including those laden with toxic metals, and metalloids, for example, arsenic (Devi et al. 2020; Raj et al. 2022; Yadav et al. 2022). In some cases, the cement stabilized or solidified spent adsorbents can then be disposed of in sanitary landfills (Malviya and Chaudhary 2006). Cement stabilization or solidification is also reported to have a high environmental impact, possibly due to the use of lime and cement, which have a high energy and climate footprints (Malviya and Chaudhary 2006). Although long-term data are still limited, there is also a possibility that the cement stabilized or solidified spent adsorbents may undergo weathering, and release contaminants into the environment. Hence, the long-term behavior and fate of cement stabilized/solidified spent adsorbents require further investigation.

12.4 Emerging Technologies

The limitations of current disposal methods, coupled with their human health and environmental risks, motivated research on the development of novel or emerging technologies. These technologies entail the upcycling and recycling of the spent adsorbents in the development of novel industrial materials (Figure 12.2). These novel technologies include the use of spent adsorbents in the development of catalysts, construction materials, ceramics, and solid fuels.

12.4.1 Novel Catalysts

Metal/metal oxides of transition elements and rare earth elements and their oxides are known to possess catalytic activities. Indeed, several metal/metal oxide adsorbents and those based on rare earth elements and their compounds have been converted into catalysts for organic synthesis (He et al. 2019; Baskar et al. 2022; Rial and Ferreira 2022). For example, spent adsorbents enriched in chromium (Cr) have been used for the catalytic elimination of methyl mercaptan (He et al. 2019). In another study, a spent mercury (Hg(II)) adsorbent comprising novel magnetic mesoporous silica or chitosan composite was shown to possess excellent catalytic activity. Duarte et al. (2013) also showed that spent activated carbon/Fe-Fenton-like catalysts facilitated the degradation of orange II in aqueous solution. These studies point to the potential to use spent adsorbents as catalysts, especially those derived from metal/metal oxides and rare earth elements and their compounds. However, given the diversity of adsorbents developed to date, scope exists to investigate the catalytic activity of several other spent adsorbents.

12.4.2 Novel Construction Materials

Similar to stabilized or solidified wastes, there is a possibility of applying spent adsorbents in novel construction material containing industrial wastes and in this case, they can be utilized as filler materials (Konadu-Amoah et al., 2022; Rial and Ferreira 2022). In this regard, cement stabilizes contaminants, while converting the spent adsorbents into novel materials that reduce demand for pristine conventional construction materials. The use of spent adsorbents in cement-based construction materials is similar to the process of cement stabilization of spent adsorbents reported earlier in the present chapter. The application of spent adsorbents in construction materials is also similar to the use of metal-laden industrial solid wastes such as coal ash as fillers in construction materials (Gwenzi and Mupatsi 2016). To date, a number of studies have reported the potential to use spent adsorbents as fillers (Rathore and Mondal 2017; Rial and Ferreira 2022). For example, Rathore and Mondal (2017) investigated the use of spent adsorbents for arsenic and fluoride as filler materials in clay bricks. In the same study, the authors discussed the preparation, characterization, and leaching behavior of contaminants from such bricks. In a recent review paper, the use of spent polyurethane adsorbents as fillers in construction materials with high thermal and acoustic insulation has been discussed (Selvasembian et al. 2021). In one study, Karanac et al. (2018) reported the elimination of trace metal and metalloid ions (Zn^{2+}, Pb^{2+}, As [V]) using lime fly ash (activated) as an adsorbent. The spent adsorbent was valorized as a potential filler in construction materials. However, most of these studies evaluating the use of spent adsorbents in construction materials are still limited to laboratory-scale experiments, while pilot and large-scale applications and investigations are still lacking. Similar to other remediation methods, data on long-term fate and behavior of pollutants and spent adsorbents are still lacking for construction materials containing spent adsorbents.

12.4.3 Solid Fuels

Carbon-rich or carbonaceous adsorbents, such as those developed from live or dead biomass, including activated carbon, biochars, chitosan, and alginate, are widely used as adsorbents. Thus, spent carbonaceous adsorbents can be converted into solid fuels such as briquettes and used as fuel in stand-alone or co-combustion waste-to-energy systems. However, to date, no data exist on potential application of spent carbonaceous adsorbents as a precursor material for solid fuels or briquettes. This lack of studies may point to putative concerns about the human health and environmental risks of such applications, especially the risk of volatilization of contaminants during combustion process. The large-scale use of activated carbon in wastewater and drinking water treatment, and recent interest in biochar-based water and wastewater treatment systems indicate the need to explore the potential to use spent carbonaceous adsorbents as energy materials. Such studies should investigate the energy, mechanical and emission characteristics of the solid fuel. In addition, such studies should also investigate the contaminant leaching characteristics of the resultant combustion ash relative to that of conventional biomass or carbonaceous solid fuels such as wood, charcoal, and coal.

12.4.4 Re-Engineered Adsorbents

Scope also exists to develop novel adsorbents through the re-engineering of spent or exhausted adsorbents (Zhao et al. 2007, 2021a, 2021b). In this regard, spent adsorbents may be precursor materials for the production of re-purposed or re-engineered adsorbents. For example, Abd Aziz et al. (2014) developed a water adsorbent from spent bleaching earth used as an adsorbent in the refining of crude palm oil. In the same study, the adsorbent was then applied to dehydrate or remove water from an ethanol-water mixture. This re-engineering of spent adsorbents is similar to the regeneration and subsequent reuse of spent adsorbents. For example, adsorbents laden with metal/metal oxides can be re-engineered or repurposed to develop novel adsorbents to remove other contaminants, such as F and As in aqueous systems (Jo et al. 2021). For example, Jo et al. (2021) developed an alum sludge adsorbent in the form of pellets from water and wastewater treatment sludge and bentonite and utilized it in removing arsenic from aqueous systems. Alum sludge derived from wastewater and water treatment plants has also been used in the removal of phosphate in aqueous environments (Zhao et al. 2007; Muisa et al. 2020). In another study, Jeon et al. (2018) re-engineered alum sludge via calcination and used the resultant adsorbent for the efficient removal of arsenic. These few studies focusing on alum sludge point to the potential to re-engineer spent or exhausted adsorbents and use them for other remediation applications. Therefore, there is a need for further work to understand the scope to re-engineer and reuse other spent adsorbents in environmental remediation.

12.4.5 Novel Raw Materials

Metal/metal oxide adsorbents and their adsorbed industrial metals and those developed from rare earth elements may serve as raw materials or novel ores for the enriched elements. These elements can be recovered through various hydrometallurgical, pyrometallurgical, and biorecovery methods similar to those used for other ores and solid waste material (Selvasembian et al. 2021). Some spent adsorbents, especially those rich in metal oxides and clays, have been used as raw materials in ceramics (Verbinnen et al. 2015; Rial and Ferreira 2022). For example, Verbinnen et al. (2015) used a spent adsorbent used for the adsorption of oxyanions and toxic metals to develop ceramics. However, literature on the recovery of critical elements and use of spent adsorbents as industrial raw materials are still limited, pointing to potential health concerns and stakeholders' knowledge, attitudes, and perceptions about such practices.

Recent interest in urban mining could create opportunities for the recovery of critical elements and use of spent adsorbents as industrial raw materials. In this regard, spent adsorbents can be integrated into the concept of urban mining, where waste repositories such as landfills are mined and processed to recover industrial metals and other critical elements (Glombitza and Reichel 2013; Tejaswini et al. 2022). However, it is unclear whether concentrations of industrial metals and other critical elements are high enough to justify recovery. In addition, limited data exist on whether or not quantities justify recovery of metals and rare earth elements. Several urban mining studies and pilot projects exist in a few countries, such as Switzerland (Simoni et al. 2015) and China (Hu and Poustie 2018).

12.5 Looking Ahead: Future Perspectives and Research Directions

12.5.1 Opportunities and Challenges

The various potential applications that can be developed from spent adsorbents point to potential opportunities to valorize or develop value-added products from spent adsorbents. If proven scalable, such applications may lead to the development of specific value chains, including novel construction materials, catalysts, solid fuels, and even soil amendment to improve soil quality and productivity. Such value chains may be integrated into existing industrial production and applications to close the material loops or create opportunities for stand-alone waste-to-product enterprises for the collection, reuse, and recycling of spent adsorbents. Similar to other waste-to-product systems, the transformation of used adsorbents into advanced products will serve a dual function. First, it will solve the challenges associated with disposing of spent adsorbents and their associated environmental health concerns. Second, it will potentially create waste-to-product value chains entailing research and development, production, marketing, and application of goods and services derived from spent adsorbents.

Despite the potential opportunities, the valorization of spent adsorbents still faces several challenges. These include the costs and labor associated with the logistics of collection, sorting, transport, and storage of the spent adsorbents. In addition, data on the feasibility of scaling up the various applications of the spent adsorbents are still limited. Lack of a supportive regulatory and policy framework coupled with stakeholders' knowledge, attitudes, and perceptions may also affect the potential to upscale such technologies. Finally, as discussed below, several gaps still exist in the development and application of technologies and materials derived from spent adsorbents.

12.5.2 Knowledge Gaps and Future Research Directions

1) *Human and ecological exposure, and health risks of spent adsorbents.*
   Limited data exist on the environmental and human health risks associated with the application and subsequent disposal of spent adsorbents. Therefore, further research is required to investigate the ecotoxicity, human exposure, and health risks of using spent adsorbents and final disposal. Specifically, there is a need to investigate how the application of spent

adsorbents in construction materials has led to human exposure and health risks. Regarding the environmental health problems, the leaching of contaminants and ecotoxicity of spent adsorbents on soil and aquatic ecosystems require further investigation.

2) *Long-term environmental fate and behavior of spent adsorbents and derivatives*
The bulk of existing studies on the removal of metals and metalloids by adsorption tends to be short-term batch and column experiments. Besides studies focusing on regeneration and recycling of adsorbents, limited attention has been paid to the long-term fate and behavior of the spent adsorbents and their derivatives. In addition, there is a need to have a greater understanding of the behavior and fate of the contaminants and their degradation by-products after a long period of time. Such studies should also investigate the biogeochemical controls of the behavior and fate of contaminants in various environmental matrices including soils and aquatic systems.
3) *Stakeholders' knowledge, attitudes, and practices on spent adsorbents and their derivatives*
Stakeholders, including consumers, practitioners, regulators, and decision- and policy-makers may have knowledge, attitudes, and perceptions toward spent adsorbents and associated derivatives and materials. These knowledge, attitudes, and perceptions may affect the production, adoption, and application of spent adsorbents, their derivatives, and associated materials. In a recent paper, these stakeholders' knowledge, attitudes, and perceptions have been collectively referred to as the "human factor" (Gwenzi 2022). Yet limited research exists on the effect of stakeholders' knowledge, attitudes, and perceptions on spent adsorbents and associated derivatives and materials. Therefore, the role of the human factor in the production, uptake, and application of spent adsorbents warrants further research.
4) *Pilot-scale testing of emerging/novel applications of spent adsorbents*
Several potential applications of spent adsorbents and their derivatives were discussed in the present chapter. However, research on these various applications is still in its infancy, hence has been limited to small-scale laboratory experiments. To date, pilot and large-scale production and evaluation of the applications of spent adsorbents and their derivatives are still lacking. Thus, the development and pilot testing of the various applications of spent adsorbents and their derivatives should be the next research step toward the large-scale application of such technologies.
5) *Technical, social, and economic viability of applications of spent adsorbents*
The large-scale application and production of technologies using spent adsorbents as precursor materials require comprehensive data on the technical and socio-economic feasibility and viability of such technologies. Data for evaluation of the socio-economic and technical feasibility can be obtained from pilot-scale studies. Lacking pilot- and large-scale studies, information on the technical and socio-economic feasibility and viability of technologies based on spent adsorbents is still missing. Therefore, this aspect requires further investigation to demonstrate technical and socio-economic feasibility and viability at scales relevant to industrial production and applications.
6) *LCA of environmental and climate footprints of spent adsorbents*
Public and research interest in the environmental and climate footprints of goods and services has increased in recent years due to awareness about the environment, including climate change and its potential impacts. Life cycle assessment (LCA) tools provide a critical tool for quantifying the environmental and climate change footprints of spent adsorbents and their derivatives relative to competing materials and technologies. Yet limited evidence exists on the environmental and climate footprints of the production

and applications of spent adsorbents and their derivatives. In this regard, LCA-based studies are required to quantify the environmental and climate footprints of spent adsorbents and their derivatives relative to competing technologies and pristine materials.

12.6 Conclusions and Outlook

Adsorption is a prominent low-cost and efficient method for the remediation of organic and inorganic contaminants in soils and aquatic media. However, adsorption generates contaminant-laden spent adsorbents. This points to the need to develop and evaluate safe disposal and recycling technologies for spent adsorbents. Yet until now, compared to other aspects of adsorption, few studies including reviews exist on the technologies for the safe disposal and recycling of spent adsorbents. Therefore, the present paper investigated the nature and health risks of spent adsorbents, current and emerging disposal and recycling technologies, their limitations, and future research directions. In the present paper, the nature of adsorbents, including carbonaceous materials, metals, metal oxide/hydroxides, and MOFs, was presented. The spent adsorbents and their adsorbed contaminants, and their potential health risks were discussed. Specifically, depending on prevailing physicochemical and biogeochemical conditions, adsorbed contaminants may undergo desorption from the spent adsorbents and be released back into the environment. The desorption and release of contaminants from spent adsorbents make them bioavailable and bioaccessible. In turn, this poses environmental, ecological and human exposure and health risks. The current disposal technologies for spent adsorbents, including landfilling, soil application, and cement stabilization/solidification, and their limitations were discussed. Emerging applications of spent adsorbents include: (1) catalysis in the case of metal/metal oxide-based adsorbents, (2) as fillers in construction materials, (3) solid fuels for carbonaceous or biomass-based adsorbents, (4) as novel raw materials for industrial metals, metalloids, and high-technology rare earth elements, and (5) as precursor materials for the development of re-engineered or repurposed adsorbents and ceramics. Future perspectives, including opportunities, challenges, and future research needs, were highlighted. Specifically, further research should address the following knowledge gaps: (1) ecological and human exposure and human health risks of spent adsorbents and their derivatives, (2) long-term environmental fate and behavior of spent adsorbents and derivatives, (3) stakeholders' knowledge, attitudes, and practices on spent adsorbents and their derivatives, (4) pilot-scale testing of emerging/novel applications of spent adsorbents, (5) technical, social, and economic viability of large-scale industrial applications of spent adsorbents, and (6) LCA of environmental and climate footprints of spent adsorbents.

Acknowledgments

This paper is an output of a broad environmental and climate footprinting project titled "*Environmental and climate footprints of a biochar and biogas technology versus current practices in sub-Saharan Africa: A life cycle analysis.*" The project is supported by the Georg Forster Fellowship for Experienced Researchers, awarded to Willis Gwenzi, and funded by the Alexander von Humboldt Stiftung/Foundation, Germany.

References

Abd Aziz, Z.S., Nour, A.H., and Rosli, M.Y. (2014). Development of water adsorbents synthesized from spent bleaching earth to dehydrate ethanol water mixture. In: *Advanced Materials Research*, vol. vol. 970, 187–191. Trans Tech Publications Ltd.

Abdollahi, N., Moussavi, G., and Giannakis, S. (2022). A review of heavy metals' removal from aqueous matrices by metal-organic frameworks (MOFs): state-of-the art and recent advances. *Journal of Environmental Chemical Engineering* 10: 107394.

Adegoke, K.A., Adegoke, O.R., Araoye, A.O. et al. (2022). Engineered raw, carbonaceous, and modified biomass-based adsorbents for Rhodamine B dye removal from water and wastewater. *Bioresource Technology Reports* 18: 101082.

Adil, H.I., Thalji, M.R., Yasin, S.A. et al. (2022). Metal–organic frameworks (MOFs) based nanofiber architectures for the removal of heavy metal ions. *RSC Advances* 12 (3): 1433–1450.

Anand, N. and Palani, S.G. (2022). A comprehensive investigation of toxicity and pollution potential of municipal solid waste landfill leachate. *Science of the Total Environment* 838: 155891.

Badiei, M., Asim, N., Mohammad, M. et al. (2021). New graphene nanocomposites-based adsorbents. In: *Handbook of Nanomaterials for Wastewater Treatment*, 367–416. Elsevier.

Baskar, A.V., Bolan, N., Hoang, S.A. et al. (2022). Recovery, regeneration and sustainable management of spent adsorbents from wastewater treatment streams: a review. *Science of the Total Environment* 822: 153555.

Batchelor, B. (2006). Overview of waste stabilization with cement. *Waste Management* 26 (7): 689–698.

Chaukura, N., Gwenzi, W., Tavengwa, N., and Manyuchi, M.M. (2016). Biosorbents for the removal of synthetic organics and emerging pollutants: opportunities and challenges for developing countries. *Environment and Development* 19: 84–89.

Dassanayake, K.B., Jayasinghe, G.Y., Surapaneni, A., and Hetherington, C. (2015). A review on alum sludge reuse with special reference to agricultural applications and future challenges. *Waste Management* 38: 321–335.

Dehmani, Y., Dridi, D., Lamhasni, T. et al. (2022). Review of phenol adsorption on transition metal oxides and other adsorbents. *Journal of Water Process Engineering* 49: 102965.

Devi, P., Kothari, P., and Dalai, A.K. (2020). Stabilization and solidification of arsenic and iron contaminated canola meal biochar using chemically modified phosphate binders. *Journal of Hazardous Materials* 385: 121559.

Du, C., Zhang, Z., Yu, G. et al. (2021). A review of metal organic framework (MOFs)-based materials for antibiotics removal via adsorption and photocatalysis. *Chemosphere* 272: 129501.

Duarte, F., Maldonado-Hódar, F.J., and Madeira, L.M. (2013). New insight about orange II elimination by characterization of spent activated carbon/Fe Fenton-like catalysts. *Applied Catalysis B: Environmental* 129: 264–272.

Elgarahy, A.M., Elwakeel, K.Z., Akhdhar, A., and Hamza, M.F. (2021). Recent advances in greenly synthesized nanoengineered materials for water/wastewater remediation: an overview. *Nanotechnology for Environmental Engineering* 6 (1): 1–24.

Fatimah, I., Fadillah, G., Yanti, I., and Doong, R.A. (2022). Clay-supported metal oxide nanoparticles in catalytic advanced oxidation processes: a review. *Nanomaterials* 12 (5): 825.

Ghosh, N., Das, S., Biswas, G., and Haldar, P.K. (2022). Review on some metal oxide nanoparticles as effective adsorbent in wastewater treatment. *Water Science and Technology* 85: 3370–3395.

Glombitza, F. and Reichel, S. (2013). Metal-containing residues from industry and in the environment: Geobiotechnological urban mining. *Geobiotechnology* I: 49–107.

Gorbounov, M., Taylor, J., Petrovic, B., and Soltani, S.M. (2022). To DoE or not to DoE? A technical review on & roadmap for optimisation of carbonaceous adsorbents and adsorption processes. *South African Journal of Chemical Engineering* 41: 111–128.

Gwandu, T., Blake, L.I., Nezomba, H. et al. (2021). Waste to resource: use of water treatment residual for increased maize productivity and micronutrient content. *Environmental Geochemistry and Health* 44: 1–18.

Gwenzi, W. (2022). Emerging contaminants: a handful of conceptual and organizing frameworks. In: *Emerging Contaminants in the Terrestrial-Aquatic-Atmosphere Continuum*, 3–15. Elsevier https://doi.org/10.1016/B978-0-323-90051-5.00014-6.

Gwenzi, W. and Mupatsi, N.M. (2016). Evaluation of heavy metal leaching from coal ash-versus conventional concrete monolithics and debris. *Waste Management* 49: 114–123.

Gwenzi, W. and Chaukura, N. (2018). Organic contaminants in African aquatic systems: current knowledge, health risks, and future research directions. *Science of the Total Environment* 619: 1493–1514. https://doi.org/10.1016/j.scitotenv.2017.11.121.

Gwenzi, W., Chaukura, N., Noubactep, C., and Mukome, F.N. (2017). Biochar-based water treatment systems as a potential low-cost and sustainable technology for clean water provision. *Journal of Environmental Management* 197: 732–749.

Gwenzi, W., Chaukura, N., Mukome, F.N. et al. (2015). Biochar production and applications in sub-Saharan Africa: opportunities, constraints, risks and uncertainties. *Journal of Environmental Management* 150: 250–261.

Gwenzi, W., Chaukura, N., Wenga, T., and Mtisi, M. (2021). Biochars as media for air pollution control systems: contaminant removal, applications and future research directions. *Science of the Total Environment* 753: 142249.

Gwenzi, W., Makuvara, Z., and Marumure, J. (2022). Rare earth elements: human exposure, risk factors, and health risks. In: *Emerging Contaminants in the Terrestrial-Aquatic-Atmosphere Continuum*, 273–290. Elsevier.

He, D., Zhang, Y., Yang, S. et al. (2019). Development of a strategy to reuse spent Cr adsorbents as efficient catalysts: from the perspective of practical application. *ACS Sustainable Chemistry & Engineering* 7 (3): 3251–3257.

Hu, Y. and Poustie, M. (2018). Urban mining demonstration bases in China: a new approach to the reclamation of resources. *Waste Management* 79: 689–699.

Hu, R., Gwenzi, W., Sipowo-Tala, V.R., and Noubactep, C. (2019). Water treatment using metallic iron: a tutorial review. *Processes* 7 (9): 622.

Hu, R., Yang, H., Tao, R. et al. (2020). Metallic iron for environmental remediation: starting an overdue progress in knowledge. *Water* 12 (3): 641.

Hu, R., Ndé-Tchoupé, A.I., Cao, V. et al. (2021). Metallic iron for environmental remediation: the fallacy of the electron efficiency concept. *Frontiers in Environmental Chemistry* 2: 677813.

Huang, W.H., Lee, D.J., and Huang, C. (2021). Modification on biochars for applications: a research update. *Bioresource Technology* 319: 124100.

Jeon, E.K., Ryu, S., Park, S.W. et al. (2018). Enhanced adsorption of arsenic onto alum sludge modified by calcination. *Journal of Cleaner Production* 176: 54–62.

Jo, J.Y., Choi, J.H., Tsang, Y.F., and Baek, K. (2021). Pelletized adsorbent of alum sludge and bentonite for removal of arsenic. *Environmental Pollution* 277: 116747.

Karanac, M., Đolić, M., Veljović, Đ. et al. (2018). The removal of Zn2+, Pb2+, and As(V) ions by lime activated fly ash and valorization of the exhausted adsorbent. *Waste Management* 78: 366–378.

Kasiuliene, A., Carabante, I., Bhattacharya, P., and Kumpiene, J. (2019). Treatment of metal (loid) contaminated solutions using iron-peat as sorbent: is landfilling a suitable management option for the spent sorbent? *Environmental Science and Pollution Research* 26 (21): 21425–21436.

Khan, N.A., Hasan, Z., and Jhung, S.H. (2013). Adsorptive removal of hazardous materials using metal-organic frameworks (MOFs): a review. *Journal of Hazardous Materials* 244: 444–456.

Konadu-Amoah, B., Hu, R., Ndé-Tchoupé, A.I. et al. (2022). Metallic iron (Fe0)-based materials for aqueous phosphate removal: a critical review. *Journal of Environmental Management* 315: 115157. https://doi.org/10.1016/j.jenvman.2022.115157.

Kumar, A., Yeshwanth, M., Kumar, K. et al. (2022). Functionalized Cu-based metal oxide nanoparticles with enhanced Cd+ 2 adsorption capacity and their ecotoxicity assessment by molecular docking. *Journal of Environmental Management* 307: 114523.

Kundu, S. and Gupta, A.K. (2008). Immobilization and leaching characteristics of arsenic from cement and/or lime solidified/stabilized spent adsorbent containing arsenic. *Journal of Hazardous Materials* 153 (1–2): 434–443.

Mahesh, N., Balakumar, S., Shyamalagowri, S. et al. (2022). Carbon-based adsorbents as proficient tools for the removal of heavy metals from aqueous solution: a state of art-review emphasizing recent progress and prospects. *Environmental Research* 213: 113723.

Malhotra, P. and Jain, A. (2021). Graphene oxide-based nanocomposites for adsorptive removal of water pollutants. In: *Contamination of Water*, 431–448. Academic Press.

Malviya, R. and Chaudhary, R. (2006). Factors affecting hazardous waste solidification/ stabilization: a review. *Journal of Hazardous Materials* 137 (1): 267–276.

Mohan, D., Sarswat, A., Ok, Y.S., and Pittman, C.U. Jr. (2014). Organic and inorganic contaminants removal from water with biochar, a renewable, low cost and sustainable adsorbent–a critical review. *Bioresource Technology* 160: 191–202.

Muisa, N., Nhapi, I., Ruziwa, W., and Manyuchi, M.M. (2020). Utilization of alum sludge as adsorbent for phosphorus removal in municipal wastewater: a review. *Journal of Water Process Engineering* 35: 101187.

Naseem, T. and Waseem, M. (2021). A comprehensive review on the role of some important nanocomposites for antimicrobial and wastewater applications. *International journal of Environmental Science and Technology* 19: 1–26.

Ndé-Tchoupé, A.I., Hu, R., Gwenzi, W. et al. (2020). Characterizing the reactivity of metallic iron for water treatment: H2 evolution in H2SO4 and uranium removal efficiency. *Water* 12 (6): 1523.

Qian, Y., Zhang, F., and Pang, H. (2021). A review of MOFs and their composites-based photocatalysts: synthesis and applications. *Advanced Functional Materials* 31 (37): 2104231.

Raj, S., Sinha, U., Singh, H., and Bhattacharya, J. (2022). Novel GO/Fe–Mn hybrid for the adsorptive removal of Pb(II) ions from aqueous solution and the spent adsorbent disposability in cement mix: compressive properties and leachability study for circular economy benefits. *Environmental Science and Pollution Research* 29: 1–19.

Rathi, B.S. and Kumar, P.S. (2021). Application of adsorption process for effective removal of emerging contaminants from water and wastewater. *Environmental Pollution* 280: 116995.

Rathore, V.K. and Mondal, P. (2017). Stabilization of arsenic and fluoride bearing spent adsorbent in clay bricks: preparation, characterization and leaching studies. *Journal of Environmental Management* 200: 160–169.

Ren, J., Dyosiba, X., Musyoka, N.M. et al. (2017). Review on the current practices and efforts towards pilot-scale production of metal-organic frameworks (MOFs). *Coordination Chemistry Reviews* 352: 187–219.

Rial, J.B. and Ferreira, M.L. (2022). Potential applications of spent adsorbents and catalysts: re-valorization of waste. *Science of the Total Environment* 823: 153370.

Salehi, E., Khajavian, M., Sahebjamee, N. et al. (2022). Advances in nanocomposite and nanostructured chitosan membrane adsorbents for environmental remediation: a review. *Desalination* 527: 115565.

Selvasembian, R., Gwenzi, W., Chaukura, N., and Mthembu, S. (2021). Recent advances in the polyurethane-based adsorbents for the decontamination of hazardous wastewater pollutants. *Journal of Hazardous Materials* 417: 125960.

Silva, M.A., Mater, L., Souza-Sierra, M.M. et al. (2007). Small hazardous waste generators in developing countries: use of stabilization/solidification process as an economic tool for metal wastewater treatment and appropriate sludge disposal. *Journal of Hazardous Materials* 147 (3): 986–990.

Simoni, M., Kuhn, E.P., Morf, L.S. et al. (2015). Urban mining as a contribution to the resource strategy of the Canton of Zurich. *Waste Management* 45: 10–21.

Styszko, K., Proctor, K., Castrignanò, E., and Kasprzyk-Hordern, B. (2021). Occurrence of pharmaceutical residues, personal care products, lifestyle chemicals, illicit drugs and metabolites in wastewater and receiving surface waters of Krakow agglomeration in South Poland. *Science of the Total Environment* 768: 144360.

Tao, R., Yang, H., Cui, X. et al. (2022). The suitability of hybrid Fe0/aggregate filtration systems for water treatment. *Water* 14 (2): 260.

Tejaswini, M.S.S.R., Pathak, P., and Gupta, D.K. (2022). Sustainable approach for valorization of solid wastes as a secondary resource through urban mining. *Journal of Environmental Management* 319: 115727.

Vaitsis, C., Sourkouni, G., and Argirusis, C. (2019). Metal organic frameworks (MOFs) and ultrasound: a review. *Ultrasonics Sonochemistry* 52: 106–119.

Vakili, M., Deng, S., Shen, L. et al. (2019). Regeneration of chitosan-based adsorbents for eliminating dyes from aqueous solutions. *Separation and Purification Reviews* 48 (1): 1–13.

Verbinnen, B., Block, C., Van Caneghem, J., and Vandecasteele, C. (2015). Recycling of spent adsorbents for oxyanions and heavy metal ions in the production of ceramics. *Waste Management* 45: 407–411.

Xiao, M., Hu, R., Ndé-Tchoupé, A.I. et al. (2022). Metallic iron for water remediation: plenty of room for collaboration and convergence to advance the science. *Water* 14 (9): 1492.

Yadav, M.K., Saidulu, D., Ghosal, P.S. et al. (2022). A review on the management of arsenic-laden spent adsorbent: insights of global practices, process criticality, and sustainable solutions. *Environmental Technology and Innovation* 27: 102500.

Yang, H., Hu, R., Ndé-Tchoupé, A.I. et al. (2020). Designing the next generation of Fe0-based filters for decentralized safe drinking water treatment: a conceptual framework. *Processes* 8 (6): 745.

Zhang, Y., Chu, W., Chen, X. et al. (2019). Recovery of rare earth metals and synthesis of Ni0. 6Co0. 2Mn0. 2 (OH) 2 from spent asymmetric-capacitance power batteries. *Journal of Cleaner Production* 235: 1295–1303.

Zhang, H., Hu, X., Li, T. et al. (2022). MIL series of metal organic frameworks (MOFs) as novel adsorbents for heavy metals in water: a review. *Journal of Hazardous Materials* 429: 128271.

Zhao, Y.Q., Razali, M., Babatunde, A.O. et al. (2007). Reuse of aluminum-based water treatment sludge to immobilize a wide range of phosphorus contamination: equilibrium study with different isotherm models. *Separation Science and Technology* 42 (12): 2705–2721.

Zhao, Y., Nzihou, A., Ren, B. et al. (2021a). Waterworks sludge: an underrated material for beneficial reuse in water and environmental engineering. *Waste and Biomass Valorization* 12 (8): 4239–4251.

Zhao, W., Xie, H., Li, J. et al. (2021b). Application of alum sludge in wastewater treatment processes:"science" of reuse and reclamation pathways. *Processes* 9 (4): 612.

Index

Note: *Italicized* and **bold** page numbers refer to figures and tables, respectively.

Remediation of Heavy Metals: Sustainable Technologies and Recent Advances, First Edition. Edited by Rangabhashiyam Selvasembian, Binota Thokchom, Pardeep Singh, Ali H. Jawad, and Willis Gwenzi.

b

d

o

p

q

t

u

v